Statistical Definitions, Formulas and Tables

K.R. Gupta

ATLANTIC
PUBLISHERS & DISTRIBUTORS (P) LTD

Published by

ATLANTIC

PUBLISHERS & DISTRIBUTORS (P) LTD

7/22, Ansari Road, Darya Ganj,
New Delhi-110002
Phones : +91-11-40775252, 23273880, 23275880, 23280451
Fax : +91-11-23285873
Web : www.atlanticbooks.com
E-mail : orders@atlanticbooks.com

Branch Office
5, Nallathambi Street, Wallajah Road,
Chennai-600002
Phones : +91-44-64611085, 32413319
E-mail : chennai@atlanticbooks.com

Printed in India at Glorious Printers, A-13, D.S.I.D.C.,
Jhilmil Industrial Area, Delhi-110095

PREFACE

The book *Statistical Definitions, Formulas and Tables* provides a collection of statistical definitions, formulas and tables which will be valuable to students of statistics, mathematics, economics, physics, engineering and other sciences, and researchers in these fields. Care has been taken to include only those formulas and tables which are most likely to be needed in practice, rather than highly specialised results which are rarely used.

Topics covered range from elementary to advanced statistics. Elementary topics include those from Collection, Classification and Tabulation of Data; Measure of Central Tendency; Measures of Dispersion; Moments, Skewness and Kurtosis; and Curve Fitting and Method of Least Squares. Advanced topics include those from Correlation Analysis; Regression Analysis; Multiple and Partial Correlation-Regression; Probability; Sampling; Theory of Estimation; Theory of Test of Significance; and Analysis of Variance. This wide coverage of topics provides within a single volume, most of the important statistical results needed by students and researchers, regardless of their particular field of interest or level of attainment.

The book has been divided into two parts—1st part contains statistical definitions and formulas and the 2nd part covers statistical tables. An important feature of this book is that the figures in the tables have been approximated, which is basic to the application of statistical models in the real world.

The book will be very useful for the students of Science, Economics, Commerce, Engineering, Psychology, Business Management and Finance. To those preparing for various competitive examinations it will be of great help for quick revision of definitions, formulas and statistics.

K.R. Gupta

CONTENTS

DEFINITIONS & FORMULAS

COLLECTION, CLASSIFICATION AND TABULATION OF DATA

Statistics has two different meanings. In the plural sense, it implies a set of numerical figures, usually obtained by measurement or counting. These are also collectively known as data. And in the singular sense, it refers to the subject of scientific activity which deals with the theories and methods of collection, analysis and interpretation of such data.

The character of statistical data collected from a group of individuals or objects, is of two types—quantitative and qualitative. The quantitative character is technically called variable and the qualitative character is called attribute. The variable can take different values which can be measured numerically in suitable units. On the other hand, an attribute cannot be measured, but can only be classified under different heads or categories. Further variables are of two types—continuous and discrete or discontinuous. A variable is said to be continuous, when it can take any value within a specified interval. A variable is said to be discrete, when it takes only some isolated values.

Primary and Secondary Data: Statistical data may be of two types, *viz.*, primary and secondary. Primary data are those which are collected for a specific purpose directly from the field of enquiry and are, thus, original in character, whereas secondary data are such numerical information which have previously been collected by some agency for one purpose and are merely complied from that source for use in a different connection. Primary data are in the shape of raw materials to which statistical methods are applied for the purpose of analysis and interpretation. Secondary data are usually in the shape of finished goods since they have been treated statistically in some form or the other.

Primary data are generally collected using following methods:

(a) Direct personal observations,

(b) Indirect oral investigation,

(c) Questionnaires sent by mail, and

(d) Schedules sent through investigators.

Statistical Enquiry: It refers to some investigation wherein relevant information is collected, analysed and interpreted by the application of statistical methods. Different stages of a statistical enquiry are:

(a) Planning the enquiry—at this stage following points are considered: (i) objective and scope of the enquiry, (ii) source of data, whether primary or secondary, (iii) type of enquiry, whether census or sample survey, (iv) definition of statistical units and (v) degree of accuracy.

(b) Collection of data—the method of collection depends to a large extent on the nature, object and scope of the enquiry on one hand and the availability of time and money on the other.

(c) Editing the returns and presentation of data.

(d) Analysis of data and interpretation of results.

(e) Preparation of report.

Law of Statistical Regularity: It states that a moderately large number of items selected at random from a large group are almost sure, on the whole, to possess the characteristics of the larger group, *i.e.*, if a random sample of large size is taken, the sample will to a fairly accurate degree exhibit the characteristics of the entire population. Here two points should be noted—(a) the sample must be drawn at random, and (b) the number of items included in the sample should be sufficiently large.

Statistical Unit: It is that thing in terms of which requisite information is collected and expressed during a statistical inquiry. It is required to chose a suitable statistical unit for effective comparison and valid conclusions.

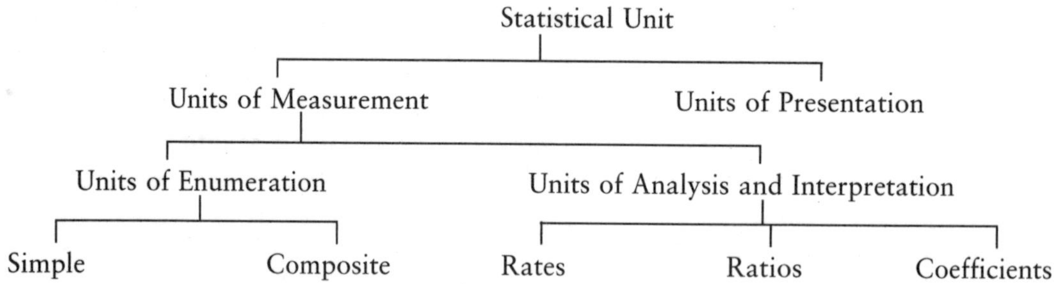

Statistical Error: It shows the extent to which the observed value of a quantity exceeds (or falls short of) the true value, *i.e.*, Error = Observed value − True value. It may be positive, negative or zero and can be classified into:

(a) **Biased errors**—arise due to personal bias or prejudices of investigators and informants, and defects in the measuring instrument. They are also known as cumulative errors because the total effect of biased errors goes on increasing as the number of observations increases.

(b) **Unbiased errors**—arise due to chance causes. They are compensating as they are sometimes in one direction and sometimes in the other; and therefore, with increase in the number of observations, the total effect of unbiased errors diminishes.

Classification: It is the process of arranging the collected statistical information under different categories or classes according to some common characteristic possessed by the individual members. It has following characteristics—(a) it should be unambiguous, (b) it should be stable, and (c) it should be flexible.

Broadly, there are four types of classification—(a) on qualitative basis, (b) on quantitative basis, (c) on time basis, and (d) on geographical basis.

Tabulation: It is defined as the logical and systematic organisation of statistical data in rows and columns, designed to simplify the presentation and facilitate comparisons.

Basically, there are two types of tabulation—(a) simple tabulation, and (b) complex tabulation. A simple table contains data in respect of one characteristic only, information relating to other characteristic being left out. A complex table contains figures relating to several characteristics.

The different parts of a table are:

(a) *Title*—a brief description of the contents shown at the top of the table.

(b) *Stub*—extreme left part of the table showing the descriptions of rows.

(c) *Caption*—upper part of the table showing description of columns and sub-columns.

(d) *Box-head*—whole of the upper part, including caption, units of measurement and column-numbers.

(e) *Body*—part of the table showing figures.

(f) *Footnote*—part below the body showing the source of data and explanations if any.

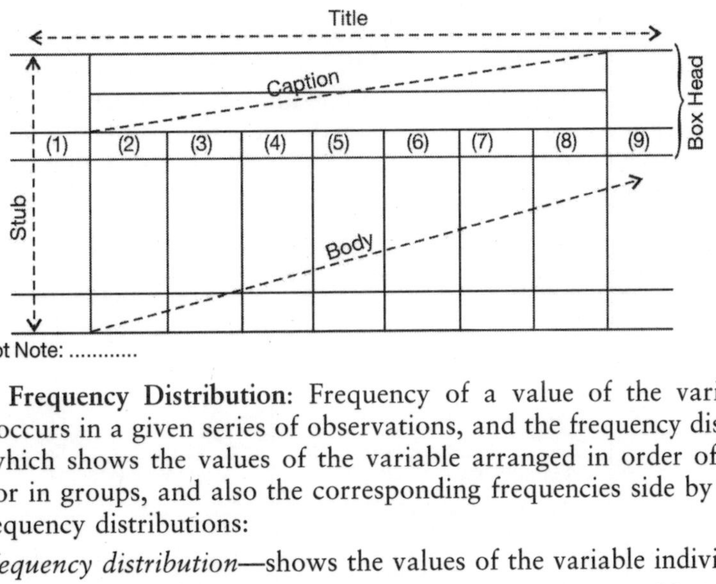

Foot Note: ....

Frequency and Frequency Distribution: Frequency of a value of the variable is the number of times it occurs in a given series of observations, and the frequency distribution is a statistical table which shows the values of the variable arranged in order of magnitude either individually or in groups, and also the corresponding frequencies side by side. There are two types of frequency distributions:

(a) *Simple frequency distribution*—shows the values of the variable individually.

(b) *Grouped frequency distribution*—shows the values of the variable in groups or intervals.

Terms associated with grouped frequency distributions are—(a) class interval, or class, (b) class frequency, and total frequency, (c) class limits, (d) class boundaries, (e) class mark, or mid-value, or mid-point of class interval, (f) width, or size of class interval, and (g) frequency density.

Cumulative Frequency Distribution: Cumulative frequency corresponding to a specified value of the variable may be defined as the number of observations smaller than (or greater than) that value. The number of observations 'upto' a given value is called less-than cumulative frequency; and the number of observations greater than a value is called the more-than cumulative frequency.

A table showing the cumulative frequencies against values of the variable systematically arranged in increasing (or decreasing) order is called cumulative frequency distribution. If cumulative percentages are shown, instead of cumulative frequencies, the table is called cumulative percentage distribution.

Diagrammatic Representation of Frequency Distribution: The diagrams commonly used to depict statistical data, given in the form of a frequency distribution, are:

(a) Dimensional diagrams—(i) one-dimensional diagrams, (ii) two-dimensional diagrams, and (iii) three-dimensional diagrams. Line diagram, bar diagram, pie diagram, histogram, frequency polygon, and ogive (or cumulative frequency polygon) are some of the commonly used dimensional diagrams.

(b) Pictograms—figures are represented by pictures. The size or the number of pictures is in proportion to the given figures.

(c) Cartograms—maps are drawn and the figures representing the phenomena at various places are shown by signs or symbols.

Frequency Curve: If the widths of classes be made smaller and smaller and at the same time the total frequency be also increased indefinitely, then the histogram and the frequency polygon will closely approach a smooth curve, known as frequency curve, which shows the probability distribution of the variable in the population and its area bounded by the ordinates at two specified points on the horizontal axis represents the probability that a value of the variable lies between those two limits.

Commonly, there are four types of frequency curves—(a) symmetrical bell-shaped, (b) asymmetrical single-humped, (c) J-shaped, and (d) U-shaped.

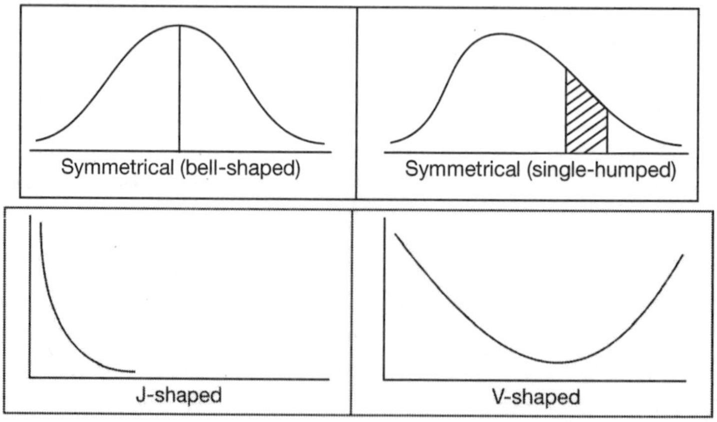

Zee-Chart or Z-Curve: In this type of graph, three curves are drawn which appear somewhat in the shape of the letter Z. The three curves drawn are—(a) curve of original data, (b) curve of cumulative data, and (c) moving total curve, which is plotted from the moving annual total, if figures are monthly or weekly. If the figures are daily, then the curve is plotted from moving monthly totals.

Ratio Scale: It is meant to study relative changes in values. It tells us about the rate or ratio of change, *i.e.*, it measures proportionate changes. Natural scale is always in arithmetic progression, whereas the ratio scale is in geometric progression.

CLASS INTERVAL (OR CLASS)

When a large number of observations varying in a wide range are available, these are usually classified in several groups according to the size of values. Each of these groups, defined by an interval, is called class interval, or simply class.

When one end of a class is not specified, the class is called an open-end class.

RANGE

It is the difference between the highest and the lowest value of the variable in question.

MAGNITUDE OF CLASS INTERVALS

Magnitude of class intervals is given by Sturges' formula as

$$i = \frac{Range}{1 + 3.322 \log N}$$

where, i is the magnitude of the class interval and N is the total number of observations.

CLASS FREQUENCY AND TOTAL FREQUENCY

The number of observations falling within a class is called its class frequency, or simply frequency. The sum of all the class frequencies is called total frequency.

CLASS LIMITS

The two numbers used to specify the limits of a class interval for the purpose of tallying the original observations into the various classes, are called class limits. The smaller of the pair is known as *Lower Class Limit* and the larger as *Upper Class Limit* with reference to the particular class.

CLASS BOUNDARIES

The most extreme values which would ever be included in a class interval are called class boundaries. Class boundaries are the real limits of a class interval. The lower extreme point is called *Lower Class Boundary* and the upper extreme point is called *Upper Class Boundary* with reference to any particular class.

Class boundaries are calculated from class limits by applying the following rule:

$$\text{Lower class boundary} = \text{Lower class limit} - d/2$$
$$\text{Upper class boundary} = \text{Upper class limit} + d/2$$

where d is the common difference between the upper class limit of any class interval and the lower class limit of the next class interval.

CLASS MARK (OR MID-VALUE OR MID-POINT)

The value exactly at the middle of a class interval is called class mark or mid-value. It lies half-way between the class limits or between the class boundaries.

$$\begin{aligned}\text{Class mark} &= (\text{Lower class limit} + \text{Upper class limit})/2 \\ &= (\text{Lower class boundary} + \text{Upper class boundary})/2\end{aligned}$$

WIDTH (OR SIZE) OF CLASS

Width or size of a class is the difference between the lower and the upper class boundaries (not class limits).

$$\text{Width of class} = \text{Upper class boundary} - \text{Lower class boundary}$$

NUMBER OF CLASSES

An ideal number of classes for any frequency distribution would be that which gives the maximum information in the clearest fashion. According to Sturges, the number of classes is given by

$$k = 1 + 3.322 \log N$$

where k is the number of classes and $\log N$ is the logarithm of the total number of observations.

FREQUENCY DENSITY

Frequency density of a class is its frequency per unit width. It represents the concentration of frequency in a class and is given by

$$\text{Frequency density} = \text{Class frequency}/\text{Width of the class}$$

AREA OF EACH RECTANGLE OF HISTOGRAM

$$\begin{aligned}\text{Area of each rectangle} &= \text{width} \times \text{height} \\ &= \text{width of class} \times \text{frequency density} \\ &= \text{width of class} \times \frac{\text{class frequency}}{\text{width of class}} = \text{class frequency}\end{aligned}$$

MEASURES OF CENTRAL TENDENCY

Statistic deals with a large number of big figures which are not only confusing to mind but also difficult to analyse. Therefore, condensation of data is necessary in statistical analysis to reduce the size of the data and enable us to make comparative studies of related variables. To reduce the complexity of data and to make them comparable, it is essential that the various phenomena which are being compared are reduced to one figure each. A figure which is used to represent a whole series should neither have the lowest value in the series nor the highest value but a value somewhere between these two limits, possibly in the centre, where most of the items of the series cluster. Such figures are called *measures of central tendency* or averages. Thus, the word 'average' denotes a 'representative' or 'typical value' of a whole set of observations. It is a single figure which describes the entire series of observations with their varying sizes and is, generally, located in the centre or middle of the distribution.

The measures of central tendency have been defined by various authors as:

Lawrence J. Kaplan: One of the most widely used set of summary figures is known as measures of location, which are often referred to as averages, central tendency or central location. The purpose for computing an average value for a set of observations is to obtain a single value which is representative of all the items and which the mind can grasp in simple manner quickly. The single value is the point of location around which the individual items cluster.

Simpson and Kafka: A measure of central tendency is a typical value around which other figures congregate.

Ya-Lun Chou: An average is a typical value in the sense that it is, sometimes, employed to represent all the individual values in a series or of a variable.

The objectives of measures of central tendency are: (a) to get a single value which is representative of the characteristics of the entire mass of data, and (b) to facilitate comparison.

Characteristics of a Good Average: According to G.U. Yule, the desirable characteristics of a satisfactory average are:

(a) It should be rigidly defined.

(b) It should be based on all observations.

(c) It should be easy to calculate and simple to follow.

(d) It should be easily calculated.

(e) It should be affected as little as possible by sampling fluctuations.

(f) It should readily lend itself to algebraic treatment.

Measures of Various Orders: Statistical series may differ from each other in the following three ways and accordingly three measures are designed to study the differences.

(a) They may differ in the values of the variable around which most of the items cluster—measures of the first order or measures of central tendency or averages are used.

(b) They may differ in the extent to which items are dispersed round the central value—measures of the second order, or measures of dispersion are used.

(c) They may differ in the extent of departure from a normal distribution—measures of the third order or skewness, kurtosis, *etc.*, are used.

In this chapter we shall study the measures of the first order, or measures of central tendency.

Measures of Central Tendency

- Mathematical Averages
  - Arithmetic Mean (A.M.)
  - Geometric Mean (G.M.)
  - Harmonic Mean (H.M.)
- Positional Averages
  - Median
  - Mode

ARITHMETIC MEAN

It is the figure obtained by dividing the sum of the values of the various items by their number.

Given n observations $y_1, y_2, y_3, \ldots, y_n$, their A.M., denoted by the symbol $\bar{y}$, is

$$\bar{y} = (y_1 + y_2 + y_3 + \ldots + y_n)/n = \frac{1}{n}\sum_{i=1}^{n} y_i$$

If $y_1, y_2, y_3, \ldots, y_n$ have frequencies $f_1, f_2, f_3, \ldots, f_n$ respectively, *i.e.*, y_1 occurs f_1 times, y_2 occurs f_2 times and so on, then the sum of all the $f_1 + f_2 + f_3 + \ldots + f_n$ observations is

$$\underbrace{y_1 + y_1 + y_1 + \ldots + y_1}_{f_1 \text{ terms}} + \underbrace{y_2 + y_2 + y_2 + \ldots + y_2}_{f_2 \text{ terms}} + \ldots + \underbrace{y_n + y_n + y_n + \ldots + y_n}_{f_n \text{ terms}}$$

$$= f_1 y_1 + f_2 y_2 + f_3 y_3 + \ldots + f_n y_n$$

Hence, the A.M. in this case is

$$\bar{y} = \frac{f_1 y_1 + f_2 y_2 + f_3 y_3 + \ldots + f_n y_n}{f_1 + f_2 + f_3 + \ldots + f_n} = \frac{\sum_{i=1}^{n} f_i y_i}{\sum_{i=1}^{n} f_i} = \frac{\sum_{i=1}^{n} f_i y_i}{N}$$

This is known as the *weighted arithmetic mean*.

A.M. OF GROUPED FREQUENCY

The A.M. from a grouped frequency distribution with a common width is given by

$$A.M. = 1st\ mid\text{-}point + \frac{Common\ width \times Total\ of\ more\text{-}than\ cumulative\ frequency,\ except\ the\ 1st}{Total\ frequency}$$

i.e., $A.M. = y_1 + hS/N$

where, y_1 = mid-point of the first class, h = common width or class interval, S = total of more-than cumulative frequency, except the 1st, and N = total frequency.

SIMPLIFIED CALCULATION FOR A.M.

(a) If $x_1, x_2, x_3, \ldots, x_n$ represent the deviations of $y_1, y_2, y_3, \ldots, y_n$ from an arbitrary constant c, then

$$Mean\ of\ y = c + Mean\ of\ x$$

i.e., if $x = y - c$, then $\bar{y} = c + \bar{x}$.

(b) If $x_1, x_2, x_3, \ldots, x_n$ represent the deviations of $y_1, y_2, y_3, \ldots, y_n$ from an arbitrary constant c, in units of another constant d, then

$$Mean\ of\ y = c + d(Mean\ of\ x)$$

i.e., if $x = (y - c)/d$, then $\bar{y} = c + d\bar{x}$.

MEAN OF COMPOSITE GROUP

If two groups contain n_1 and n_2 observations with means $\bar{y}_1$ and $\bar{y}_2$ respectively, then the mean $(\bar{y})$ of the composite group of $n_1 + n_2$ observations is given by the relation

$$\bar{y} = (n_1\bar{y}_1 + n_2\bar{y}_2)/N$$

where, $N = n_1 + n_2$.

In general, if there are several groups, the i-th group containing n_i observations with mean $\bar{y}_i$, then the mean $(\bar{y})$ of the composite group is given by the relation

$$\bar{y} = \Sigma n_i\bar{y}_i/\Sigma n_i = \Sigma n_i\bar{y}_i/N$$

where, $N = \Sigma n_i$.

CHARLIER'S ACCURACY CHECK

The accuracy of the calculation of the arithmetic average can be checked with the help of the following formula given by Charlier.

$$\Sigma f dy = \Sigma[f(d+1)] - \Sigma f$$

where, Σfdy = total of the products of the deviations from the assumed average and the respective frequencies of the items.

Σf = total of the frequencies of the items.

d = deviations from the assumed average.

GEOMETRIC MEAN (G.M.)

Geometric mean of a group of n observations is the n-th root of their product. It is defined only when all observations have the same sign, and none of them is zero.

If $y_1, y_2, y_3, ..., y_n$ are n observations, then

$$G.M. = \sqrt[n]{y_1 \times y_2 \times y_3 \times ... \times y_n} \text{ or } (y_1 \times y_2 \times y_3 \times ... \times y_n)^{1/n}$$

If $y_1, y_2, y_3, ..., y_n$ have frequencies $f_1, f_2, f_3, ..., f_n$ respectively, the product of all the N $(= f_1 + f_2 + f_3 + ... + f_n)$ observations is

$$\underbrace{y_1 \cdot y_1 \cdot y_1 \cdots y_1}_{f_1 \text{ terms}} \cdot \underbrace{y_2 \cdot y_2 \cdot y_2 \cdots y_2}_{f_2 \text{ terms}} \cdots \underbrace{y_n \cdot y_n \cdot y_n \cdots y_n}_{f_n \text{ terms}} = y_1^{f_1} \cdot y_2^{f_2} \cdot y_3^{f_3} \cdots y_n^{f_n}$$

so that

$$G.M. = \sqrt[N]{y_1^{f_1} \cdot y_2^{f_2} \cdot y_3^{f_3} \cdots y_n^{f_n}} \text{ or } \left(y_1^{f_1} \cdot y_2^{f_2} \cdot y_3^{f_3} \cdots y_n^{f_n}\right)^{1/N}$$

where, $N = \Sigma f_i$ is the total frequency.

Thus,

$$\textit{Simple G.M.}(g) = (y_1 \cdot y_2 \cdot y_3 \cdots y_n)^{1/n}$$

$$\textit{Weighted G.M.}(G) = \left(y_1^{f_1} \cdot y_2^{f_2} \cdot y_3^{f_3} \cdots y_n^{f_n}\right)^{1/N}$$

These formulae cannot be applied directly. Taking logarithms of both sides, we get

$$\log g = (1/n)(\log y_1 + \log y_2 + \log y_3 + ... + \log y_n) = (1/n)\Sigma \log y_i$$

$$\log G = (1/N)[f_1(\log y_1) + f_2(\log y_2) + f_3(\log y_3) + ... + f_n(\log y_n)] = (1/N)\Sigma f_i(\log y_i)$$

HARMONIC MEAN (H.M.)

Harmonic mean of a set of observations is the reciprocal of the arithmetic mean of their reciprocals. It is defined only when no observation is zero.

$$\textit{Simple H.M.} = \frac{n}{1/y_1 + 1/y_2 + 1/y_3 + ... + 1/y_n} = \frac{n}{\Sigma(1/y_i)}$$

$$\textit{Weighted H.M.} = \frac{f_1 + f_2 + f_3 + ... + f_n}{f_1/y_1 + f_2/y_2 + f_3/y_3 + ... + f_n/y_n} = \frac{\Sigma f_i}{\Sigma(f_i/y_i)} = \frac{N}{\Sigma(f_i/y_i)}$$

They are equal only when all weights are equal.

RELATIONS BETWEEN A.M., G.M. AND H.M.

(a) For any given set of observations, A.M. is greater than or equal to G.M., and G.M. is greater than or equal to H.M., *i.e.*,

$$A.M. \geq G.M \geq H.M.$$

They are equal, only when all observations are equal.

(b) For two observations only,

$$A.M./G.M. = G.M./H.M.$$

i.e., $(G.M.)^2 = A.M. \times H.M.$ (for two observations only)

QUADRATIC MEAN

This is also known as *Root Mean Square*. It is calculated by taking the square root of the average of the squares of the numbers. It is useful when some items have negative value and other positive. Symbolically,

$$Q.M. = \sqrt{(y_1^2 + y_2^2 + y_3^2 + ... + y_n^2)/N}$$

MEDIAN

Median of a set of observations is the middle-most value when the observations are arranged in order of magnitude. Median is calculated as follows:

(a) ***From Simple Series***—The given data are arranged in order of magnitude. If the number of observations be odd, the value of the middle-most item, $\left(\frac{n+1}{2}\right)$th item, is the median. If the number of observations be even, the arithmetic mean of the two middle-most items, $\frac{n}{2}$ and $\left(\frac{n}{2}+1\right)$th items, is taken as median.

(b) ***From Simple Frequency Distribution***—The cumulative frequency (less than type) corresponding to each distinct value of the variable is calculated. If the total frequency be N, the value of the variable corresponding to cumulative frequency $(N + 1)/2$ gives the median.

(c) ***From Grouped Frequency Distribution***—Median from a grouped frequency distribution is that value which corresponds to cumulative frequency N/2. Median from a grouped frequency distribution can be calculated by any of the following methods:

(i) *By the Application of Formula for Median*—The class in which cumulative frequency N/2 lies, is called the median class.

$$Median = l_1 + \frac{N/2 - F}{f_m} \times c$$

where, l_1 = lower boundary of median class;
N = total frequency;
F = cumulative frequency below l_1;
f_m = frequency of median class; and
c = width of median class.

(ii) *By the Application of Simple Interpolation in a Cumulative Frequency Distribution*—If F_1 and F_2 be the cumulative frequencies which are just smaller than and just larger than $N/2$, and they correspond to the class boundaries l_1 and l_2 respectively, then

$$\frac{Median - l_1}{l_2 - l_1} = \frac{(N/2) - F_1}{F_2 - F_1}$$

where, l_1 and l_2 are the lower and the upper boundaries of the median class, and $N/2$ lies between F_1 and F_2.

(iii) *Graphical Method*—An approximate value of median can be obtained graphically from ogive, or cumulative frequency polygon. Draw a horizontal line from the point $N/2$ on the vertical scale showing the cumulative frequencies, until it meets the ogive (either less-than or more-than type). From the point of intersection, a perpendicular is now drawn on the horizontal axis. The position of the foot of the perpendicular is read from the horizontal scale showing values of the variable, and this gives the median. If both ogives (less-than and more-than) are available on the same graph paper, the position of the foot of the perpendicular drawn from the point of intersection of the two ogives, gives the median.

MODE

Mode of a given set of observations is that value which occurs with the maximum frequency.

From a simple series, mode can be determined by locating that value, which occurs the maximum number of times.

From a simple frequency distribution, mode can be determined by inspection only. It is that value of the variable which corresponds to the largest frequency.

From a grouped frequency distribution, if all classes are of equal width, mode is usually calculated by the formula

$$Mode = l_1 + \frac{d_1}{d_1 + d_2} \times c$$

where, l_1 = lower boundary of the modal class, *i.e.*, the class containing the largest frequency;

d_1 = difference of the largest frequency and the frequency of class just preceding the modal class;

d_2 = difference of the largest frequency and the frequency of class just following the modal class; and

c = common width of classes.

If f_0, f_{-1} and f_1 represent the frequencies of the modal class, the class just preceding and the class just following it, then

$$d_1 = f_0 - f_{-1} \text{ and } d_2 = f_0 - f_1$$

$$\therefore \qquad Mode = l_1 + \frac{f_0 - f_{-1}}{2f_0 - f_{-1} - f_1} \times c$$

The formula for the calculation of mode is based on the assumption that the distances of mode from the two boundaries of modal class are proportional to the differences in frequencies of the modal class and its two adjoin classes, *viz.*, d_1 and d_2.

RELATION BETWEEN MEAN, MEDIAN AND MODE

Mean, Median and Mode are related by the identity:

$$Mean - Mode = 3(Mean - Median)$$

PARTITION VALUES

Just as median divides the total number of observations into two equal parts, there are similar other measures which are used to divide, or partition, the observations into a fixed number of parts. These are collectively known as *partition values* or *quantiles* or *fractiles*. Some of the important types of partition values are:

(a) Median, (b) Quartiles, (c) Deciles, (d) Percentiles

Quartiles are such values which divide the total number of observations into 4 equal parts. There are three quartiles:

(i) First quartile (or Lower quartile): Q_1

(ii) Second quartile (or Middle quartiles): Q_2

(iii) Third quartile (or Upper quartile): Q_3

Corresponding to Q_1, the cumulative frequency $= N/4$

" Q_2, " " $= 2N/4$

" Q_3, " " $= 3N/4$

$Q_1 < Q_2 < Q_3$

Median $= Q_2$; *Quartile deviation* $= (Q_3 - Q_1)/2$; *Skewness* $= (Q_3 - 2Q_2 + Q_1)/(Q_3 - Q_1)$

Deciles are such values which divide the total number of observations into 10 equal parts and Percentiles are such values which divide the total number of observations into 100 equal parts.

CALCULATION OF PARTITION VALUES

The method of calculation of quartiles, deciles, percentiles is exactly the same as that employed for median, using simple interpolation in a cumulative frequency distribution.

(a) *From Simple Series*—The given data are arranged in increasing order of magnitude and a number showing the rank is attached to each observation. The smallest value is given rank 1, the next higher value rank 2, *etc*., and the largest value is given rank n. The ranks of partition values are as follows:

Rank of median = $(1/2)(n + 1)$; Rank of $Q_1 = (1/4)(n + 1)$; Rank of $Q_3 = (3/4)(n + 1)$; Rank of deciles $D_k = (k/10)(n + 1)$; Rank of percentiles $P_k = (k/100)(n + 1)$.

Using simple interpolation, the value of the variable corresponding to the appropriate rank is determined, giving the partition value.

Median = Value corresponding to rank $(1/2)(n + 1)$
Q_1 = ,, ,, ,, $(1/4)(n + 1)$
Q_3 = ,, ,, ,, $(3/4)(n + 1)$
D_k = ,, ,, ,, $(k/10)(n + 1)$
P_k = ,, ,, ,, $(k/100)(n + 1)$

(b) *From Simple Frequency Distribution*—The cumulative frequency corresponding to each distinct value of the variable is calculated. If the total frequency be N,

Median = Value corresponding to cumulative frequency $(1/2)(N + 1)$
Q_1 = ,, ,, ,, ,, ,, $(1/4)(N + 1)$
Q_3 = ,, ,, ,, ,, ,, $(3/4)(N + 1)$
D_k = ,, ,, ,, ,, ,, $(k/10)(N + 1)$
P_k = ,, ,, ,, ,, ,, $(k/100)(N + 1)$

(c) *From Grouped Frequency Distribution*—(i) *By application of simple interpolation*—A cumulative frequency distribution is constructed showing the class boundaries and the corresponding cumulative frequencies (less-than type). Using simple interpolation, we now find:

Median = Value corresponding to cumulative frequency $(1/2)N$
Q_1 = ,, ,, ,, ,, $(1/4)N$
Q_3 = ,, ,, ,, ,, $(3/4)N$
D_k = ,, ,, ,, ,, $(k/10)N$
P_k = ,, ,, ,, ,, $(k/100)N$

(ii) *Graphical method*—An ogive (less-than types) is drawn. From this ogive

Median = Abscissa corresponding to ordinate $(1/2)N$
Q_1 = ,, ,, ,, $(1/4)N$
Q_3 = ,, ,, ,, $(3/4)N$
D_k = ,, ,, ,, $(k/10)N$
P_k = ,, ,, ,, $(k/100)N$

MEASURES OF DISPERSION

The word *dispersion* implies the degree of heterogeneity in the data. It indicates the extent to which observations vary among themselves. The dispersion of a given set of observations will be zero, only when all of them are equal. The wider the discrepancy from one observation to another, the larger will be the dispersion.

The measures of dispersion have been defined by various authors as:

Dyckman and Thomas: Measures of variability are usually used to indicate how tightly bunched the sample values are around the mean.

B.C. Brooks and W.F.L. Dick: Dispersion or spread is the degree of the scatter or variation of the variable about a central value.

A.L. Bowley: Dispersion is the measure of the variations of the items.

Spiegel: The degree to which numerical data tend to spread about an average value is called the variation or dispersion of the data.

L.R. Connor: Dispersion is a measure of the extent to which the individual items vary.

W.I. King: The term dispersion is used to indicate the facts that within a given group, the items differ from one another in size or in other words, there is lack of uniformity in their sizes.

From the above definitions, it can be concluded that in a general sense the term dispersion refers to the variability in the size of items. If the variation is substantial, dispersion is said to be considerable and if the variation is very little, dispersion is insignificant.

The measures of dispersion are also called *averages of the second order* because in the calculation of measures of dispersion, the average values are derived by the use of the *averages of the first order*, *i.e.*, arithmetic mean, geometric mean, harmonic mean, median, mode, *etc.*

Objectives of Measures of Dispersion: The objectives of the measures of dispersion are:

(a) To judge the reliability of measures of central tendency.

(b) To make a comparative study of the variability of two series.

(c) To identify the causes of variability with a view to control it.

(d) To obtain other statistical measures for further analysis of data.

Characteristics for an Ideal Measure of Dispersion: The properties of a good measure of dispersion are the same as the properties of a good measure of central tendency, *viz.*:

(a) It should be rigidly defined.
(b) It should be based on all the observations of the series.
(c) It should be easy to calculate and easy to understand.
(d) It should be capable of further algebraic treatment.
(e) It should not be affected by fluctuations of sampling.
(f) It should not be affected much by extreme observations.

Types of Measures of Dispersion: A measure of dispersion is designed to state numerically the extent to which individual observations vary on the average. The various measures of dispersion are:

Measures of Dispersion

- *Absolute Measures*
  - *Range*
  - *Inter-quartile Range*
  - *Quartile Deviation*
  - *Mean Deviation*
  - *Standard Deviation*
  - *Lorenz Curve*
- *Relative Measures*
  - *Coefficient of Range*
  - *Coefficient of Variation*
  - *Coefficient of Quartile Deviation*
  - *Coefficient of Mean Deviation*

RANGE

It is the difference between the maximum and the minimum values of a set of observations, *i.e.*,

$$Range = Maximum\ value - Minimum\ value$$

COEFFICIENT OF RANGE

If the range is divided by the sum of the extreme items, the resulting figures is called the coefficient of the range or the coefficient of the scatter, *i.e.*,

$$Coefficient\ of\ range = \frac{Maximum\ value - Minimum\ value}{Maximum\ value + Minimum\ value} = \frac{Absolute\ range}{Sum\ of\ the\ extreme\ values}$$

INTER-QUARTILE RANGE

It is the difference in the values of two quartiles, *i.e.*,

$$Inter\text{-}quartile\ range = Third\ quartile - First\ quartile = Q_3 - Q_1$$

It is not affected by the values of the extreme items, therefore it is better measure of dispersion than range.

QUARTILE DEVIATION OR SEMI-INTER-QUARTILE RANGE

It is defined as half the difference between the upper and the lower quartiles, *i.e.*,

$$Quartile\ deviation = (Q_3 - Q_1)/2$$

It is dependent on the two quartiles and does not take into account the variability of the largest 25% or the smallest 25% of observations. Therefore, it is unaffected by extreme values.

CO-EFFICIENT OF QUARTILE DEVIATION

Quartile deviation is an absolute measure of dispersion. If it is divided by the average value of the two quartiles, a relative measure of dispersion, known as co-efficient of quartile deviation, is obtained, *i.e.*,

$$\textit{Coefficient of quartile deviation} = \left(\frac{Q_3 - Q_1}{2}\right)\Big/\left(\frac{Q_3 + Q_1}{2}\right) = \frac{Q_3 - Q_1}{Q_3 + Q_1}$$

MEAN DEVIATION (OR MEAN ABSOLUTE DEVIATION)

It is the arithmetic average of the absolute deviations of various items from a measure of central tendency (either mean, median or mode). Theoretically, mean deviations can be taken from any of the three averages (mean, median or mode), but in actual practice, mean deviation is calculated either from mean or from median. Mode is usually not considered, as its value is indeterminate, and it gives erroneous conclusions.

Given the observations $y_1, y_2, y_3, ..., y_n$, in order to find mean deviation about A, we first obtain the deviations $(y_1 - A), (y_2 - A), (y_3 - A), ..., (y_n - A)$. Some of these deviations may be positive and some negative. If we write $|y_i - A|$ to denote the positive value of $(y_i - A)$, whatever be the actual sign, the sum of these 'absolute deviations' is given by

$$|y_1 - A| + |y_2 - A| + |y_3 - A| + ... + |y_n - A| = \Sigma|y_i - A|$$

and A.M. of the absolute deviations is

$$\textit{Mean deviation about } A = \frac{1}{n}\Sigma|y_i - A|$$

Mean deviation is usually calculated about arithmetic mean ($\bar{y}$), and hence mean deviation only refers to mean deviation about mean.

For simple series,

$$\textit{Mean deviation} = \frac{1}{n}\Sigma|y_i - \bar{y}|$$

For frequency distribution,

$$\textit{Mean deviation} = \frac{1}{N}\Sigma f_i|y_i - \bar{y}|$$

where, $N = \Sigma f_i$.

Mean deviation is also known as first moment of dispersion and is a absolute measure of dispersion, expressed in the same units in which the original data are.

COEFFICIENT OF MEAN DEVIATION

In order to transform mean deviation into a relative measure, it is divided by the average from which it has been calculated, *i.e.*,

$$\textit{Coefficient of mean deviation} = \frac{\textit{Mean deviation}}{\textit{Average about which it is calculated}}$$

$$\textit{Coefficient of mean deviation about mean (or median)} = \frac{\textit{Mean deviation}}{\textit{mean (or median)}}$$

STANDARD DEVIATION

It is the square-root of the arithmetic mean of squares of deviations from arithmetic mean. In other words, it is the root-mean-square-deviations from mean. It is usually denoted by the Greek letter σ (sigma).

If $y_1, y_2, y_3, \ldots, y_n$ be a set of observations and $\bar{y}$ their arithmetic mean, then,

Deviations from mean: $(y_1 - \bar{y}), (y_2 - \bar{y}), (y_3 - \bar{y}), \ldots, (y_n - \bar{y})$

Square-deviations from mean: $(y_1 - \bar{y})^2, (y_2 - \bar{y})^2, (y_3 - \bar{y})^2, \ldots, (y_n - \bar{y})^2$

Mean-square-deviation from mean:

$$\frac{1}{n}\left[(y_1 - \bar{y})^2 + (y_2 - \bar{y})^2 + (y_3 - \bar{y})^2 + \ldots + (y_n - \bar{y})^2\right] = \frac{1}{n}\Sigma(y_i - \bar{y})^2$$

Root-mean-square-deviation from mean, *i.e.*,

$$\text{Standard deviation } (\sigma) = \sqrt{\frac{1}{n}\Sigma(y_i - \bar{y})^2} = \sqrt{\frac{\Sigma y^2}{n} - \left(\frac{\Sigma y}{n}\right)^2}$$

For frequency distributions,

$$\sigma = \sqrt{\frac{\Sigma f y^2}{N} - \left(\frac{\Sigma f y}{N}\right)^2}$$

The square of standard deviation, *i.e.*, σ^2, is known as variance. Thus, $\textit{Variance} = (S.D.)^2$

For simple series: $\sigma^2 = \frac{1}{n}\Sigma(y_i - \bar{y})^2$

For frequency distribution: $\sigma^2 = \frac{1}{N}\Sigma f_i(y_i - \bar{y})^2$

where, $N = \Sigma f_i$. σ is always considered as positive and therefore it is the positive square-root of variance.

Important Properties: (a) S.D. is independent of the change of origin, *i.e.*, if $y = x - c$, where c is a constant, then

$$\text{S.D. of } x = \text{S.D. of } y \quad \Rightarrow \quad \sigma_x = \sigma_y$$

Thus, same S.D. will be obtained if each of the observations is increased or decreased by a constant.

(b) If two variables x and y are so related that $y = ax + b$ for each $x = x_i$, where a and b are constants, then $\sigma_y = |a| \cdot \sigma_x$, where $|a|$ denotes the positive value of a.

In particular, if $y = (x - c)/d$, where c and d are constants (d positive), then $\sigma_x = d \cdot \sigma_y$. Thus, S.D. does not depend on origin, but depends on scale of measurement. If each observation is multiplied or divided by a constant, S.D. will also be similarly affected.

(c) S.D. is the minimum root-mean-square-deviation, *i.e.*,

$$\sqrt{(1/n)\Sigma(y_i - \bar{y})^2} \le \sqrt{(1/n)\Sigma(y_1 - A)^2}$$

whatever be the value of A.

STANDARD DEVIATION FROM CONTINUOUS SERIES

In this case, the class intervals are represented by their respective middle points. Let us represent the middle ponits by y. Now, take the deviations of each middle values from some value and represent these deviations as dy, then standard deviation is given by:

$$\text{Standard deviation } (\sigma) = \sqrt{\frac{\Sigma f(dy)^2}{\Sigma f} - \left[\frac{\Sigma f(dy)}{\Sigma f}\right]^2}$$

if the deviations are divided by some common factor, say, k defined by $d'y = dy/k$, then

$$\text{Standard deviation } (\sigma) = \sqrt{\frac{\Sigma f(d'y)^2}{\Sigma f} - \left[\frac{\Sigma f(d'y)}{\Sigma f}\right]^2} \times k$$

where 'k' stands for the common factor or the magnitude of the class interval, and the use of the values of $d'x$ is called step deviation.

STANDARD DEVIATION OF COMPOSITE GROUP

If a group of n_1 observations has mean $\bar{y}_1$ and S.D. σ_1, and another group of n_2 observations has mean $\bar{y}_2$ and S.D. σ_2, then S.D. (σ) of the composite group of $n_1 + n_2$ (= N, say) observations can be obtained by

$$N\sigma^2 = (n_1\sigma_1^2 + n_2\sigma_2^2) + (n_1 d_1^2 + n_2 d_2^2)$$

where, $d_1 = \bar{y}_1 - \bar{y}$, $d_2 = \bar{y}_2 - \bar{y}$, and $N\bar{y} = n_1\bar{y}_1 + n_2\bar{y}_2$.

The above relation can be extended to any number of groups. Thus,

$$N\sigma^2 = \Sigma n_i \sigma_i^2 + \Sigma n_i d_i^2$$

where, $d_i = \bar{y}_i - \bar{y}$, $N = \Sigma n_i$, and $\bar{y}$ is the mean of composite group, given by $N\bar{y} = \Sigma n_i \bar{y}_i$.

COEFFICIENT OF STANDARD DEVIATION

It is obtained by dividing the standard deviation by the arithmetic mean, *i.e.*,

$$\text{Coefficient of standard deviation} = \text{Standard Deviation/Mean} = \sigma/\bar{y}.$$

CHARLIER'S CHECK OF ACCURACY

As in case of arithmetic average, the accuracy of the calculation of standard deviation can be checked by a formula given by Charlier as follows:

$$\Sigma f(dy+1)^2 - \Sigma f dy^2 - 2\Sigma f dy = N$$

SHEPHERD'S CORRECTION FOR GROUPING

In case of continuous distribution, the frequency may taper off to zero in both the direction. Therefore, a correction in the value of standard deviation is usually done to remove the effects of grouping. These corrections are known as Shepherd's corrections for grouping.

$$\text{Corrected standard deviation} = \sqrt{\sigma^2 - h^2/12}$$

where, σ = standard deviation before correction, and

h = magnitude of the class interval.

CHEBYSHEV'S THEOREM

For any $k > 1$, at least $100(1 - 1/k^2)\%$ of the data values lie within k standard deviations of the mean of the data values, *i.e.*, within the limits

$$\bar{y} - k\sigma \text{ and } \bar{y} + k\sigma.$$

STANDARD SCORES OR Z-SCORES

It gives the number of standard deviations, the original value is from the mean. It is obtained on shifting the origin in the original measurement (y) to its mean and the scale to its standard deviation.

$$\text{Standard score, } Z = \frac{y - mean}{S.D.}$$

The value of Z-scores are independent of the units of measurements.

GINNI'S MEAN DIFFERENCE

It is a measure of dispersion based on the absolute values of the differences between all possible pairs of observations. Ginni's coefficient of mean difference (Δ) for a set of n observations $y_1, y_2, y_3, \ldots, y_n$ is given by

$$\Delta = \frac{1}{n(n-1)} \sum_{i=1}^{n} \sum_{\substack{j=1 \\ (i \neq j)}}^{n} |y_i - y_j|$$

Since, $y_i - y_j = -(y_j - y_i)$, we have $|y_i - y_j| = |y_j - y_i|$

If $y_i > y_j$, then $|y_i - y_j| = |y_j - y_i| = y_i - y_j$

$$\therefore \quad \Delta = \frac{1}{n(n-1)} \sum_{\substack{i,j \\ (i \neq j)}} 2(y_i - y_j); \qquad (y_i > y_j)$$

$$\Rightarrow \quad \Delta = \frac{2}{n(n-1)} \sum_{\substack{i,j \\ (i \neq j)}} (y_i - y_j); \qquad (y_i > y_j)$$

$$\Rightarrow \quad \Delta = g/m$$

where, g = sum of positive differences of all possible pairs of observations, and

$m = [2/n(n-1)]$

STANDARD DEVIATION IN TERMS OF MUTUAL DIFFERENCES

For a set of n observations $y_1, y_2, y_3, \ldots, y_n$ the sum of the squares of all ${}^nC_2 = n(n-1)/2$ mutual differences $(y_i - y_j), i \neq j = 1, 2, 3, \ldots, n$ is given by

$$\sum_{i=1}^{n} \sum_{j=1}^{n} (y_i - y_j)^2 = \sum_i \sum_j (y_i - \bar{y} + \bar{y} - y_j)^2 = \sum_i \sum_j [(y_i - \bar{y}) - (y_j - \bar{y})]^2$$

where, $\bar{y} = \frac{1}{n} \sum_{i=1}^{n} y_i$ is the arithmetic mean of the given observations.

$$\therefore \quad \sum_i \sum_j (y_i - y_j)^2 = \sum_i \sum_j [(y_i - \bar{y})^2 + (y_j - \bar{y})^2 - 2(y_i - \bar{y})(y_j - \bar{y})]$$

$$= \sum_i \sum_j (y_i - \bar{y})^2 + \sum_i \sum_j (y_i - \bar{y})^2 - 2 \sum_i \sum_j (y_i - \bar{y})(y_j - \bar{y})$$

$$= \sum_i \left[\sum_i (y_i - \bar{y})^2 \right] + \sum_j \left[\sum_i (y_i - \bar{y})^2 \right] - 2 \sum_i (y_i - \bar{y}) \sum_j ((y_i - \bar{y})$$

$$= \sum_i n(y_i - \bar{y})^2 + \sum_j n(y_j - \bar{y})^2 - 2 \times 0 \times 0 = \sum_i n(y_i - \bar{y})^2 + \sum_j n(y_j - \bar{y})^2 \ldots (**)$$

But,

$$\sigma^2 = \frac{1}{n} \sum_i (y_i - \bar{y})^2 = \frac{1}{n} \sum_j (y_j - \bar{y})^2 \Rightarrow \sum_i (y_i - \bar{y})^2 = \sum_j (y_i - \bar{y})^2 = n\sigma^2$$

Substituting in (**), we get

$$\sum_i \sum_j (y_i - y_j)^2 = n \sum_i (y_i - \bar{y})^2 + n \sum_j (y_i - \bar{y})^2 = n \cdot (n\sigma^2) + n \cdot (n\sigma^2) = 2n^2\sigma^2$$

$$\therefore \quad \sigma^2 = \frac{1}{2n^2} \sum_i \sum_j (y_i - y_j)^2 \quad \Rightarrow \quad \sigma = \frac{1}{n} \sqrt{\frac{1}{2} \sum_i \sum_{\substack{j \\ (i \neq j)}} (y_i - y_j)^2}$$

Hence, $$\sigma = \frac{1}{n}\sqrt{\frac{1}{2}(Sum\ of\ squares\ of\ all\ possible\ mutual\ differences)}$$

LORENZ CURVE

It is a graphic method of studying the dispersion in a distribution. A very distinctive feature of this curve consists in dealing with the cumulative values of the variable and the cumulative frequencies rather than its absolute values and the given frequencies. The technique of drawing the curve consists of the following steps:

(a) The size of items as well as frequencies are first cumulated.

(b) Taking the final cumulated figure as 100, percentages are calculated for all cumulative values.

(c) On the X-axis begin from 100 to 0 and let it represent frequencies.

(d) On the Y-axis begin from 0 to 100 and let it represent values.

(e) Draw a line from 0 of the X-axis to the 100 of Y-axis. This is the line of equal distribution.

(f) The curve is drawn taking various points of X and Y. If the distribution is not proportionately equal, the curve would be away from the line of equal distribution. The farther is the curve from the line of equal distribution, the greater is the variability in the series.

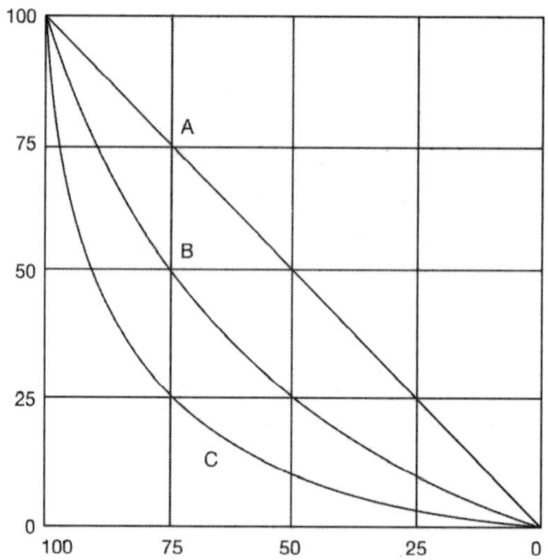

GINNI'S COEFFICIENT OF CONCENTRATION

Ginni's coefficient of concentration, also known as concentration ratio, is based on his coefficient of mean difference and is defined as:

$$G = \Delta/2\overline{y}$$

where, G = Ginni's coefficient of concentration,

Δ = Ginni's coefficient of mean difference, and

$\bar{y}$ = mean of the given distribution.

Ginni's coefficient of concentration is a pure number independent of units of measurement and lies between 0 and 1. In case of equal distribution, G is equal to 0 and its values goes up with increase in the inequality. In fact, it is the ratio of area of concentration to the total area of lower triangle below line of equal distribution.

MODULUS

It is the square root of twice the second moment of dispersion about the mean and is given by

$$C = \sqrt{2fd^2/N}$$

where, $d = y - \bar{y}$

Hence, modulus is equal to standard deviation multiplied by the square root of 2, *i.e.*,

$$C = \sqrt{2} \times \sigma$$

Like standard deviation, this measure is also based on the second moment about the mean.

PRECISION

It is the reciprocal of modulus, *i.e.*,

$$\text{Precision} = \frac{1}{\sqrt{2fd^2/N}} = \frac{1}{\sqrt{2} \times \sigma}$$

PROBABLE ERROR

It is equal to 0.67 × standard deviation.

RELATIONS BETWEEN VARIOUS MEASURES OF DISPERSION

In case of a normal distribution, the relationship between quartile deviation, mean deviation and standard deviation is fixed. The reason is that in a normal distribution, median, mode and arithmetic mean have the same value, and the two quartiles are equidistant from the median. We have the following relations between the different measures of dispersion:

(a) Mean ± Q.D. covers 50% of the observations of the distribution.

(b) Mean ± M.D. covers 57.5% of the observations of the distribution.

(c) Mean ± σ covers 68.27% of the observations of the distribution.

(d) Mean ± 2σ covers 95.45% of the observations of the distribution.

(e) Mean $\pm$ 3 σ covers 99.73% of the observations of the distribution.

(f) $Q.D. = 0.6745\sigma = (2/3)\sigma$

(g) $M.D. = \sqrt{2/\pi} \times \sigma = 0.7979\sigma = (4/5)\sigma$

(h) $Q.D. = 0.8459\, M.D. = (5/6)\, M.D.$

(i) $Q.D. : M.D. : S.D. :: (2/3)\sigma : (4/5)\sigma : \sigma \quad \Rightarrow \quad Q.D. : M.D. : S.D. :: 10 : 12 : 15$

(j) Range $= 6\, S.D. = 6\sigma$.

MOMENTS, SKEWNESS AND KURTOSIS

In the previous chapters we have studied the statistical measures which reveal clearly the salient features of a frequency distribution. The measures of central tendency show the concentration of the observations about the middle of the distribution and the measures of dispersion give an idea about the spread or scatter of the observations about some measure of central tendency. These measures of central tendency and dispersion do not reveal whether the dispersal of values on either side of an average is symmetrical or not; and the equal distance on either side of the average contains equal number of values or not.

Thus, the two measures, measures of central tendency and measures of dispersion, are inadequate to characterise a distribution completely and they must be supported and supplemented by two more measures—*skewness*, which helps us to study the shape, *i.e.*, symmetry or asymmetry of the distribution, and *kurtosis*, which refers to the flatness or peakedness of the curve which can be drawn with the help of the given data. All the four measures, *viz.*, central tendency, dispersion, skewness and kurtosis, taken together are sufficient to describe a frequency distribution completely.

Moments: It is a mechanical term for the measure of a force with reference to its tendency to produce rotation. The strength of this tendency depends on the amount of the force and the distance from the origin of the point at which the force is exerted. Thus, it involves two factors—the quantum of the force and the distance from which it is applied.

If a number of forces represented by $f_1, f_2, f_3, \ldots$ are applied with distances of $y_1, y_2, y_3, \ldots$ respectively, then f_1y_1 is the moment of the first force, f_2y_2 of the second force, f_3y_3 of the third force and so on. If these moments are added, we get a value Sfy which is the total moment. If total moment, Sfy, is divided by the total of the force of Sf, we get a value $\Sigma fy/\Sigma f$ which is called a moment.

However, the term moment as used in physics has nothing to do with the moment used in statistics, the only analogy being that in statistics we talk of moment of random variable about some point and these moments are used to describe the various characteristics of a frequency distribution *viz.*, central tendency, dispersion, skewness and kurtosis. In statistics, the term 'moment' is used with reference to frequencies and the respective values against them. The frequencies are treated as forces and the values as distances.

Skewness: A frequency distribution is said to be 'symmetrical', if the frequencies are symmetrically distributed about mean, *i.e.*, when values of the variable equidistant from mean have equal frequencies.

The word 'skewness' is used to denote the extent of asymmetry in the data. When the frequency distribution is not symmetrical, it is said to be 'skew'. The word 'skewness' literally denotes 'asymmetry', or 'lack of symmetry', and 'skew' denotes 'asymmetrical'. A symmetrical distribution has therefore zero skewness. Some of the definitions of skewness are:

Riggleman and Frisbee—Skewness is lack of symmetry. When a frequency distribution is plotted on a chart, skewness present in the items tends to the disperse the chart more on one side of the mean than on the other.

Croxton and Cowden—When a series is not symmetrical, it is said to be asymmetrical or skewed.

Garren—A distribution is said to be 'skewed' when the mean and the median fall at different points in the distribution, and balance (or centre of gravity) is shifted to one side or the other.

Simpson and Kafka—Measures of skewness tell us the direction and the extent of skewness. In symmetrical distribution, the mean, median and mode are identical. The more the mean moves away from the mode, the larger the asymmetry or skewness.

Skewness may be positive, zero or negative.

(a) In a symmetrical distribution, Mean (μ) = Median (M) = Mode (Z), and hence skewness is zero.

(b) In a positively skewed distribution, Mean (μ) > Median (M) > Mode (Z).

(c) In a negatively skewed distribution, Mean (μ) < Median (M) < Mode (Z).

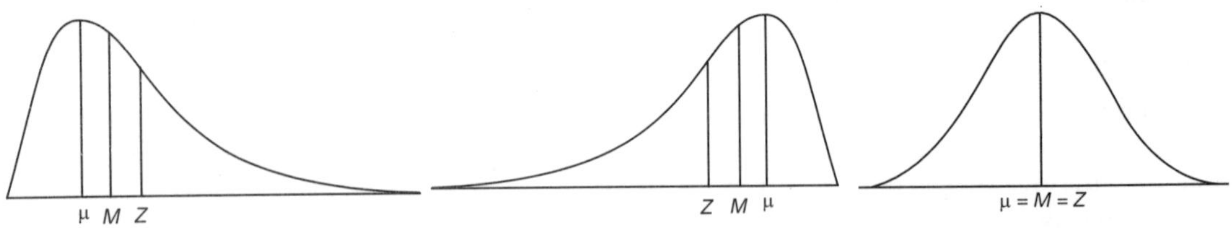

Positive Skewness Negative Skewness Zero Skewness

Kurtosis: Kurtosis refers to the degree of 'peakedness' of the frequency curve. Two distributions may have the same average, dispersion and skewness; yet, in one there may be high concentration of values near the mode, showing a sharper peak in the frequency curve than in the other. This characteristic of the frequency distribution is known as kurtosis and it indicates whether the distribution, if plotted on a graph would give us a normal curve, a curve more flat than the normal curve or a curve more peaked than the normal curve. Some of the definitions of kurtosis are:

Simpson and Kafka—The degree of Kurtosis of a distribution is measured relative to the peak of a normal curve.

Croxton and Cowden—A measure of Kurtosis indicates the degree to which a curve of a frequency distribution, is peaked or flat topped.

If a distribution is more peaked than the normal distribution, it is called *Leptokurtic*. If the distribution is more flat than the normal distribution, it is called *Platykurtic*. The normal distribution is known as *Mesokurtic*. All these three types of curve can be represented by the following diagram.

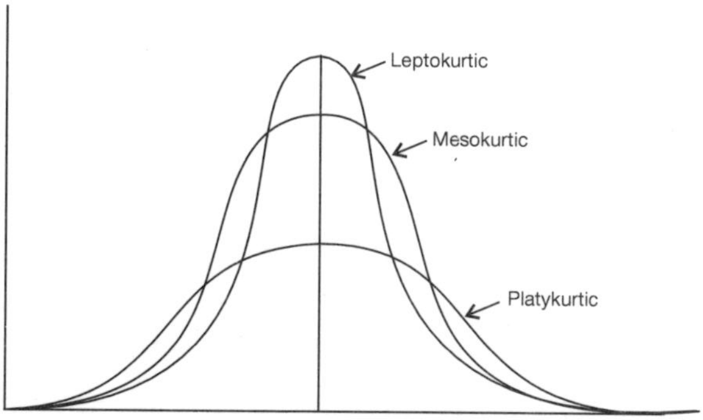

MOMENTS

Given n observations $y_1, y_2, ..., y_n$, and an arbitrary constant A,

$$\frac{1}{n}\Sigma(y_i - A) \text{ is called the } \textit{1st moment about A},$$

$$\frac{1}{n}\Sigma(y_i - A)^2 \text{ is called the } \textit{2nd moment about A},$$

$$\frac{1}{n}\Sigma(y_i - A)^3 \text{ is called the } \textit{3rd moment about A},$$

and so on. Let us denote these moments successively by m'_1, m'_2, m'_3, *etc.* (Sometimes, these are also represented by μ'_1, μ'_2, μ'_3, *etc.*)

Then, $m'_1 = \Sigma(y_i - A) / n = (\Sigma y_i - \Sigma A) / n = (\Sigma y_i - nA) / n = \bar{y} - A$

i.e., the 1st moment about A equals $(\bar{y} - A)$.

MOMENTS ABOUT ZERO (OR RAW MOMENTS)

$$\text{1st moment about zero} = \frac{1}{n}\Sigma y_i = \bar{y}$$

$$\text{2nd moment about zero} = \frac{1}{n}\Sigma y_i^2$$

$$\text{3rd moment about zero} = \frac{1}{n}\Sigma y_i^3$$

and so on. Note that the 1st moment about zero is the mean $\bar{y}$, $m'_1 = \bar{y}$.

MOMENTS ABOUT MEAN (OR CENTRAL MOMENTS)

$$\text{1st moment about mean} = \frac{1}{n}\Sigma(y_i - \bar{y}) = 0$$

$$\text{2nd moment about mean} = \frac{1}{n}\Sigma(y_i - \bar{y})^2 = \sigma^2$$

$$\text{3rd moment about mean} = \frac{1}{n}\Sigma(y_i - \bar{y})^3$$

$$\text{4th moment about mean} = \frac{1}{n}\Sigma(y_i - \bar{y})^4$$

and so on. These are usually denoted by m_1, m_2, m_3, m_4, *etc.* (Sometimes, these are represented by μ_1, μ_2, μ_3, μ_4 *etc.*). Note that the 1st central moment is always zero, and the 2nd central moment is the variance σ^2, *i.e.*, $m_1 = 0$ and $m_2 = \sigma^2$.

From the second relation, we find that the standard deviation is the square-root of the second central moment m_2.

The 3rd central moment m_3 is used to measure *skewness* and the 4th central moment m_4 to measure *kurtosis*. Higher order moments m_5, m_6 *etc.*, are seldom used.

In general, given n observations $y_1, y_2, ..., y_n$ the r-th order, moments ($r = 0, 1, 2, ...$) are defined as follows:

$$r\text{-th moment about } A: \quad m'_r = \frac{1}{n}\Sigma(y_i - A)^r$$

$$r\text{-th raw moment:} \quad m'_r = \frac{1}{n}\Sigma y_i^r$$

$$r\text{-th central moment:} \quad m_r = \frac{1}{n}\Sigma(y_i - \bar{y})^r$$

For a frequency distribution,

$$r\text{-th moment about } A: \quad m'_r = \frac{1}{N}\Sigma f_i(y_i - A)^r$$

$$r\text{-th raw moment:} \quad m'_r = \frac{1}{N}\Sigma f_i y_i^r$$

$$r\text{-th central moment:} \quad m_r = \frac{1}{N}\Sigma f_i(y_i - \bar{y})^r$$

where, $N = \Sigma f_i$.

Note: Moments about mean are written without dashes ('), but moments about any other origin, *i.e.*, non-cential moments, with dashes.

There are important relations between central and non-central moments. For example, if the non-central moments (m'_1, m'_2, m'_3 *etc.*) about any arbitrary origin A are known, the central moments can be obtained as:

$$m_2 = m'_2 - m'^2_1$$

$$m_3 = m'_3 - 3m'_2m'_1 + 2m'^3_1$$

$$m_4 = m'_4 - 4m'_3m'_1 + 6m'_2m'^2_1 - 3m'^4_1$$

In particular, using the first two moments, m'_1 and m'_2, about an arbitrary origin A, the mean and the variance may be obtained as:

$$\bar{y} = m'_1 + A \quad \text{and} \quad \sigma^2 = m'_2 - m'^2_1$$

BETA-COEFFICIENTS

Beta-coefficients are defined as follows:

$$\beta_1 = m_3^2/m_2^3 \text{ and } \beta_2 = m_4/m_2^2$$

The beta-coefficients can never be negative, and are pure numbers, independent of origin and scale of observations.

Since, for a symmetrical distribution all odd order moments are zero, $m_3 = 0$, consequently $\beta_1 = 0$.

Frequency distributions are classified as leptokurtic, platykurtic or mesokurtic, according as the value of β_2 is greater than, less than or equal to 3. Beta-coefficients are used for measuring skewness and kurtosis.

GAMMA-COEFFICIENTS

Gamma-coefficients are defined as follows:

$$\gamma_1 = \sqrt{\beta_1} \text{ and } \gamma_2 = \beta_2 - 3$$

γ_1 must have the same sign as m_3. The gamma-coefficients may be positive, negative, or zero; but are pure numbers like the beta-coefficients. These are used as measure of skewness and kurtosis as:

$$\text{Skewness } (\gamma_1) = \sqrt{\beta_1} = m_3/\sigma^2 \text{ and Kurtosis } (\gamma_2) = \beta_2 - 3 = m_4/\sigma^4 - 3$$

Distributions are said to be 'positively skew', 'negatively skew', or 'symmetrical', according as γ_1 is positive, negative, or zero. Similarly, positive, negative, or zero values of γ_2 are associated with 'leptokurtic', 'platykurtic', or 'mesokurtic' distributions.

STANDARDISED VARIABLE

In statistical theories, it is often found convenient to express deviations from mean ($\bar{y}$) in terms of standard deviation (σ) as unit.

$$z = \frac{y - \bar{y}}{\sigma} = \frac{Variable - Mean}{S.D.}$$

is called the standardised variable or standardised form of y.

The standardised variable z is independent of origin and unit of measurement. Mean of the standardised variable z is zero, and its S.D. is 1, *i.e.*, $\bar{z} = 0$ and $\sigma_z = 1$.

Since, the mean is zero, the *rth* central moment of z is the same as the *rth* raw moment of z, and is usually denoted by α_r. Thus, $\alpha_r = m_r / \sigma^r$.

Note: $\alpha_1 = 0,\ \alpha_2 = 1,\ \alpha_3 = \sqrt{\beta_1} = \gamma_1,\ \alpha_4 = \beta_2 = \gamma_2 + 3$.

CHARLIER'S CHECK

C.V.L. Charlier gave a simple and effective check on the calculations for moments from a grouped frequency distribution. Expanding $(y + 1)^4$ by binomial theorem, multiplying by f, and then summing, we have the identity:

$$\Sigma f(y + 1)^4 = \Sigma fy^4 + 4\Sigma fy^3 + 6\Sigma fy^2 + 4\Sigma fy + N$$

Along with the columns for fy, fy^2, fy^3, fy^4 needed for the calculation of moments upto the 4*th* order, an extra column for $f(y + 1)^4$ is also calculated. If it is found that the numerical values obtained from the table satisfy the above identity, then we are sure that the calculations are correct.

SHEPPARD'S CORRECTION FOR ERRORS DUE TO GROUPING

For the calculation of mean, variance and other moments from a grouped frequency distribution, the mid-value of a class interval is taken as representative of all observations falling within that class interval. However, this introduces some error in the calculated values, known as '*error due to grouping*'.

Under certain conditions it is possible to make corrections for these discrepancies by using certain formulae, known as Sheppard's corrections for errors due to grouping. So far as the first four moments are concerned:

(a) No correction is necessary for the mean m_1' and the third central moment m_3.

(b) Corrections for the 2*nd* and the 4*th* central moments are:

$$m_2 \text{ (corrected)} = m_2 \text{ (uncorrected)} - c^2/12$$

$$m_4 \text{ (corrected)} = m_4 \text{ (uncorrected)} - (c^2/2)m_2 \text{ (uncorrected)} + 7c^4/240$$

where c is the common width of class intervals.

Sheppard's corrections should be applied only when:

(a) the distribution relates to a continuous variable, and is of moderate asymmetry;

(b) the frequency becomes smaller and smaller approaching zero at each end of the distribution.

TESTS FOR SKEWNESS

The following tests are applied in order to find out whether a particular distribution is skewed.

(a) In a skew distribution, values of mean, median and mode would not coincide. The mean and mode would be pulled wide apart and median would usually lie between them. For a moderately asymmetrical distribution,

$$Mean - Mode = 3(Mean - Median)$$

$$\Rightarrow \quad 3Median = 3Mean + Mode - Mean = 2Mean + Mode$$

$$= 2Mean + Mode + 2Mode - 2Mode = 3Mode + 2(Mean - Mode)$$

$$\Rightarrow \quad Median = Mode + (2/3)(Mean - Mode)$$

(b) In a skewed distribution, the two quartiles would not be equidistant from the median or, in other words, $(Q_3 - M) - (M - Q_1)$ would not be zero.

(c) A skewed distribution, when plotted on a graph paper, would not give a symmetrical bell-shaped curve.

(d) The sum of positive deviations from the medians is not equal to the sum of negative deviations.

(e) Frequencies are not equally distributed at various points which are equidistant from the mode. In a moderately asymmetrical distribution,

$$Mean - Median = 3(Mean - Mode)$$

on either side.

MEASURES OF SKEWNESS

Skewness is measured by the following formulae:

(a) *Pearson's First Measure*—

$$Skewness = (Mean - Mode)/\text{Standard Deviation}$$

(b) *Pearson's Second Measure*—

$$Skewness = 3(Mean - Median)/\text{Standard Deviation}$$

(c) *Bowley's Measure*—

$$Skewness = \frac{(Q_3 - Q_2) - (Q_2 - Q_1)}{(Q_3 - Q_2) + (Q_2 - Q_1)} = \frac{Q_3 - 2Q_2 + Q_1}{Q_3 - Q_1}$$

where, Q_1, Q_2, Q_3 denote the first, second and third quartiles of the distribution.

(d) *Kelly's Measure*—

$$Skewness = \frac{D_9 + D_1 - 2Median}{D_9 - D_1} = \frac{P_{90} + P_{10} - 2Median}{P_{90} - P_{10}}$$

where, D is the deciles and P is the percentiles.

(e) *Measures of Skewness based on Moments—*

$$\text{Skewness } (\gamma_1) = m_3/\sigma^3 = m_3/(\sqrt{m_2})^3$$

where, m_2 and m_3 are the second and third central moments, and σ denotes the S.D.

Note: All the measures of skewness are pure numbers which do not depend upon the units of measurement, and have the value zero when the distribution is symmetrical.

MEASURES OF KURTOSIS

Kurtosis is measured by co-efficient β_2 or its derivation γ_2.

$$\beta_2 = m_4/m_2^2 = m_4/\sigma^4$$

$$\gamma_2 = \left(m_4/\sigma^4\right) - 3 = \beta_2 - 3$$

where, m_4 and σ denote the fourth central moment and S.D. respectively.

Note: m_4/σ^4 is the fourth moment α_4 of the standardised variable z, *i.e.*, $\alpha_4 = m_4/\sigma^4 = \beta_2$.

For Platykurtic distribution, $\beta_2 < 3$.

For Mesokurtic distribution, $\beta_2 = 3$.

For Leptokurtic distribution, $\beta_2 > 3$.

For the normal distribution, which is neither very peaked nor flat-topped, $\beta_2 = 3$. This important distribution is taken as a standard for measuring kurtosis, and it has become customary to use $\gamma_2 = \beta_2 - 3$ as a measure of kurtosis.

Thus, a distribution is said to be 'Platykurtic', when γ_2 is negative; it is said to be 'Mesokurtic' when $\gamma_2 = 0$, and 'Leptokurtic' when γ_2 is positive.

CURVE FITTING AND METHOD OF LEAST SQUARES

Very often there exists a relation between available observations in respect of two variables. For instance, expenditure depends on income, weight and height of a person are interdependent, agricultural production depends on the amount of rainfall, price of a commodity depends on demand of that commodity, *etc*. These relationship between variables can be expressed by means of some mathematical equation, representing a certain geometrical curve. The process of finding such a curve or its equation on the basis of a given set of observations is called *curve fitting*.

In other words, suppose $(x_1, y_1), (x_2, y_2), (x_3, y_3), \ldots, (x_n, y_n)$ are n paired-values of two variables x and y. Again, suppose that x is independent variable and y is dependent variable; and y is a function of x. Then, the device of finding a functional relationship of the form $y = f(x)$ between two variables x and y on the basis of a finite set of pairs of observations $(x_1, y_1), (x_2, y_2), (x_3, y_3), \ldots, (x_n, y_n)$ on these variables, is called curve fitting. Following are equations of some common types of curves:

Straight line:	$y = a + bx$
Parabola:	$y = a + bx + cx^2$
Cubic curve:	$y = a+bx+cx^2+dx^3$
Polynomial of n-th degree:	$y = a+bx+cx^2+\ldots+px^n$
Exponential curve:	$y = ab^x$
Modified exponential curve:	$y = a+bc^x$
Logistic curve:	$1/y = a+bc^x$
Gompertz curve:	$\log y = a + bc^x$
Geometric curve:	$y = ax^b$

The following rules help to select the appropriate equation to be fitted to a given set of observations:

(a) Plot the corresponding observations of x and y as points on a graph paper. If the pattern of points shows approximately a linear path, use the straight line curve.

(b) If $\log y$ when plotted against x shows a linear path, the exponential curve is used.

(c) If $\log y$ when plotted against $\log x$ shows a linear path, the geometric curve is used.

Curve fitting is very important tool both from the point of view of theoretical and practical statistics. In theoretical statistics, the study of regression and correlation can be regarded as fitting of linear curves (such as straight lines, planes, hyper-planes) to the given bivariate or multivariate frequency or probability distributions. In practical statistics, it enables us to get a close functional relation between the two variables x and y. In general, the relation is expressed by a polynomial, but other types of relationship, algebraic, exponential or logarithmic can also be fitted by using the principle of least squares.

Method of Least Squares: It is a device for finding the equation of a specified type of curve, which best fits a given set of observations. In other words, to find the values for a set of unknown quantities so that a set of given equations may be satisfied as nearly as possible, when the number of equations is greater than the number of unknowns and the equations not strictly compatible with each other, Legendre suggested a method, known as *Legendre Principle* or the *method of least squares*. Gauss connected the method of least squares with the theory of probability and placed the method on a logical foundation.

Let there be a number of independent linear equations in n unknowns $x_1, x_2, x_3, \ldots, x_n$:

$$\left.\begin{array}{l} a_{11}x_1 + a_{12}x_2 + a_{13}x_3 + \ldots + a_{1n}x_n = b_1 \\ a_{21}x_1 + a_{22}x_2 + a_{23}x_3 + \ldots + a_{2n}x_n = b_2 \\ \ldots\ldots\ldots\ldots\ldots\ldots\ldots\ldots\ldots\ldots\ldots \\ a_{m1}x_1 + a_{m2}x_2 + a_{m3}x_3 + \ldots + a_{mn}x_n = b_m \end{array}\right\} \quad i.e., \sum_{j=1}^{n} a_{ij}x_j = b_i,\ i = 1, 2, 3, \ldots, m$$

where a's and b's are constants.

If $m = n$, we have a unique set of values satisfying the given system of equations. If $m > n$, *i.e.*, if the number of equations is greater than the number of unknowns, no such solution exists. Therefore, we try to find those values of $x_1, x_2, x_3, \ldots, x_n$ which satisfy the given system of equations as nearly as possible.

The principle of least squares states that these values are those which make sum of squares (S) a minimum, where

$$S = \sum_{i=1}^{m} (a_{i1}x_1 + a_{i2}x_2 + a_{i3}x_3 + \ldots + a_{in}x_n - b_i)^2 = \sum_{i=1}^{m} E_i^2$$

where, $$E_i = a_{i1}x_1 + a_{i2}x_2 + a_{i3}x_3 + \ldots + a_{in}x_n - b_i$$

The extreme values of the function $f = f(x_1, x_2, x_3, \ldots, x_n)$ are given by $\partial f/\partial x_1 = \partial f/\partial x_2 = \partial f/\partial x_3 = \ldots = \partial f/\partial x_n = 0$, provided that the partial derivatives exist. Applying this concept, S will have a maximum or minimum, for those values of the unknowns x_j ($j = 1, 2, 3, \ldots, n$) which satisfy the following n equations:

$$\partial S/\partial x_j = 0, \qquad j = 1, 2, 3, \ldots, n$$

i.e., $$\sum_{i=1}^{m} a_{i1}E_i = 0,\ \sum_{i=1}^{m} a_{i2}E_i = 0,\ \sum_{i=1}^{m} a_{i3}E_i = 0,\ \ldots,\ \sum_{i=1}^{m} a_{in}E_i = 0$$

These n equations are called the normal equations and can be solved like simultaneous equations for the n unknowns $x_1, x_2, x_3, \ldots, x_n$. If the common rank of the matrix of coefficients and the augmented matrix is n, we can get a unique solution for $x_1, x_2, x_3, \ldots, x_n$.

These solutions obtained from the normal equations make the second order partial derivatives of S positive and so S is minimum.

FITTING STRAIGHT LINE

Let us consider an equation of straight line

$$y = a + bx$$

to be fitted to a given set of n pairs of observations $(x_1, y_1), (x_2, y_2), (x_3, y_3), ..., (x_n, y_n)$. Applying the method of least squares, the values of a and b are so determined as to minimise $\sum_{i=1}^{n}(y_i - a - bx_i)^2$. This leads to two normal equations:

$$\Sigma y = an + b\Sigma x \text{ and } \Sigma xy = a\Sigma x + b\Sigma x^2$$

The values of n, Σx, Σy, Σx^2 and Σxy are substituted on the basis of the given data. We have then two equations involving a and b, solving which the values of a and b are obtained.

FITTING PARABOLA

Let us consider an equation of parabola

$$y = a + bx + cx^2$$

to be fitted to a given set of n pairs of observations $(x_1, y_1), (x_2, y_2), (x_3, y_3), ..., (x_n, y_n)$. Applying the method of least squares, the values of a, b, and c are so determined as to minimise $\sum_{i=1}^{n}(y_i - a - bx_i - cx_i^2)^2$. This leads to three normal equations:

$$\Sigma y = an + b\Sigma x + c\Sigma x^2; \quad \Sigma xy = a\Sigma x + b\Sigma x^2 + c\Sigma x^3 \text{ and}$$

$$\Sigma x^2 y = a\Sigma x^2 + b\Sigma x^3 + c\Sigma x^4$$

The values of n, Σx, Σy, Σx^2, Σx^3, Σx^4, Σxy, and $\Sigma x^2 y$ are substituted on the basis of the given data. We have then three equations involving a, b and c, solving which the values of a, b and c are obtained.

FITTING EXPONENTIAL AND GEOMETRIC CURVE

To fit curves with equations of the form $y = ab^x$ and $y = ax^b$, the procedure is to take logarithms of both sides and then form normal equations.

Exponential Curve: To fit the exponential curve of the form $y = ab^x$, we take logarithms of both sides and get

$$\log y = \log a + x \log b$$

which can be written as: $Y = A + Bx$,

where, $Y = \log y$, $A = \log a$, and $B = \log b$.

The normal equations are:

$$\Sigma Y = nA + B\Sigma x \text{ and } \Sigma xY = A\Sigma x + B\Sigma x^2$$

These equations are solved for A and B, and then taking antilog, we find the values of a and b.

Geometric Curve: Similarly, the geometric curve of the form $y = ax^b$ may be written as

$$\log y = \log a + b \log x \quad \Rightarrow \quad Y = A + bX$$

where, $\quad Y = \log y$, $A = \log a$, and $X = \log x$.

The normal equations are:

$$\Sigma Y = nA + b\Sigma X \text{ and } \Sigma XY = A\Sigma X + b\Sigma X^2$$

These equations are solved for A and b, and then a = antilog A is obtained.

FITTING OF LOGISTIC CURVE

Let us consider the logistic curve of the form: $y = a/(1+be^{cx})$.

This is an S-shaped curve, as $x \to \infty$, $y \to a$. Thus, a can be estimated from the data since it is the value of y when x is reasonably large.

The given logistic curve can be written as:

$$1/y = 1/a + (b/a)e^{cx} \Rightarrow (1/y) - (1/a) = (b/a)(e^c)^x \Rightarrow Y = AB^x$$

where, $\quad Y = (1/y) - (1/a)$, $A = b/a$, and $B = e^c$.

Since, our equation has been converted to exponential curve, therefore now proceed using the method of exponential curve fitting.

Knowing x, we can plot $\log y$ against x and use the method of least squares to estimate A and B and therefore to estimate b and c.

FITTING OF CURVE THROUGH THE ORIGIN

Given a set of n paired values (x_1, y_1), (x_2, y_2), (x_3, y_3), ..., (x_n, y_n), the curve to be fitted is passing through origin, *i.e.*, $(0, 0)$ is a point on the curve; then the straight line to be fitted is simply, $y = ax$ and the quadratic parabola is $y = ax + bx^2$.

FITTING A CURVE OF THE FORM $x(y - a) = b$

$$x(y - a) = b \quad \Rightarrow \quad y = a + b/x, \quad \text{taking } x \text{ as independent variable.}$$

Sum of squares of residuals is: $R = \Sigma(y - a - b/x)^2$

$$\partial R/\partial a = 0 \quad \Rightarrow \quad \Sigma y = na + b\Sigma(1/x)$$

$$\partial R/\partial b = 0 \quad \Rightarrow \quad \Sigma(y/x) = a\Sigma(1/x) + b\Sigma(1/x^2)$$

Substituting the values of n, Σy, $\Sigma(1/x)$, $\Sigma(1/x^2)$, and $\Sigma(y/x)$ in the above equations we can find the value of a and b.

FITTING A CURVE OF THE FORM y = ax + b/x

Let the set of n paired values be (x_1, y_1), (x_2, y_2), (x_3, y_3), ..., (x_n, y_n), or (x_i, y_i), $i = 1, 2, 3, ..., n$. Then, the sum of the squares of residuals is

$$R = \sum_{i=1}^{n} (y_i - ax_i - b/x_i)^2$$

Differentiating R w.r.t. a and b separately and setting these partial derivatives equal to zero, we get the normal equations.

$$\partial R/\partial a = 0 = -2\sum_{i=1}^{n} x_i (y_i - ax_i - b/x_i)$$

$$\Rightarrow \qquad \sum_{i=1}^{n} x_i y_i = a\sum_{i=1}^{n} x_i^2 + \sum_{i=1}^{n} b = a\sum_{i=1}^{n} x_i^2 + nb$$

$$\partial R/\partial b = 0 = -2\sum_{i=1}^{n} (1/x_i)(y_i - ax_i - b/x_i)$$

$$\Rightarrow \qquad \sum_{i=1}^{n} (y_i/x_i) = \sum_{i=1}^{n} a + b\sum_{i=1}^{n} (1/x_i^2) = na + b\sum_{i=1}^{n} (1/x_i^2)$$

Substituting the values and solving the above equations, we get the values of a and b.

FITTING A CURVE OF THE FORM y = ax² + b/x

Let the set of n paired values be (x_1, y_1), (x_2, y_2), (x_3, y_3), ..., (x_n, y_n), or (x_i, y_i), $i = 1, 2, 3, ..., n$. Then, the sum of the squares of residuals is

$$R = \sum_{i=1}^{n} \left(y_i - ax_i^2 - b/x_i\right)^2$$

Differentiating R with respect to a and b respectively and setting these partial derivatives equal to zero, we get the normal equations:

$$\partial R/\partial a = 0 = -2\sum_{i=1}^{n} x_i^2 (y_i - ax_i^2 - b/x_i)$$

$$\Rightarrow \qquad \sum_{i=1}^{n} x_i^2 y_i = a\sum_{i=1}^{n} x_i^4 + b\sum_{i=1}^{n} x_i$$

$$\partial R/\partial b = 0 = -2\sum_{i=1}^{n} (1/x_i)(y_i - ax_i^2 - b/x_i)$$

$$\Rightarrow \qquad \sum_{i=1}^{n} (y_i/x_i) = a\sum_{i=1}^{n} x_i + b\sum_{i=1}^{n} (1/x_i^2)$$

Substituting the values and solving the above equations, we get the values of a and b.

FITTING A CURVE OF THE FORM $y = ae^{bx}$

Taking logarithm on both sides of the equation $y = ae^{bx}$, we get

$$\log_{10} y = \log_{10} a + bx \log_{10} e \quad \Rightarrow \quad Y = A + Bx$$

where, $Y = \log_{10} y$, $A = \log_{10} a$ and $B = b \log_{10} e$.

This is a polynomial of degree 1 (*i.e.*, a straight line). For n pairs of values (x_i, y_i), $i = 1, 2, 3, \ldots, n$ the usual normal equations are:

$$\Sigma Y_i = nA + B\Sigma x_i \quad \text{and} \quad \Sigma x_i Y_i = A\Sigma x_i + b\Sigma x_i^2$$

These equations are solved for A and B and then a and b are obtained by the relations:

$$a = \text{antilog } A \text{ and } b = B/\log_{10} e = B/0.4343 = 2.3026B.$$

CHANGE OF ORIGIN AND SCALE

(a) If we change the origin of x only, *i.e.*, write $X = x - c$, where c is an arbitrary constant, then x is replaced by X and we get

$$\Sigma y = an + b\Sigma X \quad \text{and} \quad \Sigma Xy = a\Sigma X + b\Sigma X^2$$

(b) If we change the origins of both x and y, *i.e.*, write $X = x - c$ and $Y = y - c'$, where c and c' are arbitrary constants, then x and y are replaced by X and Y and we get

$$\Sigma Y = an + b\Sigma X \quad \text{and} \quad \Sigma XY = a\Sigma X + b\Sigma X^2$$

(c) In case the successive values of the independent variable x are found to have a common difference, two special transformations are available for the cases, when (i) n is odd, and (ii) n is even.

(i) When n is odd, we write $u = \dfrac{x - \textit{central value of } x}{\textit{common difference}}$

The values of u will be 0 at the middle; 1, 2, 3, 4, ... for the successive values; and –1, –2, –3, –4, ... for the preceding values.

(ii) When n is even, we get $u = \dfrac{x - \textit{mean of two central values of } x}{(1/2) \times \textit{common difference}}$

The values of u will be successively 1, 3, 5, 7, ... for the lower half, and –1, –3, –5, –7, ... for the upper half, starting from the middle.

If we write the equation of straight line in the form $y = a + bu$, the normal equations will be

$$\Sigma y = an + b\Sigma u \quad \text{and} \quad \Sigma uy = a\Sigma u + b\Sigma u^2$$

But, since by the above transformation $\Sigma u = 0$, these equations take very simple forms:

$$\Sigma y = an \quad \text{and} \quad \Sigma uy = b\Sigma u^2$$

from which the values of a and b can easily determined.

However, in all cases, finally we have to rewrite the transformed variable X, Y, u in terms of the original variables x and y, and then simplify the result to get the equation of the fitted straight line.

CORRELATION ANALYSIS

The word "correlation" is used to denote the degree of association between variables. Two variables x and y are said to be correlated if they are so related that variations in the magnitude of one variable tend to be accompanied by variations in the magnitude of the other variable. If y tends to increase as x increases, the variables are said to be positively correlated and if y tends to decrease as x increases, the variables are negatively correlated. If the values of y are not affected by changes in the values of x, the variables are said to be uncorrelated.

The association between variables may be *linear* or *non-linear*. If the amount of change in one variable tends to bear a constant ratio to the amount of change in the other variable, then correlation is said to be linear and it is measured by *correlation coefficient*, represented by r.

The correlation coefficient (r) measures the strength of the linear association between two quantitative variables. It does not distinguish explanatory from response variables and is not affected by changes in the unit of measurement of either or both variables. To interpret the numerical value of the correlation coefficient (r), following points should be noted:

(a) The value of r always falls between –1 and 1. Positive r indicates positive association between the variables, and negative r indicates negative association.

(b) The extreme values $r = -1$ and $r = 1$ occur only in the case of perfect linear association, when the points in a scatterplot lie exactly along a straight line. Values of r close to 1 or –1 indicate that the points lie close to a straight line.

(c) The value of r is not changed when the unit of measurement of x, y, or both changes. The correlation r has no unit of measurement; it is a dimensionless number between –1 and 1.

(d) Correlation measures only the strength of linear association between two variables. Curved relationships between variables, no matter how strong, need not be reflected in the correlation.

KARL PEARSON'S COEFFICIENT OF CORRELATION

When deviations are taken from actual mean:

$$r_{xy} = \frac{\Sigma xy}{n\sigma_x \sigma_y} = \frac{\Sigma xy}{n\sqrt{\frac{\Sigma x^2}{n}}\sqrt{\frac{\Sigma y^2}{n}}} = \frac{\Sigma xy}{\sqrt{\Sigma x^2 \times \Sigma y^2}}$$

where,

r_{xy} = Karl Pearson's coefficient of correlation,

$x = (X - \bar{X})$,

$y = (Y - \bar{Y})$,

n = number of pairs of observations,

σ_x = standard deviation of X series, and

σ_y = standard deviation of Y series.

When deviations are taken from assumed mean:

$$r_{xy} = \frac{\frac{\Sigma dxdy}{n} - \frac{\Sigma dx}{n}.\frac{\Sigma dy}{n}}{\sigma_x \sigma_y} = \frac{\Sigma dxdy - n\cdot\frac{\Sigma dx}{n}.\frac{\Sigma dx}{n}}{n\sigma_x \sigma_y}$$

or,

$$r_{xy} = \frac{n\Sigma dxdy - \Sigma dx \Sigma dy}{n^2 \sigma_x \sigma_y}$$

or,

$$r_{xy} = \frac{n\Sigma dxdy - \Sigma dx \Sigma dy}{n^2\sqrt{\frac{\Sigma dx^2}{n} - \left(\frac{\Sigma dx}{n}\right)^2}\sqrt{\frac{\Sigma dy^2}{n} - \left(\frac{\Sigma dy}{n}\right)^2}}$$

$$= \frac{n\Sigma dxdy - \Sigma dx \Sigma dy}{n\sqrt{n\Sigma x^2 - (\Sigma x)^2}\sqrt{n\Sigma y^2 - (\Sigma y)^2}}$$

where, $dx = (X - A)$, $dy = (Y - A)$, and A is the assumed mean.

In grouped data of bivariate distribution:

$$r_{xy} = \frac{\frac{\Sigma fdxdy}{n} - \left(\frac{\Sigma fdx}{n}\right)\left(\frac{\Sigma fdy}{n}\right)}{n\sqrt{\frac{\Sigma fdx^2}{n} - \left(\frac{\Sigma fdx}{n}\right)^2}\sqrt{\frac{\Sigma fdy^2}{n} - \left(\frac{\Sigma fdy}{n}\right)^2}}$$

or,

$$r_{xy} = \frac{\Sigma fdxdy - \frac{\Sigma fdx \cdot \Sigma fdy}{n}}{\sqrt{\Sigma fdx^2 - \frac{(\Sigma fdx)^2}{n}}\sqrt{\Sigma fdy^2 - \frac{(\Sigma fdy)^2}{n}}}$$

When actual value of x and y are taken and no deviations are found:

$$r_{XY} = \frac{N\Sigma XY - (\Sigma X)(\Sigma Y)}{\sqrt{N\Sigma X^2 - (\Sigma X)^2}\sqrt{N\Sigma Y^2 - (\Sigma Y)^2}}$$

$$= \frac{\frac{\Sigma XY}{n} - \frac{\Sigma X}{n}\cdot\frac{\Sigma Y}{n}}{\sqrt{\frac{\Sigma X^2}{n} - \left(\frac{\Sigma X}{n}\right)^2}\sqrt{\frac{\Sigma Y^2}{n} - \left(\frac{\Sigma Y}{n}\right)^2}}$$

SPEARMAN'S RANK CORRELATION COEFFICIENT

If ranks are not repeated:

$$r_s = 1 - \frac{6\Sigma d^2}{n(n^2 - 1)} \quad or \quad 1 - \frac{6\Sigma d^2}{(n^3 - n)}$$

If ranks are repeated:

$$r_s = 1 - \frac{6\left[\Sigma d^2 + \Sigma(m^3 - m)/12\right]}{n^3 - n}$$

where,

r_s = Spearman's rank correlation,
d = difference of rank,
n = number of pairs of observation, and
m = number of times the values are repeated or the ranks are common.

COEFFICIENT OF CONCURRENT DEVIATIONS

$$r_c = \pm\sqrt{\left|\left(\frac{2c - n}{n}\right)\right|}$$

where,

r_c = coefficient of correlation by the concurrent deviation method,
c = number of concurrent deviations or the number of positive signs, and
n = number of pairs of deviation compared, *i.e.*, $(N - 1)$.

COEFFICIENT OF CORRELATION BY THE METHOD OF LEAST SQUARES

$$r_{xy} = \frac{S_y}{\sigma_y}$$

where,

S_y = the standard deviations of linear or computed value of y, and

σ_y = the standard deviation of the original value of y.

COEFFICIENT OF CORRELATION BY COVARIANCE

$$r_{xy} = \frac{Covariance\ of\ x\ and\ y}{\sqrt{Variance\ of\ x}\ \sqrt{Variance\ of\ y}}$$

$$\text{Covariance} = \frac{\Sigma xy}{n}$$

where, x and y are deviations of the series from their mean.

PROBABLE ERROR OF THE COEFFICIENT OF CORRELATION

$$P.E_r. = 0.6745\frac{1-r^2}{\sqrt{n}} \text{ and } S.E_r. = \frac{1-r^2}{\sqrt{n}}$$

where,

$P.E_r.$ = probable error,

$S.E_r.$ = standard error,

r = correlation, and

n = number of pairs of observation.

Coefficient of Determination $= r^2$

Coefficient of Non-determination $k^2 = 1 - r^2$

Coefficient of Alienation $\sqrt{k^2} = \sqrt{1 - r^2}$

REGRESSION ANALYSIS

The word 'regression' is used to denote estimation or prediction of the average value of one variable for a specified value of the other variable and the regression analysis is mathematical measure of the average relationship between two or more variables, in terms of the original units of the data. The estimation is done by means of suitable equations, derived on the basis of available bivariate data. Such an equation is known as a regression equation and its geometrical representation is called a regression curve.

In regression analysis, there are two types of variables:

(a) 'predicted' or 'dependent' or 'regressed' or 'explained' variable—the value of which can be predicted, and

(b) 'independent' or 'predictor' or 'regressor' or 'explanator' variable—on the basis of the values of this variable, the values of the other variable are predicted.

In simple regression (or linear regression) the relationship between the variables is assumed to be linear. The estimate of x (say, x') is obtained from an equation of the form

$$x' - \overline{x} = b_{xy}(y - \overline{y}) \qquad \text{... (i)}$$

and the estimate of y (say, y') from another equation (usually different from the former) of the form

$$y' - \overline{y} = b_{yx}(x - \overline{x}) \qquad \text{... (ii)}$$

Equation (i) is known as regression equation of x on y, and equation (ii) as regression equation of y on x. The coefficient b_{xy} in equation (i) is called regression coefficient of x on y, and b_{yx} in equation (ii) is called regression coefficient of y on x. The geometrical representation of linear regression equations (i) and (ii) are known as regression lines. These lines are 'best fitting' straight lines obtained by the Method of Least Squares.

The regression coefficients b_{xy} and b_{yx} poses some important properties, which can be summarised as follows:

(a) The product of the two regression coefficients, b_{xy} and b_{yx}, is equal to the square of correlation coefficient, *i.e.*, $b_{xy} \cdot b_{yx} = r^2$. In other words, the correlation coefficient is the geometric mean between the regression coefficients.

(b) r, b_{xy}, and b_{yx}, all have the same sign. If the correlation coefficient r is zero, the regression coefficients b_{xy} and b_{yx} are also zero.

(c) One of the two regression coefficient must be numerically less than unity and the other must be numerically greater than unity or both will be unity or both will be numerically less than unity. Both coefficients cannot be greater than unity.

(d) The regression lines always intersect at the point $(\bar{x}, \bar{y})$. The slopes of the regression line of y on x and the regression line of x on y are respectively b_{yx} and $1/b_{xy}$.

(e) The regression coefficients are independent of the change of origin, but not of scale.

(f) The angle between the two regression lines depends on the correlation coefficient r. When $r = 0$, the two lines are perpendicular to each other; when $r = +1$, or $r = -1$, they coincide. As r increases numerically from 0 to 1, the angle between the regression lines diminishes from 90° to 0°.

(g) Arithmetic mean of the regression coefficients is greater than the correlation coefficient, provided the correlation coefficient is positive.

(h) The two regression equations are usually different. However, when $r = \pm 1$, they became identical; and in this case, there is an exact linear relationship between the variables. When $r = 0$, the regression equations reduce to $y = \bar{y}$ and $x = \bar{x}$, and neither y nor x can be estimated from linear regression equations.

	x on y	y on x
Regression Equations	$x = a + by$	$y = a + bx$
	$x - \bar{x} = b_{xy}(y - \bar{y})$	$y - \bar{y} = b_{yx}(x - \bar{x})$
Normal Equations	$\Sigma x = na + b\Sigma y$	$\Sigma y = na + b\Sigma x$
	$\Sigma xy = a\Sigma y + b\Sigma y^2$	$\Sigma xy = a\Sigma x + b\Sigma x^2$
Regression Coefficients	$b_{xy} = r\frac{\sigma_x}{\sigma_y}$	$b_{yx} = r\frac{\sigma_y}{\sigma_x}$
When deviations are taken	$r\frac{\sigma_x}{\sigma_y} = \frac{\Sigma xy}{\Sigma y^2}$	$r\frac{\sigma_y}{\sigma_x} = \frac{\Sigma xy}{\Sigma x^2}$
from Mean	$x = (X - \bar{X}); y = (Y - \bar{Y})$	
When deviations are	$r\frac{\sigma_x}{\sigma_y} = \frac{\Sigma dxdy - \frac{(\Sigma dx)(\Sigma dy)}{n}}{\Sigma dy^2 - \frac{(\Sigma dy)^2}{n}}$	$r\frac{\sigma_y}{\sigma_x} = \frac{\Sigma dxdy - \frac{(\Sigma dx)(\Sigma dy)}{n}}{\Sigma dx^2 - \frac{(\Sigma dx)^2}{n}}$
taken from Assumed Mean	$dx = (X - A); dy = (Y - A)$	
When the original values of	$r\frac{\sigma_x}{\sigma_y} = \frac{\Sigma XY - N\bar{X}\bar{Y}}{\Sigma Y^2 - N(\bar{Y})^2}$	$r\frac{\sigma_y}{\sigma_x} = \frac{\Sigma XY - N\bar{X}\bar{Y}}{\Sigma X^2 - N(\bar{X})^2}$
***X* and *Y* Series are taken**		
Standard Error of Estimate	$S_x = \sigma_x\sqrt{(1 - r^2)}$	$S_y = \sigma_y\sqrt{(1 - r^2)}$
Coefficient of Correlation	$r = \sqrt{b_{xy} \times b_{yx}}$	

REGRESSION CURVES

The conditional mean for a continuous bivariate distribution is given by

$$E(Y|X=x) = \int_{-\infty}^{+\infty} y f(x,y)dy \Big/ \int_{-\infty}^{+\infty} f(x,y)dy = \int_{-\infty}^{+\infty} y f(x,y)dy \Big/ f_X(x)$$

and is called the regression function of Y on X. And the curve

$$y = E(Y|X=x)$$

is called the regression curve of Y on X, or in other words the regression curve for the mean of Y.

Similarly, the regression function of X on Y is defined by

$$E(X|Y=y) = \int_{-\infty}^{+\infty} x f(x,y)dx \Big/ f_Y(y)$$

and the curve

$$x = E(X|Y=y)$$

is called the regression curve of X on Y.

Note: (a) If one of the regression curve is a straight line, so that the regression is linear, it does not follow that the other is also linear.

(b) For a discontinuous bivariate distribution, the regression curve of Y on X will be a disconnected set of points.

(c) If $E(Y|X) = a + bx$, *i.e.*, regression curve of Y on X is linear, then

$$a = \bar{y} - \frac{\mu_{11}}{\sigma_x^2}\bar{x} \text{ and } b = \frac{\mu_{11}}{\sigma_x^2} = \frac{r\sigma_y}{\sigma_x}$$

and thus

$$E(Y|X) = \bar{Y} + \frac{r\sigma_y}{\sigma_x}(X - \bar{X}).$$

MULTIPLE AND PARTIAL CORRELATION-REGRESSION

In case of bivariate distributions, the values of one variable are influenced by those of another and in that case the simple correlation coefficient r provides a measure of the degree of association between these two variables, and the appropriate linear regression equation is used to estimates the values of the variable. But, in many cases the values of a variable are influenced not only by another single variable, but by several others. In that case we are concerned with the joint distribution of three or more variables, called multivariate distribution.

Multivariate distribution has three types of problems:

(a) The problem of *multiple regression*—To estimate the value of one variable from those of several others.

(b) The problem of *multiple correlation*—To measure the extent of combined influence of a group of variables upon another variable.

(c) The problem of *partial correlation*—To examine the influence of one variable upon another, after eliminating the effects of all other variables.

Partial Correlation: In partial correlation, we study the effect of one independent variable on a dependent variable by excluding the effect of other independent factors. It is also called net correlation.

Since in simple correlation no factor is held constant, correlation between two variables is called the *Zero order coefficient*. However, if a partial correlation is studied between two variables by keeping a third variable constant, it is called a *first order coefficient* as one variable is kept constant. In case, a partial correlation is studied between two variables by keeping two other variable constant, it is called *second order coefficient* as two variables are kept constant.

A first order partial correlation coefficient is represented by $r_{12.3}$, which means that a partial correlation is being studied between variables (1) and (2) while the third variable is kept constant.

Multiple Correlation: In multiple correlation, we study the effects of all the independent variables simultaneously on a dependent variable. Here we study three or more variables at a time.

The coefficient of multiple linear correlation is represented by R and the necessary subscripts are added to it. For example, if there are three variables then $R_{1.23}$ represents the multiple correlation coefficient with variable (1) as dependent variable and variables (2) and (3) as independent variables.

Multiple Regression: In multiple regression, we study the effect of two or more independent variables on one dependent variable. For example, if there are 3 or more variables, say, Y_1, Y_2, and Y_3, we take Y_1 as the dependent variable and try to find out its relative movement for movements in both Y_2 and Y_3, which are independent variables.

The multiple regression is studied in the same way as the simple regression, with the difference that the other variables are added in the regression equation. The multiple regression for the three variables Y_1, Y_2, and Y_3 is given by

$$Y_1 = a_{1.23} + b_{12.3}Y_2 + b_{13.2}Y_3$$

In this equation, there are three constants and the subscript after the dot indicates the variable which are held constant. For instance, $b_{12.3}$ indicates the slope of the regression line of Y_1 on Y_2 when Y_3 is held constant. Similarly, $b_{13.2}$ indicates the slope of the regression line of Y_1 on Y_3 when Y_2 is held constant.

Partial Regression: Since, the changes in Y_1 are due partially to changes in Y_2 and partially to changes in Y_3, therefore $b_{12.3}$ and $b_{13.2}$ are called partial regression coefficients of Y_1 on Y_2 keeping Y_3 constant and of Y_1 on Y_3 keeping Y_2 constant.

PARTIAL CORRELATION COEFFICIENT

If there are three variables, only then there are three partial correlation coefficients which are given by

$$r_{12.3} = \frac{r_{12} - r_{13}\,r_{23}}{\sqrt{(1 - r_{13}^2)(1 - r_{23}^2)}}, \quad r_{13.2} = \frac{r_{13} - r_{12}\,r_{23}}{\sqrt{(1 - r_{12}^2)(1 - r_{23}^2)}}, \text{ and}$$

$$r_{23.1} = \frac{r_{23} - r_{12}\,r_{13}}{\sqrt{(1 - r_{12}^2)(1 - r_{13}^2)}}.$$

These correlation coefficients are calculated on the basis of zero order coefficient or simple coefficient of correlation where no variable is kept constant. Thus, r_{12} is simple correlation coefficient between variables (1) and (2). These correlation coefficient lies between –1 and +1.

Case of four variables: In the case of four variables, two of them have to be kept constant and the correlation is studied between the remaining two. Thus, if correlation is studied between variables (1) and (2), keeping the remaining two variables (3) and (4) constant, the partial correlation is symbolically represented by $r_{12.34}$, and this is a second order partial correlation coefficient as two variables have to be kept constant. There will be in all, six partial correlation coefficients, *viz.*, $r_{12.34}$, $r_{13.24}$, $r_{14.23}$, $r_{23.14}$, $r_{24.13}$, and $r_{34.12}$. The first partial correlation coefficients of the second order are given by

$$r_{12.34} = \frac{r_{12.4} - r_{13.4}\, r_{23.4}}{\sqrt{(1 - r_{13.4}^2)(1 - r_{23.4}^2)}}, \quad r_{13.24} = \frac{r_{13.4} - r_{12.4}\, r_{23.4}}{\sqrt{(1 - r_{12.4}^2)(1 - r_{23.4}^2)}}, \text{ and}$$

$$r_{14.23} = \frac{r_{14.3} - r_{12.3}\, r_{24.3}}{\sqrt{(1 - r_{12.3}^2)(1 - r_{24.3}^2)}}.$$

Similarly, the remaining 3 partial correlation coefficients can be derived.

STANDARD ERROR OF PARTIAL CORRELATIONS

Z-transformation is used to study the reliability of the partial coefficients of correlation. The standard error of $Z = 1/\sqrt{n-3}$. In case of partial correlation like $r_{12.3}$ the standard error of Z would be $1/\sqrt{n-3-1}$ or $1/\sqrt{n-4}$ as one more degree of freedom is lost when one independent variable is made constant. On this basis, if there are four variables under study and two of them have been made constant, two more degrees of freedom are lost. Thus, for $r_{12.34}$ the degrees of freedom would be $n - 3 - 2$ and the standard error of Z would be $1/\sqrt{n-5}$.

MULTIPLE CORRELATION COEFFICIENT

It is given by

$$R_{1.23} = \sqrt{\frac{r_{12}^2 + r_{13}^2 - 2r_{13} - 2r_{12}\, r_{13}\, r_{23}}{1 - r_{23}^2}}, \quad R_{2.13} = \sqrt{\frac{r_{21}^2 + r_{23}^2 - 2r_{12}\, r_{13}\, r_{23}}{1 - r_{23}^2}}, \text{ and}$$

$$R_{3.12} = \sqrt{\frac{r_{31}^2 + r_{32}^2 - 2r_{12}\, r_{13}\, r_{23}}{1 - r_{13}^2}}.$$

Following are some of the properties of multiple correlation coefficient:

(a) It is a non-negative coefficient. Its value lies between 0 and 1, *i.e.*, $0 \le R_{1.23} \le 1$.

(b) $R_{1.23} \ge r_{12}$ and $R_{1.23} \ge r_{13}$.

(c) If $R_{1.23} = 0$ then $r_{12} = 0$ and $r_{13} = 0$.

(d) The position of the subscript to the right of the dot does not make a difference, *i.e.*, $R_{1.23}$ is the same thing as $R_{1.32}$.

(e) The coefficient of multiple determination is obtained by squaring the coefficient of multiple correlation.

MULTIPLE REGRESSION EQUATIONS

The multiple regression equation of Y_1 on Y_2 and Y_3 is given by

$$Y_1 = a_{1.23} + b_{12.3} Y_2 + b_{13.2} Y_3$$

The values of $b_{12.3}$ and $b_{13.2}$ are determined by solving simultaneously the following three normal equations.

$$\Sigma Y_1 = n a_{1.23} + b_{12.3}\Sigma Y_2 + b_{13.2}\Sigma Y_3$$

$$\Sigma Y_1 Y_2 = a_{1.23}\Sigma Y_2 + b_{12.3}\Sigma Y_2^2 + b_{13.2}\Sigma Y_2 Y_3$$

$$\Sigma Y_1 Y_3 = a_{1.23}\Sigma Y_3 + b_{12.3}\Sigma Y_2 Y_3 + b_{13.2}\Sigma Y_3^2$$

The multiple regression equation of Y_2 on Y_1 and Y_3 is given by

$$Y_2 = a_{2.13} + b_{21.3}Y_1 + b_{23.1}Y_3$$

The values of $b_{21.3}$ and $b_{23.1}$ are determined by solving simultaneously the following three normal equations.

$$\Sigma Y_2 = n a_{2.13} + b_{21.3}\Sigma Y_1 + b_{23.1}\Sigma Y_3$$

$$\Sigma Y_1 Y_2 = a_{2.13}\Sigma Y_1 + b_{21.3}\Sigma Y_1^2 + b_{23.1}\Sigma Y_1 Y_3$$

$$\Sigma Y_2 Y_3 = a_{2.13}\Sigma Y_3 + b_{21.3}\Sigma Y_1 Y_3 + b_{23.1}\Sigma Y_3^2$$

The multiple regression equation of Y_3 on Y_1 and Y_2 is given by

$$Y_3 = a_{3.12} + b_{31.2}Y_1 + b_{32.1}Y_2$$

The values of $b_{31.2}$ and $b_{32.1}$ are determined by solving simultaneously the following three normal equations.

$$\Sigma Y_3 = n a_{3.12} + b_{31.2}\Sigma Y_1 + b_{32.1}\Sigma Y_2$$

$$\Sigma Y_1 Y_3 = a_{3.12}\Sigma Y_1 + b_{31.2}\Sigma Y_1^2 + b_{32.1}\Sigma Y_1 Y_3$$

$$\Sigma Y_2 Y_3 = a_{3.12}\Sigma Y_2 + b_{31.2}\Sigma Y_1 Y_2 + b_{32.1}\Sigma Y_2^2$$

In Case of Deviations from the Mean

The multiple regression equation of Y_1 on Y_2 and Y_3 is given by

$$Y_1 - \bar{Y}_1 = \frac{\sigma_1}{\sigma_2}\left[\frac{r_{12} - r_{13}r_{23}}{1 - r_{23}^2}\right](Y_2 - \bar{Y}_2) + \frac{\sigma_1}{\sigma_3}\left[\frac{r_{13} - r_{12}r_{23}}{1 - r_{23}^2}\right](Y_3 - \bar{Y}_3)$$

Similarly, the multiple regression equation of Y_2 on Y_3 and Y_1 is given by

$$Y_2 - \bar{Y}_2 = \frac{\sigma_2}{\sigma_3}\left[\frac{r_{23} - r_{12}r_{13}}{1 - r_{13}^2}\right](Y_3 - \bar{Y}_3) + \frac{\sigma_2}{\sigma_1}\left[\frac{r_{12} - r_{23}r_{31}}{1 - r_{31}^2}\right](Y_1 - \bar{Y}_1)$$

Lastly, the multiple regression equation of Y_3 on Y_1 and Y_2 is given by

$$Y_3 - \bar{Y}_3 = \frac{\sigma_3}{\sigma_2}\left[\frac{r_{23} - r_{12}r_{13}}{1 - r_{12}^2}\right](Y_2 - \bar{Y}_2) + \frac{\sigma_3}{\sigma_1}\left[\frac{r_{13} - r_{23}r_{12}}{1 - r_{12}^2}\right](Y_1 - \bar{Y}_1)$$

STANDARD ERROR OF THE ESTIMATES

The standard error of the estimate of Y_1 on Y_2 and Y_3 is given by

$$S_{1.23} = \sqrt{\frac{\Sigma(Y_1 - \hat{Y}_1)^2}{n - 3}}$$

where $S_{1.23}$ is the standard error of the estimate of Y_1 on Y_2 and Y_3. Y_1 is the original value of Y and $\hat{Y}_1$ is the estimated value on the basis of the regression equations.

In terms of the correlation coefficients r_{12}, r_{13}, r_{23} the standard error of the estimate is given by

$$S_{1.23} = \sigma_1 \sqrt{\frac{1 - r_{12}^2 - r_{13}^2 - r_{23}^2 + 2r_{12}r_{13}r_{23}}{1 - r_{23}^2}}$$

If the estimated values of Y for all data points and the partial correlation coefficients are not available, the standard error of the regression estimate is given by

$$S_{1.23} = \sqrt{\frac{\Sigma Y_1^2 - a\Sigma Y_1 - b_1 \Sigma Y_1 Y_2 - b_2 \Sigma Y_3 Y_1}{n - 3}}$$

Note: The standard error is an estimate of unexplained variation in the values of Y. If the total variation of Y_1 is divided into two parts, the standard error would represent the unexplained variation. The remainder or the explained variation would be due to independent variables.

PARTIAL REGRESSION COEFFICIENT

Partial regression coefficient of

Y_1 on Y_2 and Y_3 is given by $$b_{12.3} = \frac{\sigma_1}{\sigma_2} \times \frac{r_{12} - r_{13}r_{23}}{1 - r_{23}^2}$$

Y_2 on Y_1 and Y_3 is given by $$b_{21.3} = \frac{\sigma_2}{\sigma_1} \times \frac{r_{12} - r_{13}r_{23}}{1 - r_{13}^2}$$

Y_3 on Y_1 and Y_2 is given by $$b_{31.2} = \frac{\sigma_3}{\sigma_1} \times \frac{r_{13} - r_{12}r_{23}}{1 - r_{23}^2}$$

SIGNIFICANCE OF REGRESSION COEFFICIENT

The SE of regression coefficient is given by

$$\sigma_{12.34...m} = \frac{\sigma_{1.234...m}}{\sigma_{2.34....m}\sqrt{n - m}}$$

where, m = number of variables correlated, n = size of the sample, and $(n - m)$ = degree of freedom.

SIGNIFICANCE OF MULTIPLE CORRELATION

Standard error of R is given by

$$SE_R \text{ or } \sigma_R = \frac{1 - R^2}{\sqrt{n - k}}$$

where, R = multiple correlation, n = size of sample, and k = number of variables.

CORRECTED MULTIPLE CORRELATION

When n is small and number of variables is quite large, the value of R can be corrected by Shrinkage formula to have a better measure of the population R. The formula is given as

$$R_c^2 = 1 - \frac{(n-1)K^2}{(n-m)}$$

where, n = size of the sample, m = number of variables, and $K^2 = 1 - R^2$.

MULTIPLE REGRESSION IN TERMS OF β COEFFICIENTS

$$R_{1.23}^2 = \beta_{12.3}\, r_{12} + \beta_{13.2}\, r_{13}$$

where,

$$\beta_{12.3} = \frac{r_{12} - r_{13}\, r_{23}}{1 - r_{23}^2} \quad \text{and} \quad \beta_{13.2} = \frac{r_{13} - r_{12}\, r_{23}}{1 - r_{23}^2}$$

F-TEST

$$F = \frac{R^2}{1 - R^2} \times \frac{n - k - 1}{k}$$

where, R = multiple correlation, n = size of the sample, and k = number of variables.

INTERPOLATION AND EXTRAPOLATION

Interpolation is defined as the art of reading between the lines of a statistical table and it usually denotes the process of finding the intermediate value of a function from a set of given values of that function. In other words, it is a technique of obtaining the most likely estimate of certain quantity under certain assumptions.

According to W.H. Harper, interpolation consists in reading a value which lies between two extreme points and extrapolation means reading a value that lies outside the two extreme points.

According to Hirach, interpolation is the estimation of a most likely estimate in given conditions. The technique of estimating a past figure is termed interpolation, while that of estimating a probable figure for the future is called extrapolation.

In interpolation, the value of a variable X for which the corresponding value of variable Y is to be estimated would always be within the lowest and highest value of X-series. In extrapolation, the values of X for which the corresponding values of Y is to be estimated would always be outside the minimum and maximum values of X-series.

The technique of interpolation is based on two assumptions:

(a) There are no sudden jumps in figures from one period to another, *i.e.*, the data should be in the shape of continuous or smoothed curve.

(b) The rate of change of the figures is uniform, *i.e.*, the data should not be of highly fluctuating nature.

Methods of Interpolation: (a) Graphic methods—use different types of graphs in different situations such as graphs on natural scale or logarithmic or semi-logarithmic scale for continuous series, for series showing periodicity or series showing positive or negative correlation.

(b) Algebraic methods—use a large number of formulae under different conditions.

(1) When X-series advances by equal intervals following methods are used—(i) Newton's method of advancing differences, (ii) Newton-Gauss forward method, (iii) Newton-Gauss backward method, (iv) Newton's method of backward differences, and (v) Binomial expansion method.

(2) When X-series advances by unequal intervals following methods are used—(i) Newton's divided difference method, (ii) Lagrange's method, and (iii) Parabolic curve fitting.

Methods of Extrapolation: Extrapolation is a futuristic estimate because it refers to estimating a value beyond the given values of the variable. There is no special formula for extrapolation. All methods used in interpolation, including the graphic methods, can be used for extrapolation. The choice of the method depends on the nature of the data given, and the conditions applicable with different formulae for interpolation hold good for extrapolation also.

Finite Differences: Δ and *E* Operators: In interpolation, the independent variable x is often known as '*argument*', and the dependent variable or the function $y = f(x)$ is known as '*entry*'.

Let us consider a equidistant values of the argument denoted by $x_0, x_1, x_2, \ldots, x_n$, *i.e.*,

$$x_1 - x_0 = x_2 - x_1 = x_3 - x_2 = \ldots = x_n - x_{n-1} = h$$

where h is a constant; and the corresponding values of the entry by $y_0, y_1, y_2, \ldots, y_n$.

Differences of the successive values of the entry, y, *i.e.*,

$$(y_1 - y_0), (y_2 - y_1), (y_3 - y_2), \ldots, (y_n - y_{n-1})$$

are called finite differences of the first order or simply first differences and are denoted by: $\Delta y_0, \Delta y_1, \Delta y_2, \Delta y_3, \ldots, \Delta y_{n-1}$ respectively. The differences of the successive first order differences Δy, namely

$$(\Delta y_1 - \Delta y_0), (\Delta y_2 - \Delta y_1), (\Delta y_3 - \Delta y_2), \ldots, (\Delta y_{n-1} - \Delta y_{n-2})$$

are known as finite differences of the second order or simply second differences and are denoted by: $\Delta^2 y_0, \Delta^2 y_1, \Delta^2 y_2, \Delta^2 y_3, \ldots, \Delta^2 y_{n-2}$ respectively. Similarly, the third differences $\Delta^3 y$, the fourth differences $\Delta^4 y$, and differences of higher order may be defined. In this way, we can obtain the following difference table:

x	y	Δy	$\Delta^2 y$	$\Delta^3 y$	$\Delta^4 y$	$\Delta^5 y$
x_0	y_0	$y_1 - y_0$				
		$= \Delta y_0$	$\Delta y_1 - \Delta y_0$			
x_1	y_1	$y_2 - y_1$	$= \Delta^2 y_0$	$\Delta^2 y_1 - \Delta^2 y_0$		
		$= \Delta y_1$	$\Delta y_2 - \Delta y_1$	$= \Delta^3 y_0$	$\Delta^3 y_1 - \Delta^3 y_0$	
x_2	y_2	$y_3 - y_2$	$= \Delta^2 y_1$	$\Delta^2 y_2 - \Delta^2 y_1$	$= \Delta^4 y_0$	$\Delta^4 y_1 - \Delta^4 y_0$
		$= \Delta y_2$	$\Delta y_3 - \Delta y_2$	$= \Delta^3 y_1$	$\Delta^3 y_2 - \Delta^3 y_1$	$= \Delta^5 y_0$
x_3	y_3	$y_4 - y_3$	$= \Delta^2 y_2$	$\Delta^2 y_3 - \Delta^2 y_2$	$= \Delta^4 y_1$	
		$= \Delta y_3$	$\Delta y_4 - \Delta y_3$	$= \Delta^3 y_2$		
x_4	y_4	$y_5 - y_4$	$= \Delta^2 y_3$			
		$= \Delta y_4$				
x_5	y_5					

The initial term y_0 of the entry is called the *leading term* and the initial terms in the difference columns, *viz.* $\Delta y_0, \Delta^2 y_0, \Delta^3 y_0$ *etc.* are called *leading differences.*

Note: Δ^2 does not imply the square of Δ, but the repetition of operation by Δ twice.

Another symbolic operator E is defined as

$$Ey_0 = y_1,\ Ey_1 = y_2,\ Ey_2 = y_3,\ \ldots$$

Both the operators Δ and E can be applied repeatedly. The repeated operations can be indicated by $\Delta^2, \Delta^3, \Delta^4, \ldots$ and $E^2, E^3, E^4, \ldots$ *etc.* Thus, we have

$$\Delta^2 y_0 = \Delta(\Delta y_0) = \Delta y_1 - \Delta y_0 = (y_2 - y_1) - (y_1 - y_0) = y_2 - 2y_1 + y_0$$

$$\Delta^3 y_0 = \Delta(\Delta^2 y_0) = \Delta^2 y_1 - \Delta^2 y_0 = (y_3 - 2y_2 + y_1) - (y_2 - 2y_1 + y_0) = y_3 - 3y_2 + 3y_1 - y_0 \text{; etc.}$$

Similarly, $E^2 y_0 = E(Ey_0) = E(y_1) = y_2$

$$E^3 y_0 = E(E^2 y_0) = E(y_2) = y_3 \text{; etc.}$$

Thus, in general we have:

$$\Delta y_r = y_{r+1} - y_r \quad \text{and} \quad Ey_r = y_{r+1}$$

Also, operators Δ and E can be interpreted as follows:

(a) Δ when prefixed to y_r implies that y_r is to be subtracted from the next value of the entry y_{r+1}.

(b) E when prefixed to y_r denotes the next value of the entry y_{r+1}.

Thus, we have

$$Ey_r = y_r + \Delta y_r \Rightarrow Ey_r = (1 + \Delta)y_r \Rightarrow E = 1 + \Delta$$

This does not mean that 1 when added to Δ gives E, but that the operation by E is equivalent to the operation by $(1 + \Delta)$. The above relation follows certain algebraic rules:

$$\Delta y_0 = y_1 - y_0 \Rightarrow \Delta y_0 = Ey_0 - y_0 \Rightarrow \Delta = E - 1$$

$$\Delta^2 y_0 = y_2 - 2y_1 + y_0 \Rightarrow \Delta^2 y_0 = E^2 y_0 - 2Ey_0 + y_0 \Rightarrow \Delta^2 = E^2 - 2E + 1 = (E - 1)^2$$

$$\Delta^3 y_0 = y_3 - 3y_2 + 3y_1 - y_0 \Rightarrow \Delta^3 y_0 = E^3 y_0 - 3E^2 y_0 + 3Ey_0 - y_0 \Rightarrow \Delta^3 = E^3 - 3E^2 + 3E - 1 = (E - 1)^3$$

$$\Delta^4 y_0 = y_4 - 4y_3 + 6y_2 - 4y_1 + y_0 \Rightarrow \Delta^4 y_0 = E^4 y_0 - 4E^3 y_0 + 6E^2 y_0 - 4Ey_0 + y_0$$

$$\Rightarrow \Delta^4 = E^4 - 4E^3 + 6E^2 - 4E + 1 = (E - 1)^4$$

N.B.: It should be noted that the interpolated figures are only best possible estimates under certain assumptions because they are obtained by methods which are entirely different from those by which actual figures were obtained. Thus, the interpolated figures are not perfect substitutes of the original figures, they are only best possible substitutes based on certain hypothesis.

NEWTON'S METHOD OF ADVANCING DIFFERENCES

This method is used when the figure to be interpolated is in the beginning of the table because it takes only leading differences into account and the leading differences are always in the beginning.

Let $y_0, y_1, y_2, ..., y_n$ be some tabulated values of a function $y = f(x)$ corresponding to the equidistant values $x = x_0, x_1, x_2, ..., x_n$ and $x_1 - x_0 = x_2 - x_1 = ... = x_n - x_{n-1} = h$. The generalised formula for this method is

$$y = y_0 + u\Delta + \frac{u(u-1)}{1\times 2}\Delta^2 + \frac{u(u-1)(u-2)}{1\times 2\times 3}\Delta^3 + ... + \frac{u(u-1)(u-2)...(u-n+1)}{1\times 2\times 3\times ...\times n}\Delta^n$$

where $u = (x - x_0)/h$.

NEWTON-GAUSS FORWARD METHOD

This method of interpolation is applied when: (a) the independent variable advances by equal interval, and (b) the dependent variable is to be interpolated for such a value of independent variable which is in the middle of the independent variable series.

Let $y_0, y_1, y_2, ..., y_n$ be some tabulated values of a function $y = f(x)$ corresponding to the equidistant values $x = x_0, x_1, x_2, ..., x_n$ and $x_1 - x_0 = x_2 - x_1 = ... = x_n - x_{n-1} = h$.

The generalised formula for this method is

$$y = y_0 + u\Delta y_0 + \frac{u(u-1)}{1\times 2}\Delta^2 y_0 + \frac{u(u-1)(u-2)}{1\times 2\times 3}\Delta^3 y_0 + ... + \frac{u(u-1)(u-2)...(u-n+1)}{1\times 2\times 3\times ...\times n}\Delta^n y_0$$

where $u = (x - x_0)/h$.

NEWTON-GAUSS BACKWARD METHOD

This method of interpolation is applied when: (a) the independent variable advances by equal interval, and (b) the figure to be interpolated is near the end of the series.

Let $x_0, x_1, x_2, ..., x_n$ be some equidistant values of the argument and $y_0, y_1, y_2, ..., y_n$ the corresponding entries.

The generalised formula for this method is

$$y = y_n + v\Delta y_{n-1} + \frac{v(v+1)}{1\times 2}\Delta^2 y_{n-2} + \frac{v(v+1)(v+2)}{1\times 2\times 3}\Delta^3 y_{n-3} + ... + \frac{v(v+1)(v+2)...(v+n-1)}{1\times 2\times 3\times ...\times n}\Delta^n y_0$$

where $v = (x - x_n)/h$.

NEWTON'S METHOD OF BACKWARD DIFFERENCE

This method is used when the figure to be interpolated is at the end of the table. In this method, the last differences are the leading differences and the differences are indicated by ∇.

In this formula, the value of w is the difference between the unit of interpolation and the last unit of the series divided by the difference between adjoining values of independent series.

The generalised formula for this method is

$$y = y_0 + w\nabla + \frac{w(w-1)}{1\times 2}\nabla^2 + \frac{w(w-1)(w-2)}{1\times 2\times 3}\nabla^3 + ... + \frac{w(w-1)(w-2)...(w-n+1)}{1\times 2\times 3\times ...\times n}\nabla^n$$

where $w = (x - x_n)/h$.

CENTRAL DIFFERENCE FORMULA

They are used when the figure to be interpolated is near the middle of the set of tabulated values. These formulae utilise differences near the central part of the difference table.

Let the function $y = f(x)$ be tabulated for some equidistant values of the argument

$$(x_0 - nh), \dots (x_0 - 2h), (x_0 - h), x_0, (x_0 + h), (x_0 + 2h), \dots (x_0 + nh)$$

the common difference being h; and let the corresponding entries be denoted by

$$y_{-n}, \quad \dots \quad y_{-2}, \quad y_{-1}, \quad y_0, \quad y_1, \quad y_2, \quad \dots \quad y_n.$$

We want to interpolate the value of the function $y = f(x)$ for a value of x in the interval $(x_0 - h)$ to $(x_0 + h)$.

Stirling's Interpolation Formula

$$y = y_0 + u.\frac{(\Delta y_0 + \Delta y_{-1})}{2} + \frac{u^2}{2!}.\,\Delta^2 y_{-1} + \frac{u(u^2 - 1^2)}{3!}.\frac{(\Delta^3 y_{-1} + \Delta^3 y_{-2})}{2}$$

$$+\frac{u^2(u^2 - 1^2)}{4!}.\,\Delta^4 y_{-2} + \frac{u(u^2 - 1^2)(u^2 - 2^2)}{5!}.\frac{(\Delta^5 y_{-2} + \Delta^5 y_{-3})}{2} + \dots$$

where $u = (x - x_0)/h$.

Stirling's formula is appropriate when the value of u lies in the interval –0.25 to +0.25. It uses y_0 and even order differences on the horizontal line through y_0, but the average of odd order differences above and below that line.

Bessel's Interpolation Formula

$$y = \frac{(y_0 + y_1)}{2} + v\Delta y_0 + \frac{\left(v^2 - 1/4\right)}{2!}.\frac{(\Delta^2 y_0 + \Delta^2 y_{-1})}{2} + \frac{v\left(v^2 - 1/4\right)}{3!}.\Delta^3 y_{-1}$$

$$+\frac{\left(v^2 - 1/4\right)\left(v^2 - 9/4\right)}{4!}.\frac{(\Delta^4 y_{-1} + \Delta^4 y_{-2})}{2} + \frac{v\left(v^2 - 1/4\right)\left(v^2 - 9/4\right)}{5!}.\Delta^5 y_{-2} + \dots$$

where $u = (x - x_0)/h$ and $v = u - ½$.

Bessel's formula is approriate for values of u lying in the interval +0.25 to +0.75, *i.e.*, for interpolation in the middle-half of two tabulated values. It uses the average of y_0 and y_1, and odd order differences on the horizontal line mid-way between them, but the average of even order differences above and below that line.

Bessel's formula is especially found useful for interpolating the value of the function exactly at the middle of two given values, *i.e.*, when $u = ½$ so that $v = 0$. Every alternate term in Bessel's formula then vanish. This special case of Bessel's formula is known as "*Formula for Interpolating to Halves*".

BINOMINAL EXPANSION METHOD

This method is used where it is possible to find the value of missing figure by directly raising a binomial without finding out the differences. The possible cases are those where x-series advances by equal intervals and the value of x for which y has to be interpolated is one of the class limits of x-series.

The generalised formula for this method is

$$(y-1)^n = y_n - ny_{n-1} + \frac{n(n-1)}{2!}y_{n-2} + \frac{n(n-1)(n-2)}{3!}y_{n-3} + \dots + y_0$$

LAGRANGE'S INTERPOLATION FORMULA

Let $y_0, y_1, y_2, ..., y_n$ denote the tabulated values of a function $y = f(x)$ corresponding to the values of the argument $x_0, x_1, x_2, ..., x_n$ (which may not be equidistant). We want to find the value of y corresponding to a specified value of x lying in between the given values. This can be obtained by:

$$y = \frac{(x - x_1)(x - x_2)...(x - x_n)}{(x_0 - x_1)(x_0 - x_2)...(x_0 - x_n)} y_0 + \frac{(x - x_0)(x - x_2)...(x - x_n)}{(x_1 - x_0)(x_1 - x_2)...(x_1 - x_n)} y_1 + ...$$
$$+ \frac{(x - x_0)(x - x_1)...(x - x_{n-1})}{(x_n - x_0)(x_n - x_1)...(x_n - x_{n-1})} y_n$$

NEWTON'S DIVIDED DIFFERENCE METHOD

In this method we use the formula

$$y = y_0 + (x - x_0)\,\Delta^1 + (x - x_0)(x - x_1)\,\Delta^2 + (x - x_0)(x - x_1)(x - x_2)\,\Delta^3 + ...$$

where Δ^1, Δ^2 and Δ^3 are the first, second and third leading divided differences, respectively.

The divided differences are obtained as follows:

x	y	*Divided Differences* Δ^1	Δ^2	Δ^3
x_0	y_0	$(y_1 - y_0)/(x_1 - x_0) = \Delta_0^1$		
x_1	y_1		$(\Delta_1^1 - \Delta_0^1)/(x_2 - x_0) = \Delta_0^2$	
		$(y_2 - y_1)/(x_2 - x_1) = \Delta_1^1$		$(\Delta_1^2 - \Delta_0^2)/(x_3 - x_0) = \Delta_0^3$
x_2	y_2		$(\Delta_2^1 - \Delta_1^1)/(x_3 - x_1) = \Delta_1^2$	
		$(y_3 - y_2)/(x_3 - x_2) = \Delta_2^1$		
x_3	y_3			

INVERSE INTERPOLATION

Given a set of tabulated values of a function $y = f(x)$ corresponding to some values of the argument x, the process of finding the value of the argument for an intermediate value of the function is called *'Inverse Interpolation'*.

Let $y_0, y_1, ..., y_n$ be a set of tabulated values of a function $y = f(x)$ corresponding to some given values (usually equidistant) of the argument $x = x_0, x_1, ..., x_n$. In some cases it may be required to determine the value of x for another value of y lying in between $y_0, y_1, ..., y_n$. The process of finding such a value of x is known as 'inverse interpolation'. If y is a function of x, say $y = f(x)$, in many cases it is possible to treat x also as a function of y, say $x = g(y)$. Since, Lagrange's interpolation formula is applicable for unequal intervals, we can use this formula for inverse interpolation, interchanging the role of the argument x and the function y.

$$x = \frac{(y - y_1)(y - y_2)(y - y_3) \dots (y - y_n)}{(y_0 - y_1)(y_0 - y_2)(y_0 - y_3) \dots (y_0 - y_n)} x_0 + \frac{(y - y_0)(y - y_2)(y - y_3) \dots (y - y_n)}{(y_1 - y_0)(y_1 - y_2)(y_1 - y_3) \dots (y_1 - y_n)} x_1$$
$$+ \frac{(y - y_0)(y - y_1)(y - y_3) \dots (y - y_n)}{(y_2 - y_0)(y_2 - y_1)(y_2 - y_3) \dots (y_2 - y_n)} x_2 + \dots + \frac{(y - y_0)(y - y_1)(y - y_2) \dots (y - y_{n-1})}{(y_n - y_0)(y_n - y_1)(y_n - y_2) \dots (y_n - y_{n-1})} x_n$$

PARABOLIC CURVE METHOD

Let y depends on the values of x in such a manner that when x is given, y can be estimated. In a parabolic curve, the relationship is of the type:

$$y = a + bx + cx^2 + dx^3 + \dots$$

where a, b, c and d are constants to be determined. The equation $y = a + bx + cx^2 + dx^3$ is a parabola of the third order. If we are given four values of y, we can fit a parabola of the third order to the series. Similarly, if 6 values of y are available, we can fit a parabola of the 5th order to such a series.

To interpolate the value of y, the following steps have to taken:

(a) The power for which the parabolic curve will be used is determined. If there are n number of known values, the power will be $n - 1$.

(b) The value of x for which y is to be interpolated is taken as the origin and from this value, the deviations of other values of x are found out. If possible, the deviations are divided by a common factor.

(c) Different values of x and y are then substituted in the parabolic equations. One of the equations would be $Y_0 = a$. It is the value of a which has to be found out from the parabolic equations. This will be the desired value.

PROBABILITY

The probability theory is that part of mathematics that aims to provide insight into happenings that depend on chance or on uncertainty. The roots of probability theory are not as ancient as those of many parts of mathematics. Only in the sixteenth and seventeenth centuries does one find the first glimmerings of the theory in the investigations made by Gerolamo Cardano, Pierre de Fermat and Blaise Pascal into games of chance. Despite the luminous reputations of these famous mathematicians and philosophers, the subject of probability theory remained on the periphery of respectability, and for a long time development was halting and lugubrious. Through the first third of the twentieth century, the eighteenth century works of Jakob Bernoulli and Abraham de Moivre continued to be viewed as the nearly definitive treatises of probability theory.

The concept of *Inverse Probability* was introduced by Thomas Bayes (1702-61). After an extensive research, Pierre Simon Laplace (1749-1827) published his monumental work, *Theory of Analytical Probability* (*Theorie Analytique des Probabilities*), in 1812 which constitutes the classical theory of probability. R.A. Fisher and Von Mises introduced empirical approach to probability through the notion of sample space.

Today, the theory of probability has been very extensively developed and there is hardly any discipline—physical or social—where it is not being extensively used. In the field of Business and Economics, it is very widely used for quantitative analysis of various problems and it forms the very basis of the modern Theory of Decision-making, *i.e.*, decision-making under conditions of uncertainty with calculated risks.

In short, probability measures the degree of uncertainty and such measures are positive and without any scale of measurement like percentages. The probability of a sample point is a number between 0 and 1 that measures the likelihood that the outcome will occur when the experiment is performed.

Broadly, probability can be classified as:

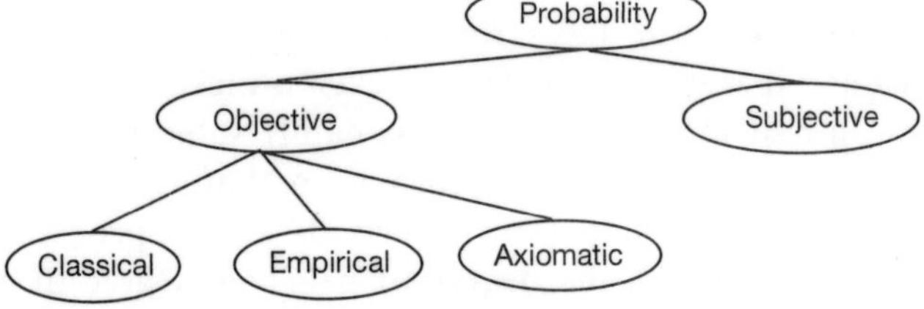

Classical approach is aslo known as mathematical or *a priori* approach to probability. And empirical approach is also known as statistical or relative frequency or *a postpriori* approach to probability.

Classical: Classical (or mathematical or *a prior*) definition of probability was proposed by Jacob Bernoulli (1654-1705) and later adopted by Laplace (1749-1827) and others. It was the earliest approach to the theory of probability. This approach assumes that various outcomes of an event are equally likely and the device with which the experiment is performed is supposed to be a fair device and therefore the probability of their happening is also equal. In this way, this approach determines the probability of an event before the event has actually taken place, and that is why this type of probability is called *a priori* probability.

Empirical: The empirical (or relative frequency or statistical or *a posteriori*) approach offers the empirical frequency of an event in past observations as a definition of its probability. This approach defines an event's probability as the limit of its relative frequency in a large number of trials. The development of this approach was motivated by the problems and paradoxes of the previously dominant viewpoint, *i.e.*, the classical interpretation. This shift from the classical view to the empirical view represents a paradigm shift in the progression of statistical thought. According to this approach, if trials be repeated a great number of times, under essentially the same conditions, then the limit of the ratio of the number of times that an event happens to the total number of trials, as the number of trials increases indefinitely, is called the probability of the happening of the event. This approach is associated with the names of Jerzy Neyman and Egon Pearson who described the logic of statistical hypothesis testing. Some other names associated with this approach are John Venn, R.A. Fisher, and Richard von Mises.

Axiomatic: A Russian mathematician Andrey Nikolaevich Kolmogorov developed the axiomatic definition of probability in the year 1933 in his book *Foundations of Probability*. His approach is entirely mathematical in character and is based on set theory. To begin with, some concepts are laid down and certain properties or postulates commonly known as axioms are defined and from these axiom alone the entire theory of probability is derived by deductive logic. This definition of probability includes the concepts of both the classical as well as the empirical definitions of probability.

Subjective: The subjective approach to probability was introduced in the year 1926 by Frank Ramsey in his book, *The Foundation of Mathematics and Other Logical Essays*. The concept was further developed by Bernard Koopman, Richard Good and Leonard Savage. This approach of probability is used when the assumption used in the classical method are not applicable and the past statistics that can be used for the relative frequency method are unavailable. In such cases, the basis for assigning probability to experimental outcomes depends on previous experiences, beliefs, and even on feelings. Since, subjective approach of probability relies on individual judgment, it is highly subjective. Nonetheless, the method is quite common in the absence of any information. Later, when the information becomes available, the assigned probability can be revised. This approach emphasises the fact that since probability of an event is the degree of belief or degree of confidence placed in the occurrence of an event by a particular individual based on the evidence available to him, different individuals may differ in their degrees of confidence even when offered the same evidence, which may consist of relative frequency of occurrence of data and any other quantitative or non-quantitative information. Persons might arrive at different probability assignments because of differences in values, experience and attitudes, *etc.*

CLASSICAL DEFINITION OF PROBABILITY

If a random experiment has n possible outcomes, which are *mutually exclusive*, *exhaustive* and *equally likely*, and m of these are favourable to an event A, then the probability of the event is defined as the ratio m/n and is denoted by

$$P(A) = p = \frac{m}{n} = \frac{\textit{number of favourable cases to A}}{\textit{total number of possible cases}}$$

Thus, the probability is a ratio of two numbers, in which the numerator (m) is the number of favourable cases and the denominator (n) is the total number of possible outcomes satisfying certain conditions. Probability of an event always lies between 0 and 1, *i.e.*, $0 \le p \le 1$.

EMPIRICAL DEFINITION OF PROBABILITY

If a random experiment under consideration is repeated n times and an event A occurs m times, then the relative frequency $m(A)/n$ of the occurrence of A approaches a definite and unique limit p as the number of trials is increased to infinity. This limit value of the relative frequency of the occurrence of the event A is called the probability of the occurrence of event A and is denoted by $P(A)$. Thus,

$$P(A) = \lim_{n \to \infty} \frac{m(A)}{n} = p$$

AXIOMATIC DEFINITION OF PROBABILITY

Let S be the simple space consisting of n simple events a_1, a_2, a_3, ..., a_n, *i.e.*,

$$S = \{a_1, a_2, a_3, \ldots, a_n\}$$

then the probability of each simple events $a_i \in S$, denoted by $P(a_i)$ is equal to a definite number and satisfy the following two conditions:

(a) The number representing the probability of any event is always positive , *i.e.*,

$$P(a_i) \ge 0, \qquad i = 1, 2, 3, \ldots, n.$$

(b) The sum of the probabilities of each event is equal to 1, *i.e.*,

$$P(a_1) + P(a_2) + P(a_3) + \ldots + P(a_n) = 1.$$

Hence, if p denotes the probability of an event then $0 \le p \le 1$.

The above two conditions are the axioms of probability, which give us the restrictions upon the probability measure. But, they do not clearly state what would be probability of an event. The probability of an event is obtained by intuition, mathematical reasoning or by any other method.

In case of equally likely simple events, the probability of happening any one simple event is $1/n$. Now let A be an event of S containing m elements of S, *i.e.*, $n(A) = m$. If $P(A)$ is the probability of happening, then

$$P(A) = \frac{1}{n} + \frac{1}{n} + \frac{1}{n} + \ldots \textit{ upto } m \textit{ terms} = \frac{m}{n}$$

or,

$$P(A) = \frac{n(A)}{n(S)}$$

In short, let E be a random experiment described by a discrete sample space S. Then, probability of an event A of S is defined as a real-valued function $P(\cdot)$ such that for any $a_1, a_2, a_3, \ldots, a_n \in A$ following basic axioms are satisfied.

Axiom 1: The probability of an event ranges from zero to one. If the event cannot take place its probability will be zero and if it is certain, *i.e.*, bound to occur, its probability will be one. Thus,

$$0 \le P(A) \le 1$$

for each event of S.

Axiom 2: The probability of entire sample space is 1, *i.e.*,

$$P(S) = 1$$

Axiom 3: If $A_1, A_2, A_3, \ldots A_n$ is a infinite sequence of disjoint events of S, then

$$P(A_1 \cup A_2 \cup A_3 \cup \ldots \cup A_n) = P(A_1) + P(A_2) + P(A_3) + \ldots + P(A_n)$$

In particular, for two mutually exclusive events A and B

$$P(A \cup B) = P(A) + P(B)$$

Here, it should be noted that we can find probability only if the event is a subset of a specified sample space S, and to each subset of S a real number, satisfying the axioms, can be assigned. a sample space on which probability has been defined is called a *probability space*.

THEOREM OF TOTAL PROBABILITY

If two events A and B are mutually exclusive, then the probability of occurrence of either A or B is given by the sum of their probabilities, *i.e.*,

Probability of (A or B) = Probability of A + Probability of B.

In symbols,

$$P(A + B) = P(A) + P(B)$$

This is also known as *Addition Theorem of probability*. In general, if $A_1, A_2, A_3, \ldots, A_n$ are mutually exclusive events defined on a sample space S, then

$$P(A_1 \cup A_2 \cup A_3 \cup ... \cup A_n) = P(A_1) + P(A_2) + P(A_3) + ... + P(A_n)$$

or, if $p_1, p_2, p_3, ..., p_n$ be separate probabilities of n mutually exclusive events, then the probability, p, that any of these events will happen is given by

$$p = p_1 + p_2 + p_3 + ... + p_n$$

or, the probability of the occurrence of any one of n mutually exclusive events is the sum of their individual probabilities.

THEOREMS BASED ON ADDITION THEOREM

(a) If $A_1, A_2, A_3, ..., A_n$ are mutually exclusive and exhaustive events, then $P(A_1) + P(A_2) + P(A_3) + ... + P(A_n) = 1$.

(b) For every event A, $P(A) + P(A') = 1$.

(c) $P(\phi) = 0$, where $\phi = \{\}$.

(d) For every two events A and B, $P(A \cap B') = P(A) - P(A \cap B)$.

(e) If $A \subset B$, then $P(B - A) = P(B) - P(A)$.

(f) For n arbitrary events $A_1, A_2, A_3, ..., A_n$

$$P(A_1 \cup A_2 \cup A_3 \cup ... \cup A_n) = 1 - P(A_1' \cap A_2' \cap A_3' \cap ... \cap A_n')$$

(g) The probability of happening exactly one of the two events A and B is

$$P(A) + P(B) - 2P(A \cap B)$$

(h) For any two arbitrary events A and B

$$P(A \cup B) = P(A \cap B') + P(A' \cap B) + P(A \cap B)$$

(i) For any two arbitrary events A and B

$$P(A \cap B) + P(A \cup B) = P(A) + P(B)$$

(j) The probability of occurrence of at least one of the rwo events A and B, which may not be mutually exclusive, is given by

$$P(A + B) = P(A) + P(B) - P(AB)$$

(k) If A, B, and C are any three events in a sample space S, then

$$P(A + B + C) = P(A) + P(B) + P(C) - P(AB) - P(AC) - P(BC) + P(ABC)$$

or,

$$P(A \cup B \cup C) = P(A) + P(B) + P(C) - P(A \cap B) - P(A \cap C) - P(B \cap C) + P(A \cap B \cap C)$$

GENERALISATION OF THE THEOREM OF TOTAL PROBABILITY

If the events $A_1, A_2, A_3, ..., A_n$, are not mutually exclusive, then the probability of the happening of at least one of the n events is given by $P(A_1 + A_2 + A_3 + ... + A_n)$ and we have

$$P\left(\bigcup_{i=1}^{n} A_i\right) = \sum_{i=1}^{n} P(A_i) - \sum_{\substack{i,j=1 \\ i<j}}^{n} P(A_i A_j) + \dots + (-1)^{n-1} P(A_1 A_2 \dots A_n)$$

THE INCLUSION-EXCLUSION FORMULA

(a) For any two events A and B, the probability of occurrence of at least one of the two given events is given by

$$P(A \cup B) = P(A) + P(B) - P(A \cap B)$$

(b) For any three events A, B, and C, the probability of occurrence of at least one of the three given events is given by

$$P(A \cup B \cup C) = P(A) + P(B) + P(C) - P(A \cap B) - P(B \cap C) - P(A \cap C) + P(A \cap B \cap C)$$

BOOLE'S INEQUALITY

(a) For any two events, A and B

$$P(A \cup B) \le P(A) + P(B)$$

(b) For any three events A, B, and C

$$P(A \cup B \cup C) \le P(A) + P(B) + P(C)$$

(c) For n events $A_1, A_2, A_3, \dots, A_n$

(i) $P\left(\bigcap_{i=1}^{n} A_i\right) \ge \sum_{i=1}^{n} P(A_i) - (n-1)$ and (ii) $P\left(\bigcup_{i=1}^{n} A_i\right) \le \sum_{i=1}^{n} P(A_i)$

(d) For any events $A_1, A_2, A_3, \dots, A_n$

$$P(A_1) + P(A_2) + P(A_3) + \dots + P(A_n) \ge 1 - P(A_1' \cap A_2' \cap A_3' \cap \dots \cap A_n')$$

CONDITIONAL PROBABILITY

Let B be an event in an arbitrary sample space with positive probability. Let A be any event in the same sample space. The probability that the event A will occur, it being known that B has occurred, is called the conditional probability of A and is defined by the relation,

$$P(A|B) = \frac{P(A \cap B)}{P(B)} = \frac{n(A \cap B)}{n(B)}, \qquad P(B) > 0,\ n(B) \ge 1$$

where,

$n(A \cap B)$ = number of sample points favourable to the event $A \cap B$, and

$n(B)$ = number of sample points favourable to the event B.

The conditional probability concept can be extended to the case where $A_1, A_2, A_3, \ldots, A_n$ is a sequence of mutually exclusive events given that an event B has occurred. Thus, the probability of the event $\bigcup_{i=1}^{n} A_i$, when the event B has occurred is,

$$P\left(\bigcup_{i=1}^{n} A_i \middle| B\right) = \frac{P\left(\bigcup_{i=1}^{n} A_i \cap B\right)}{P(B)} = \frac{\sum_{i=1}^{n} P(A_i \cap B)}{P(B)}, \qquad P(B) > 0$$

AXIOMS OF CONDITIONAL PROBABILITY

Let B be an event for which $P(B) > 0$. The conditional probability function $P(*|B)$ satisfies the following axioms:

Axiom 1: For any event A, $0 \le P(A|B) \le 1$.

Axiom 2: For a certain event A, $P(A|B) = 1$.

Axiom 3: If $A_1, A_2, A_3, \ldots, A_n$ is a sequence of disjoint events, then

$$P(A_1 \cup A_2 \cup A_3 \cup \ldots \cup A_n | B) = P(A_1|B) + P(A_2|B) + P(A_3|B) + \ldots + P(A_n|B)$$

Notes: (a) $P(A|B) = \dfrac{P(A)}{P(B)} \ge P(A) \quad if \quad A \subset B$

(b) $P(A|B) = \dfrac{P(B)}{P(B)} = 1 \ge P(A) \quad if \quad A \supset B$

Any statement about the comparison of $P(A|B)$ and $P(A)$ cannot be made if $A \not\subset B$.

(c) The concept of conditional probability is not valid if A and B are disjoint events i.e., $A \cap B = \phi$.

MULTIPLICATIVE LAW OF PROBABILITY

The probability of the simultaneous occurrence of two events A and B is equal to the probability of A multiplied by the conditional probability of B given that A has occurred (or it is equal to the probability of B, multiplied by the conditional probability of A given that B has occurred), *i.e.*,

Probability of (A and B) = Probability of A × Conditional probability of B, assuming A has occurred.

$$P(AB) = P(A)P(B|A) = P(B)P(A|B)$$

or,

$$P(A \cap B) = P(A)P(B|A) = P(B)P(A|B) \qquad \text{for } P(A) > 0,\ P(B) > 0$$

In case of three arbitrary events: Let A, B, and C be the three arbitrary events then the compound probability for the three events is given by

$$P(A \cap B \cap C) = P(A)P(B|A)P(C|A \cap B)$$

Generalisation: For any n events $A_1, A_2, A_3, \ldots, A_n$, we have

$$P(A_1 \cap A_2 \cap A_3 \cap \ldots \cap A_n) = P(A_1)P(A_2|A_1)P(A_3|A_1 \cap A_2)\ldots P(A_n|A_1 \cap A_2 \cap \ldots \cap A_{n-1})$$

provided $P(A_1) \neq 0, P(A_1 \cap A_2) \neq 0, \ldots, P(A_1 \cap A_2 \cap \ldots \cap A_{n-1}) \neq 0$

INDEPENDENT EVENTS

If events A and B are independent, the probability of the intersection of A and B equals the product of the probabilities of A and B, *i.e.*,

$$P(A \cap B) = P(A)P(B)$$

The converse is also true: If $P(A \cap B) = P(A)P(B)$, then A and B are independent.

THEOREM OF COMPOUND PROBABILITY FOR N INDEPENDENT EVENTS

For n independent events, the probability of the simultaneous occurrence (occurring together or in succession) is equal to the product of the probabilities of their occurring separately, *i.e.*,

$$P(A_1 \cap A_2 \cap A_3 \cap \ldots \cap A_n) = P(A_1)P(A_2)P(A_3)\ldots P(A_n)$$

BAYES' THEOREM

An event A can be explained by a set of n exhaustive and mutually exclusive hypotheses $H_1, H_2, H_3, \ldots, H_n$. Given

(a) '*a priori*' probabilities $P(H_1), P(H_2), P(H_3), \ldots, P(H_n)$; $P(H_i) \neq 0$, $i = 1, 2, 3, \ldots, n$ corresponding to a total absence of a knowledge regarding the occurrence of A, and

(b) conditional probabilities $P(A|H_1), P(A|H_2), P(A|H_3), \ldots, P(A|H_n)$

(i) It is required to form '*a posteriori*' probabilities $P(H_1|A), P(H_2|A)$, $P(H_3|A), \ldots, P(H_n|A)$.

(ii) Further, find the probabilities of materialisation of another event B, given the probabilities

$$P(B|AH_1), P(B|AH_2), P(B|AH_3), \ldots, P(B|AH_n)$$

Remarks: (a) The probabilities $P(H_i)$ is called '*a priori*' probabilities, because they exist before we have any information from the experiment itself.

(b) The probabilities $P(H_i|A)$, $i = 1, 2, 3, \ldots, n$ are called the '*posteriori*' probabilities.

(c) In case the materialisation of H_i makes B independent of A, then $P(B|AH_i) = P(B|H_i)$.

DISTRIBUTION FUNCTION

Let X be a random variable on (S, A, P). Then the function

$$F(x) = P(X \leq x), \qquad -\infty < x < \infty$$

is called the distribution function (d.f.) or cumulative distribution function of X, where S is the sample space, *viz.*, space of outcomes, A is the field of subsets in S and P is a probability function on A. For any value x, it adds up, or accumulates, all the probability assigned to that value and smaller values.

Following are some of the important properties of distribution function:

(a) If F is the distribution function of the random variable X and if $a < b$, then

$$P(a < X \leq b) = F(b) - F(a)$$

(b) If F is the distribution function of one-dimensional random variable X, then

(i) $0 \leq F(x) \leq 1$, and

(ii) $F(x) \leq F(y)$, if $x < y$

(c) If F is distribution function of one-dimensional random variable X, then

$$F(-\infty) = \lim_{x \to \infty} F(x) = 0 \text{ and } F(\infty) = \lim_{x \to \infty} F(x) = 1$$

DISCRETE PROBABILITY DISTRIBUTION OR PROBABILITY MASS FUNCTION

Let a discrete random variable X assume the values $x_1, x_2, x_3, ..., x_n$ with probabilities $p_1, p_2, p_3, ..., p_n$ respectively, where $\Sigma p_i = 1$. The specification of the set of values x_i together with their probabilities p_i $(i = 1, 2, 3, ..., n)$ defines the probability distribution of the discrete random variable X.

The probability (p) that the random variable X assumes a specified value (x) can be expressed in terms of a general mathematical expression $f(x)$ as

$$f(x) = \text{probability that } X \text{ assumes the value } x$$
$$= P(X = x)$$

The function $f(x)$ is known as 'probability mass function' (p.m.f.), or the 'probability function' of the discrete random variable X. If $f(x)$ is known, it is unnecessary to write the probabilities $p_1, p_2, p_3, ..., p_n$, because p_i may be obtained from $f(x)$ on putting $x = x_i$, *i.e.*, $p_i = f(x_i)$. The discrete probability distribution is then defined simply by stating the mathematical expression for $f(x)$, along with the set of possible values $x_1, x_2, x_3, ..., x_n$ as

x :	x_1	x_2	x_3	...	x_n	*Total*
$p = f(x)$:	$f(x_1)$	$f(x_2)$	$f(x_3)$	...	$f(x_n)$	1

Discrete probability distribution or probability mass function must satisfy the following two conditions:

(a) $p(x) \geq 0$, *or* $f(x) \geq 0 \qquad \forall\, x$

(b) $\Sigma p(x) = 1$, *or* $\Sigma f(x) = 1$

where, the summation of $p(x)$ or $f(x)$ is over all possible values of x.

EXPECTED VALUES OF DISCRETE RANDOM VARIABLES

Let a discrete random variable X assume the values $x_1, x_2, x_3, ..., x_n$ with probabilities $p_1, p_2, p_3, ..., p_n$ respectively. Then the *expectation* or *expected value* of X, written as $E(X)$, is defined as the sum of products of the different values of X and the corresponding probabilities. Thus,

$$E(X) = \sum_{i=1}^{n} p_i x_i$$

Similarly, the expected value of X^2 is defined as the sum of products of the squares of values and the corresponding probabilities, *i.e.*,

$$E(X^2) = \sum_{i=1}^{n} p_i x_i^2$$

In general, the expected value of any function $g(X)$ is defined as

$$E[g(X)] = \sum_{i=1}^{n} p_i \cdot g(x_i)$$

Now let us perform the experiment for a very large number of times under identical conditions. Let f_i gives x_i, $(i = 1, 2, ..., n)$ where $f_1 + f_2 + f_3 + ... + f_n = N$, then the mean of X is given by

$$\frac{f_1x_1 + f_2x_2 + f_3x_3 + ... + f_nx_n}{N} = \frac{f_1}{N}x_1 + \frac{f_2}{N}x_2 + \frac{f_3}{N}x_3 + ... + \frac{f_n}{N}x_n$$

When $N \to \infty$, using the statistical definition of probability, we have the limiting value for the mean given by

$$\bar{X} = p_1x_1 + p_2x_2 + p_3x_3 + ... + p_nx_n = \sum_{i=1}^{n} p_i x_i$$

which is known as the mathematical expectation of X and is denoted by $E(X)$. Thus, $E(X)$ may be regarded as the limiting value of the average value of X realised in N random experiments as $N \to \infty$.

Hence, mean of a probability distribution is the expected value of X.

$$Mean(\mu) = E(X)$$

Variance is the expected value of $(X - \mu)^2$, where μ is the mean.

$$Variance\ (\sigma^2) = E(X - \mu)^2 = E(X^2) - \{E(X)\}^2$$

CAUCHY-SCHWARZ'S INEQUALITY

If X and Y are two jointly distributed random variables with finite variances taking real values, then

$$[E(XY)]^2 \leq E(X^2)\cdot E(Y^2)$$

CONDITIONAL EXPECTATION AND VARIANCE

Let, subject to the occurrence of the event Y, random variable X takes the value x_i with probability $p_i(Y)$, *i.e.*,

$$P_i(Y = y_j) = P(X = x_i | Y = y_j), \qquad i = 1, 2, 3, \ldots, n$$

$$= \frac{P\{(X = x_i) \cap (Y = y_j)\}}{P(Y = y_j)}$$

then the conditional expectation of X, given Y, is

$$E(X|Y = y_j) = \frac{\sum x_i P(X = x_i | Y = y_j)}{P(Y = y_j)}$$

and the conditional variance of X, given Y, is

$$Var(X|Y = y_j) = E[\{X - E(X|Y = y_j)\}^2 | Y = y_j]$$

and the conditional variance of Y, given X, is

$$Var(Y|x = x_i) = E[\{Y - E(Y|X = x_i)\}^2 | X = x_i].$$

MOMENT GENERATING FUNCTION

Let X be a discrete random variable, then the moment generating function (m.g.f.) of probability distribution of X about the point A is defined by

$$M_{X=A}(t) = E\left[e^{t(X-A)}\right] = \sum_i e^{t(x_i - A)} p_i$$

where t is a real parameter and A is the point about which the m.g.f. is found out.

Expanding the exponential in the above equation, we have

$$M_{X=A}(t) = \sum_i p_i + t\sum_i p_i(x_i - A) + \frac{t^2}{2!}\sum_i p_i(x_i - A)^2 + \ldots + \frac{t^r}{r!}\sum_i p_i(x_i - A)^r + \ldots$$

$$= 1 + t\mu_1' + \frac{t^2}{2!}\mu_2' + \ldots + \frac{t^r}{r!}\mu_r' + \ldots$$

where μ_r' is the moment of order r about the point A.

FACTORIAL MOMENT GENERATING FUNCTION

If X be the discrete random variable, then the factorial moment generating function $w(t)$ of the probability distribution of X about the origin is given by

$$w(t) = E[(1 + t)^X]$$

Since,

$$(1 + t)^X = \sum_{r=0}^{\infty} \frac{X^r t^r}{r!}, \qquad X^r = X(X - 1)(X - 2) \dots (X - \overline{r - 1})$$

$$\therefore \qquad w(t) = 1 + t\mu_1' + \frac{t^2 \mu_2'}{2!} + \dots + \frac{t^r \mu_r'}{r!} + \dots$$

μ_r' is the coefficient of $t^r/r!$ in the expansion of $w(t)$ in ascending power of t.

CUMULANT GENERATING FUNCTION

The logarithm of m.g.f. about origin of the random variable X is called the second moment generating function or cumulant generating function. In other words, if the logarithm of the m.g.f. of a distribution is expanded as a convergent series in powers of t *viz.*,

$$K_X(t) = \log_e M_X(t) = \sum_{r=1}^{n} k_r \cdot \frac{t^r}{r!} \equiv k_1 t + k_2 \frac{t^2}{2!} + \dots + k_r \frac{t^r}{r!} + \dots$$

then the coefficients of $t^r/r!$, k_r is called the rth cumulant of the random variable X and $K(t)$ is called the cumulative function.

PROBABILITY GENERATING FUNCTION

If a discrete random variable X takes integral values 0, 1, 2, 3, ... only, then the probability generating function $P(s)$ is defined as

$$P(s) = p_0 + p_1 s + p_2 s^2 + \dots = \sum_{n=0}^{\infty} p_n s^n = E(s^X)$$

In the expansion of $P(s)$, the coefficient of s^n in powers of s gives $P(X = n)$.

SPECIAL DISCRETE DISTRIBUTIONS

Bernoulli $\quad f(x) = p^x(1 - p)^{1-x}, \quad x = 0, 1$

$0 < p < 1$ $\quad M(t) = 1 - p + pe^t$

$\mu = p, \qquad \sigma^2 = p(1 - p)$

Binomial $\quad f(x) = \dfrac{n!}{x!(n-x)!} p^x (1-p)^{n-x}, \quad x = 0,1,2,\dots,n$

$$M(t) = (1 - p + pe^t)^n$$

$$\mu = np, \qquad \sigma^2 = np(1 - p)$$

Geometric

$$f(x) = (1 - p)^{x-1}p, \quad x = 1, 2, 3, \ldots$$

$$M(t) = \frac{pe^t}{1-(1-p)e^t}, \quad t < -\ln(1-p)$$

$$\mu = \frac{1}{p}, \quad \sigma^2 = \frac{1-p}{p^2}$$

Hypergeometric

$$f(x) = \frac{\binom{N_1}{x}\binom{N_2}{n-x}}{\binom{N}{n}}, \quad x \le n, x \le N_1, n-x \le N_2$$

$$\mu = n\left(\frac{N_1}{N}\right), \quad \sigma^2 = n\left(\frac{N_1}{N}\right)\left(\frac{N_2}{N}\right)\left(\frac{N-n}{N-1}\right)$$

Negative Binomial

$$f(x) = \binom{x-1}{r-1}p^r(1-p)^{x-r}, \quad x = r, r+1, r+2, \ldots$$

$$M(t) = \frac{(pe^t)^r}{[1-(1-p)e^t]r}, \quad t < -\ln(1-p)$$

$$\mu = r\left(\frac{1}{p}\right), \quad \sigma^2 = \frac{r(1-p)}{p^2}$$

Poisson

$$f(x) = \frac{\lambda^x e^{-\lambda}}{x!}, \quad x = 0, 1, 2, \ldots$$

$$M(t) = e^{\lambda(e^t - 1)}$$

$$\mu = \lambda, \quad \sigma^2 = \lambda$$

Uniform

$$f(x) = \frac{1}{m}, \quad x = 1, 2, \ldots, m$$

$$\mu = \frac{m+1}{2}, \quad \sigma^2 = \frac{m^2-1}{12}.$$

PROBABILITY OF CONTINUOUS RANDOM VARIABLE

The probability that the random variable will take on a value in the interval from a to b is given by

$$P(a \le X \le b) = \sum_{i=1}^{m} f(x_i)\cdot \Delta x$$

If $f(x_i)$ is an integrable function defined for all values of the random variable, we shall define the probability that the value of the random variable falls between a and b by letting $\Delta x \to 0$, namely, as

$$P(a \le X \le b) = \int_a^b f(x)dx.$$

CONTINUOUS PROBABILITY DISTRIBUTION

A continuous probability distribution is defined by stating the mathematical expression for the p.d.f. $f(x)$ and specifying the range (a, b). The range of the variable may be finite or infinite, but even when the range is finite, it is convenient to consider the range as infinite, by simply postulating that the density function is zero outside the specified range. Thus, if X takes values in the closed interval $[a, b]$ or $a \le X \le b$ only and its density function is $\phi(x)$, then its distribution is equivalent to the variate with density function defined by

$$f(x) = \begin{cases} 0 & x < a \\ \phi(x) & a \le x \le b \\ 0 & x > b \end{cases}$$

The curve represented by the equation

$$y = f(x)$$

is known as the probability curve. Geometrically, the integral of the p.d.f. represents the area under the probability curve, and hence

$P(c \le x \le d) =$ Area under probability curve between the vertical lines c and d.

The cumulative distribution function (c.d.f.) $F(x)$ represents the probability that the random variable takes a value less than or equal to a specified value x. Geometrically, $F(x)$ shows the area under probability curve to the left of the ordinate at any specified value. For example,

$F(c) = P(x \le c)$

$\quad$ = Area under probability curve to the left of the vertical line at c.

If the value of $F(x)$ are known for all values of x, then the probability distribution is completely given. Because then the probabilities of intervals can be obtained from the relation

$P(c \le x \le d) =$ Area between the ordinates at c and d.

$\quad$ = (Area to the left of the ordinate at d) – (Area to the left of the ordinate at c)

$\quad = F(d) - F(c)$

Putting $c = d$, we get

$P(c \le x \le c) =$ Area between the ordinates at c and c, which is zero.

i.e., $P(x = c) = 0$

Thus, for a continuous probability distribution, the probability that a random variables takes any specific value is zero and the probability function $f(x)$ possesses the following properties:

(a) $f(x) \geq 0 \quad \forall x$ *i.e., for* $-\infty < x < \infty$

(b) $\int_{-\infty}^{+\infty} f(x)dx = 1$

(c) If $(a,b) \subseteq S$, the probability of the event $\{a < X < b\}$ is

$$P(a < X < b) = \int_a^b f(x)dx$$

Since negative probability has no meaning, property (a) is obviously necessary. Property (b) corresponds to the requirement that the probability of an event that is certain to occur should be equal to unity. The geometric interpretation of the property (c) is that it denotes an area bounded by the curve $y = f(x)$, the ordinates $x = a$, $x = b$ and the axis of x, while

$$\int_{-\infty}^{+\infty} f(x)dx = 1$$

means that total area under the curve $y = f(x)$ where $f(x)$ is the p.d.f. should be unity.

Let, $F(x) = \int_{-\infty}^{x} f(x)dx = P(X \leq x)$ then the function $F(x)$ is probability that $X \leq x$. Also,

$$F(b) = P(X \leq b) \text{ and } F(b) - F(a) = \int_a^b F(x)dx = P(a \leq X \leq b)$$

$F(x)$ is called the cumulative distribution function (c.d.f.) with following properties:

(a) Since $f(x) \geq 0$, $\int_a^x f(x)dx \geq 0$ when $x \geq a$. In particular, $\int_{-\infty}^{x} f(x)dx \geq 0$ or $F(x) \geq 0$.

(b) Since $\int_a^x f(x)dx$ is a continuous of $f(x)$, therefore $F(x)$ is a continuous and R-integral function so that

$$F(a - 0) = F(a) = F(a + 0),\ F(b - 0) = F(b) = F(b + 0)$$

and $$P(a < X < b) = P(a \leq X < b) = P(a < X \leq b) = P(a \leq X \leq b)$$

(c) $F(x)$ is a non-decreasing function of x because $f(x) \geq 0$

(d) If $F(x)$ is a continuous function, then $F'(x) = f(x)$

(e) $F(-\infty) = \lim_{x \to -\infty} \int_{-\infty}^{x} f(x)dx = 0$

(f) $F(\infty) = \lim_{x \to \infty} \int_{-\infty}^{x} f(x)dx = \int_{-\infty}^{\infty} f(x)dx = 1.$

SPECIAL CONTINUOUS DISTRIBUTIONS

Beta: $(0 < \alpha; 0 < \beta)$

$$f(x) = \frac{\Gamma(\alpha+\beta)}{\Gamma(\alpha)\Gamma(\beta)} x^{\alpha-1}(1-x)^{\beta-1}, \quad 0 < x < 1$$

$$\mu = \frac{\alpha}{\alpha+\beta}, \quad \sigma^2 = \frac{\alpha\beta}{(\alpha+\beta+1)(\alpha+\beta)^2}$$

Chi-square: $(\chi^2(r); r = 1, 2, \ldots)$

$$f(x) = \frac{1}{\Gamma(r/2)2^{r/2}} x^{r/2-1} e^{-x/2}, \quad 0 \le x < \infty$$

$$M(t) = \frac{1}{(1-2t)^{r/2}}, \quad t < \frac{1}{2}$$

$$\mu = r, \quad \sigma^2 = 2r$$

Exponential: $(0 < \theta)$

$$f(x) = \frac{1}{\theta} e^{-x/\theta}, \quad 0 \le x < \infty$$

$$M(t) = \frac{1}{1-\theta t}, \quad t < \frac{1}{\theta}$$

$$\mu = \theta, \quad \sigma^2 = \theta^2$$

Gamma: $(0 < \alpha, 0 < \theta)$

$$f(x) = \frac{1}{\Gamma(\alpha)\theta^\alpha} x^{\alpha-1} e^{-x/\theta}, \quad 0 \le x < \infty$$

$$M(t) = \frac{1}{(1-\theta t)^\alpha}, \quad t < \frac{1}{\theta}$$

$$\mu = \alpha\theta, \quad \sigma^2 = \alpha\theta^2$$

Normal: $\left(N(\mu, \sigma^2), -\infty < \mu < \infty, 0 < \sigma\right)$

$$f(x) = \frac{1}{\sigma\sqrt{2\pi}} e^{-(x-\mu)^2/2\sigma^2}, \quad -\infty < x < \infty$$

$$M(t) = e^{\mu t + \sigma^2 t^2/2}$$

$$E(X) = \mu, \quad \mathrm{Var}(X) = \sigma^2$$

Uniform: $\left(U(a,b), -\infty < a < b < \infty\right)$

$$f(x) = \frac{1}{b-a}, \quad a \le x \le b$$

$$M(t) = \frac{e^{tb} - e^{ta}}{t(b-a)}, \quad t \ne 0; \quad M(0) = 1$$

$$\mu = \frac{a+b}{2}, \quad \sigma^2 = \frac{(b-a)^2}{12}.$$

SAMPLING THEORY

The idea of sampling is to study a part in order to gain information about the whole. The aggregate of statistical information on a particular character of all the members covered by an investigation, is called Population or Universe, and the selected part, which is used to ascertain the characteristics of population is called sample. The population consist of individual units or members, some of which are included in the sample. The total number of members of the population and the number included in the sample, are called population size and sample size respectively.

Population may be finite or infinite depending on its size. It is said to be finite when the number of members of the population can be expressed as a definite quantity, otherwise it is infinite. A complete list of all the units in a finite population, properly numbered for identification, is called a sampling frame. The population of concrete individuals is called existent population and the collection of all possible ways in which an event can materialise is called hypothetical population.

Sampling is used to get as much information as possible, at most the whole of the information, about the population from which the sample has been drawn. However, the logic of the theory of sampling is the logic of induction, *i.e.*, we pass from particular (*i.e.*, sample) to general (*i.e.*, population) and hence all results are expressed in terms of probability.

There are certain important laws on which sampling theory is based. They are:

(a) *Law of Statistical Regularity*—A group of objects chosen at random from a particular group tends to possess characteristics of that group. In other words, a large number of items taken at random from each group are almost sure on an average to possess characteristics of each group. This law implies that the sample size should be large and the sample on the basis of which inference is drawn about the universe must be selected at random.

(b) *Principle of Inertia of Large Numbers*—As sample size increases, results would be more reliable. It implies that large numbers are relatively more stable in their characteristics than small numbers.

(c) *Principle of Persistence*—If some items of the universe possess some specific characteristics, these characteristics would be found in the sample also and even if the sample size is increased or the population is increased, these characteristics would be reflected in the same manner as in the previous case.

(d) *Principle of Optimisation*—It aims at obtaining a desired level of efficiency at the minimum cost and, at the same time, of obtaining maximum possible efficiency with a given level of costs. Therefore, the size of the sample is maintained in such a way that the results are optimised in terms of cost and efficiency.

(e) *Principle of Validity*—A sample design is called valid only when the inferences drawn from it are valid for the universe from which the sample has been taken. For this, not only the sample has to be random, but its size has to be adequate and the collection of data and its analysis have to be scientifically done.

Types of Sampling: Some of the important types of sampling are:

(a) *Simple Random Sampling*—It is the process of selection of a group of units in such a manner that every unit of the population has an equal chance of being included in the sample. There are two ways of drawing a simple random sample—(i) Simple Random Sampling With Replacement (SRSWR) and (ii) Simple Random Sampling Without Replacement (SRSWOR).

(b) *Purposive Sampling*—A sample which is selected on the basis of individual judgement of the sampler, is called a purposive sample.

(c) *Stratified Sampling*—In this method, the population is subdivided into several parts, called strata; and then a sub-sample is chosen from each of them. All the sub-samples combined together give the stratified sample. If the selection from strata is done by random sampling, the method is known as stratified random sampling.

(d) *Systematic Sampling*—It involves the selection of sample units at equal intervals, after all the units in the population have been arranged in some order. If the population size is finite, the units may be serially numbered and arranged. From the first k of these, a single unit is chosen at random. This unit and every k-th unit thereafter constitutes a systematic sample.

(e) *Multi-stage Sampling*—It refers to a sampling procedure which is carried out in several stages. The population is first divided into large groups, called first-stage units. These first-stage units are again divided into smaller units, called second-stage units–the second-stage units into third-stage units, and so on, until we reach the ultimate units.

Multi-stage sampling enables existing divisions and sub-divisions of the population to be used as units at various stages and permits the field work to be concentrated, although a large area is covered.

Sampling Distribution: Any statistical measure calculated on the basis of sample observations is called a *statistic*, such as, the proportion of defectives observed in the sample, sample mean, sample s.d., *etc*. Any statistical measure based on all units in the population is called a *parameter*, such as proportion of defectives in the whole lot, population mean, population s.d., *etc*.

If many samples of a fixed size are drawn from a given population, the group of units constituting the sample also varies from one sample to another. Thus, the value of a statistic varies from sample to sample, but the parameter remains a constant. This variation in the value of a statistic is known as *sampling fluctuation*. The parameter has no sampling fluctuation. Usually parameters are unknown and statistics are used as estimates of parameters. The probability distribution of a statistic is called its sampling distribution and the standard deviation in the sampling distribution is called *standard error* of the statistic.

Errors in Sampling: In statistics, error means the difference between the true value and the estimated value. It can be classified in two categories—(a) sampling errors, and (b) non-sampling errors.

Sampling errors have their origin in sampling and they arise on account of the fact that sample has been used to estimate parameters or population values. Generally, they arise due to—(i) improper selection of the sample, (ii) substitution, (iii) faulty demarcation of statistical units, and (iv) variability of population and wrong method of estimation.

Non-sampling errors generally arise when data are not properly observed, approximated and processed. They, generally, arise due to—(i) improper or ambiguous definition of the various terms, (ii) incomplete questionnaire and defective methods of interviewing, (iii) personal bias of the investigator, (iv) lack of trained and qualified investigators, (v) improper coverage and inadequate or incomplete response, and (vi) errors in compilation and tabulation.

Classes of Errors: Broadly speaking, errors may be either biased or unbiased. Biased errors are those which arise on account of some bias in the mind of the investigator or the informant or in the instruments of measurement. Whereas unbiased errors are those which arise just on account of chance.

SIMPLE RANDOM SAMPLING

Let a simple random sample of n members be drawn from a finite population consisting of N members, *i.e.*, sample size = n, and population size = N. Then,

total number of possible samples with distinct permutations of members

(a) in SRSWR = N^n

(b) in SRSWOR = ${}^NP_n = N(N-1)(N-2) \dots (N-n+1)$

Among these, the number of cases favourable to each group (ignoring the order of drawing) or distinct combination of members

(a) in SRSWR is not the same.

(b) in SRSWOR is the same, *viz.*, $n!$

Therefore, we consider

(a) in SRSWR, N^n possible samples with distinct permutations each with a probability $1/N^n$.

(b) in SRSWOR, ${}^NC_n = N!/n!(N-n)!$ possible samples with distinct combinations each with a probability $1/{}^NC_n$.

STANDARD ERROR (S.E.)

(a) For random samples of size n,

$$\text{S.E. of sample mean } (\bar{y}) = \sigma/\sqrt{n}$$

$$\text{S.E. of sample proportion } (p) = \sqrt{PQ/n}$$

where, σ is population standard deviation (s.d.) and P is the population proportion ($P + Q = 1$).

If the random sample is drawn without replacement from a finite population of size N, then the above formulae can be modified on multiplication by the correction factor $\sqrt{(N - n)/(N - 1)}$.

(b) If $\bar{y}_1$ and $\bar{y}_2$ represent the means calculated from independent random samples of sizes n_1 and n_2 drawn from two populations with standard deviations σ_1 and σ_2 respectively, then

$$\text{S.E. of } (\bar{y}_1 - \bar{y}_2) = \sqrt{\sigma_1^2/n_1 + \sigma_2^2/n_2}$$

If the populations have the same s.d. σ, then

$$\text{S.E. of } (\bar{y}_1 - \bar{y}_2) = \sigma\sqrt{1/n_1 + 1/n_2}$$

(c) If p_1 and p_2 denotes the proportions calculated from independent random samples of sizes n_1 and n_2 and drawn from two populations with proportions P_1 and P_2 respectively, then

$$\text{S.E. of } (p_1 - p_2) = \sqrt{P_1Q_1/n_1 + P_2Q_2/n_2}$$

If the two population proportions P_1 and P_2 are equal, say $P_1 = P_2 = P$, then

$$\text{S.E. of } (p_1 - p_2) = \sqrt{PQ(1/n_1 + 1/n_2)}$$

(d) For random samples from a normal population with s.d. σ,

$$\text{S.E. of sample s.d. } (S) = \sigma/\sqrt{2n}$$

$$\text{S.E. of sample variance } (S^2) = \sigma^2\sqrt{2/n}$$

These two formulae are approximate and used only when the sample size n is large (greater than 50).

PROBABLE ERROR (P.E.)

It is defined to be a quantity such that exactly half of the possible values of the statistic differ from its expected value by more than that quantity. If t is a statistic and θ its expected value, then the probability that the difference of t and θ exceeds the probable error (P.E.) is exactly 1/2.

$$P\{|t - \theta| > \text{P.E.}\} = 1/2, \text{ and consequently, } P\{|t - \theta| < \text{P.E.}\} = 1/2$$

In random samples of size n from a normal population with mean μ and s.d. σ,

$$\text{P.E. of mean} = 0.6745\sigma/\sqrt{n} = 0.6745\times\text{S.E. of mean}$$

In case of large samples, S.E. of correlation coefficient (r) is $(1 - r^2)/\sqrt{n}$, so that

$$\text{P.E. of correlation coefficient} = 0.6745\times(1 - r^2)/\sqrt{n}.$$

DISTRIBUTIONS USED IN SAMPLING THEORY

In sampling theory, probability distributions which are derived from Normal distribution are used.

(a) *Standard Normal Distribution*: If a random variable y is normally distributed with mean μ and standard deviation σ, then

$$z = (y - \mu)/\sigma = (\textit{Normal variable} - \textit{Mean})/\textit{s.d.}$$

is called a standard normal variate. The probability distribution of z is called standard normal distribution, and is defined by the p.d.f.

$$p(z) = \left(1/\sqrt{2\pi}\right)e^{-z^2/2}; \qquad (-\infty < z < +\infty)$$

Here, mean = 0, and s.d. = 1.

(b) *Chi-square* (χ^2) *Distribution*—A random variable y is said to follow Chi-square (χ^2) distribution if its p.d.f. is of the form

$$f(y) = K.e^{-y/2}\, y^{(n/2)-1}; \qquad (0 < y < \infty)$$

where K is a constant.

Here, mean = n, and s.d. = $\sqrt{2n}$, where n is the number of degrees of freedom of Chi-square distribution.

(c) *Student's t Distribution*—A random variable is said to follow Student's t distribution, or simply t distribution, if its p.d.f. is of the form

$$f(t) = K\left(1 + t^2/n\right)^{-(n+1)/2}; \qquad (-\infty < t < +\infty)$$

where, K is a constant and parameter n (positive integer) is called the number of degrees of freedom.

Here, mean = 0, and $s.d. = \sqrt{n/(n-2)}$; $(n > 2)$.

(d) *Snedecor's F Distribution*—A random variable is said to follow F distribution with degrees of freedom (n_1, n_2) if its p.d.f. is of the form

$$f(F) = KF^{(n_1/2)-1}\left(n_2 + n_1F\right)^{-(n_1+n_2)/2}; \qquad (0 < F < \infty)$$

where K is a constant. The percentage points of F distribution with d.f. (n_1, n_2) are denoted by $F_{p,(n_1,n_2)}$ or briefly F_p if the d.f. are understood. The lower percentage points are given by $F_{1-p,(n_1,n_2)} = 1/F_{p,(n_2,n_1)}$, *i.e.*, the lower percentage point is the reciprocal of the upper percentage point with the order of d.f. reversed.

Here, mean $= n_2/(n_2 - 2)$, mode $= n_2(n_1 - 2)/n_1(n_2+2)$, and *s.d.* $= \left(\dfrac{n_2}{n_2-2}\right)\sqrt{\dfrac{2(n_1+n_2-2)}{n_1(n_2-4)}}$.

(e) *Non-central t-distribution*—If X is distributed as $N(\mu,\sigma^2)$ and Y is an independent χ^2 variate with n d.f., then $t = (X/\sigma)/\left(\sqrt{Y}/\sqrt{N}\right)$ has a non-central t-distribution with n d.f. and with non-centrality parameter μ/σ.

(f) *Non-central F Distribution*—Let χ_1^2 be a non-central χ^2 with n_1 d.f. and non-centrality parameter ψ^2 and let χ_2^2, independent of χ_1^2, be a central χ^2 with n_2 d.f., then the ratio $\left(\chi_1^2/n_1\right)/\left(\chi_2^2/n_2\right)$ is known as a non-central F statistic with (n_1, n_2) d.f. and with non-centrality parameter ψ^2.

RELATIONS BETWEEN STANDARD NORMAL, χ^2, t AND F DISTRIBUTIONS

(a) A Chi-square variate with n degree of freedom is the sum of the squares of n independent standard normal variates, *i.e.*, $\chi^2 = \sum_{i=1}^{n} z_i^2$.

(b) A t-variate with n d.f. is the ratio of a standard normal variate and the square-root of an independent Chi-square variate when divided by its degree of freedom, *i.e.*, $t = z/\sqrt{y/n}$.

(c) F-variate with (n_1, n_2) d.f. is the ratio of two independent Chi-square variates each divided by its own d.f. (Chi-square with n_1 d.f. appears in the numerator), *i.e.*, $F = (y_1/n_1)/(y_2/n_2)$.

(d) F-variate with $(1, n)$ d.f. is equivalent to the square of a t-variate with n d.f. The relation between upper percentage points is $F_{p,(1,n)} = t^2_{p/2,\,n}$.

(e) When n d.f. is large, $\sqrt{2\chi^2} - \sqrt{2n-1}$ is approximately a standard normal variate.

(f) When n d.f. is large, a t-variate is approximately a standard normal variate. The relation between percentage points is $z_p = t_{p,\infty}$.

(g) For large degrees of freedom, both the Chi-square and the F distributions tend to normal distributions.

ANALYTIC METHOD

The distribution of the variate Y_i is $\left(1/\sqrt{2\pi}\right)\exp\left(\dfrac{-1}{2}y_i^2\right)dy_i, \quad i = 1,2,3,\ldots,n.$

Hence, the joint distribution of n independent variates Y_i is

$$dF = (2\pi)^{\frac{-1}{2}n} \exp\left[\frac{-1}{2}\left(y_1^2 + y_2^2 + \dots + y_n^2\right)\right] dy_1\, dy_2 \dots dy_n$$

The distribution function of $Z = Y_1^2 + Y_2^2 + \dots + Y_n^2$ is given by

$$F(y) = \int\int\int \dots \int (2\pi)^{\frac{-1}{2}n} \exp\left[\frac{-1}{2}\sum_{i=1}^{n} y_i^2\right] dy_1\, dy_2 \dots dy_n$$

where the domain of y's is conditioned by: $y_1^2 + y_2^2 + \dots + y_n^2 \le z$.

THE METHOD OF CHARACTERISTIC FUNCTIONS

If $y_1, y_2, y_3, \dots, y_n$ is a random sample of size n, so that, $y_1, y_2, y_3, \dots, y_n$ are independent and identically distributed and $\bar{y}$ be the sample mean, then

(a) $\phi y_1(t) + \phi y_2(t) + \dots + \phi y_n(t) = \phi y_1(t) \times \dots \times \phi y_n(t) = [\phi y(t)]^n$

(b) $\phi \bar{y}(t) = \phi y_1(t/n) \cdot \phi y_2(t/n) \cdots \phi y_n(t/n) = [\phi y(t/n)]^n$

The characteristic function (*c.f.*) of the Poisson distribution with parameter λ is $\exp\left[\lambda\left(e^{it} - 1\right)\right]$. Therefore, by the property (b) of *c.f.*, the *c.f.* of the mean is

$$\left[\exp \lambda\left(e^{it/n} - 1\right)\right]^n = \exp\left[n\lambda\left(e^{it/n} - 1\right)\right]$$

Hence, the distribution of the mean is the Poisson distribution with parameter $n\lambda$, the interval being $1/n$, instead of unity.

The *c.f.* of the normal distribution with mean μ and s.d. σ is $\exp\left[(-1/2)t^2\sigma^2 + it\mu\right]$. The *c.f.* of the distribution of the mean of n values is then

$$\exp\left\{n\left(\frac{-1}{2}\cdot\frac{t^2\sigma^2}{n^2} + \frac{it\mu}{n}\right)\right\} = \exp\left\{\frac{-1}{2}\cdot\frac{t^2\sigma^2}{n} + it\mu\right\}$$

This is the *c.f.* of the normal distribution with mean μ and s.d. $\sigma/\sqrt{n}$.

DISTRIBUTION OF QUARTILES

Consider a random sample of size n drawn from the continuous population whose density function is $f(y)$. The probability that one value of y will fall in the interval $(-\infty, u)$ is

$$\int_{-\infty}^{u} f(y)dy = F(u) = F$$

and the probability that one value of y will fall in the interval $(u + du, \infty)$ is

$$\int_{u+du}^{\infty} f(y)dy = 1 - \int_{-\infty}^{u+du} f(y)dy = 1 - F(u + du)$$

Consequently, the probability that out of a sample of n, $(l - 1)$ values will fall below the value u, one in the infinitesimal interval $(u, u + du)$ and the remaining $(n - l)$ fall above u, is

$$dp = \frac{n!F^{l-1} f(u)du[1 - F(u + du)]^{n-l}}{(l - 1)!(n - l)!} = \frac{n!}{(l - 1)!(n - l)!} F^{l-1}(1 - F)^{n-l} dF$$

where $dF = f(u)$ and infinitesimals of order higher than the first have been neglected on the right hand side.

This is the distribution of u, the number of the sample to which a proportion l/n of the members fall, *i.e.*, of the lth quartile.

THEORY OF ESTIMATION

Theory of estimation, developed by Prof. R.A. Fisher in 1930, refers to the technique and methods by which population parameters are estimated from sample studies. For a parameter, there can be a number of estimators and therefore the main part of estimation problem lies in selecting out sample functions or estimators such that their distributions are concentrated as closely as possible round the true parameter values.

Now, suppose that we have a random sample $y_1, y_2, y_3, \ldots, y_n$ on a variable y, whose distribution in the population involves an unknown parameter θ. It is required to find an estimate of θ on the basis of sample values. The estimation can be done in two different ways—(a) Point Estimation, and (b) Interval Estimation.

In point estimation, the estimated value is given by a single quantity, which is a function of sample observations, *i.e.*, statistic. This function is called the *estimator* of the parameter, and the value of the estimator in a particular sample is called an *estimate*.

In interval estimation, an interval within which the parameter is expected to lie is given by using two quantities based on sample values. This is known as *confidence interval*, and the two quantities which are used to specify the interval, are known as *confidence limits*.

An estimator is said to be a good estimator if it is—(a) unbiased, (b) consistent, (c) efficient, and (d) sufficient.

The point estimation uses two methods for finding an estimator which satisfies most of the criteria of a good estimator:

(a) Method of maximum likelihood—it consists in choosing as an estimator of θ that statistic, which when substituted for θ, maximises the likelihood function. Such a statistic is called a maximum likelihood estimator (m.l.e.).

(b) Method of moments—it consists in equating the first few moments of the population with the corresponding moments of the sample, *i.e.*, setting $\mu_r' = m_r'$, where $\mu_r' = E(y^r)$ and $m_r' = \Sigma y_i^r / n$. Since, the parameters enter into the population moments, these relations when solved for the parameters give the estimates.

The procedure of determining an interval (t_1, t_2) that will include a population parameter, say θ, with a certain probability $(1 - \alpha)$ is known as interval estimation. Here α is the probability that the interval does not include the true parameter value. Thus, we have

$$P(\theta, t_1, t_2) = P(t_1 < \theta < t_2) = 1 - \alpha$$

where t_1 and t_2 are random variables depending upon some estimator $\hat{\theta}$ of θ and respectively referred to as lower and upper limits of the interval. The interval (t_1, t_2) is called the confidence interval with $(t_2 - t_1)$ as the measure of precision of the estimate and $(1 - \alpha)$ is the measure of reliability.

UNBIASEDNESS

A statistic t is said to be an unbiased estimator of a parameter θ, if the expected value of t is θ, *i.e.*, $E(t) = \theta$.

Otherwise, the estimator is said to be biased. The bias of a statistic in estimating θ is given as: Bias $= E(t) - \theta$.

CONSISTENCY

A statistic t_n computed from a sample of n observations is said to be a consistent estimator of a parameter θ, if it converges in probability to θ as n tends to infinity. This implies that the larger the sample size (n), the less is the chance that the difference between t_n and θ will exceed any fixed value. Thus, given any arbitrary small positive quantity ε we have

$$\underset{n\to\infty}{Lt}\ P\{|t_n - \theta| > \varepsilon\} = 0$$

If $E(t_n) \to \theta$ and $Var(t_n) \to 0$ as $n \to \infty$, the statistic t_n will be a consistent estimator of θ.

Note: Consistency is a limiting property and several consistent estimators may exist for the same parameter.

EFFICIENCY

Of the two consistent estimators for the same parameter, the statistic with the smaller sampling variance is said to be more efficient. Thus, if t and t' are both consistent estimators of θ, and $Var(t) < Var(t')$ then t is more efficient than t' in estimating θ.

Let, V_0 be the variance of the most efficient estimator and V be the variance of any other estimator, then the efficiency of the estimator is given by V_0/V.

A statistic t which has the minimum variance among all estimators of θ is called the minimum variance estimator (MVE) and a statistic which is unbiased and has also the minimum variance is said to be the minimum variance unbiased estimator (MVUE).

SUFFICIENCY

A statistic t will be a sufficient estimator of θ if t exists such that the joint distribution of the sample is expressible as the product of two factors, one of which is the sampling distribution of t and contains θ, and the other factor is independent of θ. Thus, if y_1, y_2, y_3, ..., y_n is a random sample from a population whose p.m.f. or p.d.f. is $f(y,\theta)$, and t is a sufficient estimator of θ then we have

$$f(y_1,\theta)\cdot f(y_2,\theta)\cdot f(y_3,\theta)\cdot\ldots\cdot f(y_n,\theta) = g(t,\theta)\cdot h(y_1,y_2,y_3,\ldots,y_n)$$

where $g(t,\theta)$ is the sampling distribution of t and contains θ, but $h(y_1,y_2,y_3,\ldots,y_n)$ is independent of θ.

The parameter θ occurring in the joint distribution of all the sample observations can be contained in the distribution of the statistic t, it is said that t alone can provide all information about θ and is therefore sufficient for θ.

METHOD OF MAXIMUM LIKELIHOOD

Let y_1, y_2, y_3, ..., y_n be a random sample from a population with p.m.f. (for discrete case) or p.d.f. (for continuous case) $f(y,\theta)$, where θ is the parameter. Then, the likelihood function of the sample is given by the joint distribution of the sample observations as

$$L = f(y_1,\theta)\cdot f(y_2,\theta)\cdot f(y_3,\theta)\cdot\ldots\cdot f(y_n,\theta)$$

Since, log L is maximum when L is maximum, hence the maximum likelihood estimator of q is obtained by maximising log L. This can be obtained by differentiating log L partially with respect to q, and using the two relations:

$$\left[\frac{\partial(\log L)}{\partial\theta}\right]_{\theta=\theta_0} = 0, \text{ and } \left[\frac{\partial^2(\log L)}{\partial\theta^2}\right]_{\theta=\theta_0} < 0$$

Note: The maximum likelihood estimator (MLE) is invariant under functional transformations, *i.e.*, if T is an MLE of θ, and $g(\theta)$ is a function of θ, then $g(T)$ is the MLE of $g(\theta)$.

METHOD OF MOMENTS

Let the density function of a distribution be $f(y, \theta_1, \theta_2, \theta_3, \ldots, \theta_k)$, then

$$\mu_r' = \int_{-\infty}^{\infty} y^r f(y, \theta_1, \theta_2, \theta_3, \ldots, \theta_k)dy, \qquad r = 1,2,3,\ldots,k$$

so that $\mu_1', \mu_2', \mu_3', \ldots, \mu_k'$ are the functions of $\theta_1, \theta_2, \theta_3, \ldots, \theta_k$. Solving the equations by the value of $\mu_1', \mu_2', \mu_3', \ldots, \mu_k'$, we get the value of $\theta_1, \theta_2, \theta_3, \ldots, \theta_k$ in terms of $\mu_1', \mu_2', \mu_3', \ldots, \mu_k'$, the sample moments.

Note: The estimates obtained by this method are consistent but not in general efficient. However, they become efficient when $n \to \infty$.

INTERVAL ESTIMATION

Let $y_1, y_2, y_3, \ldots, y_n$ be random sample from a population of a known mathematical form which involves an unknown parameter θ. t_1 and t_2 are two functions based on sample observations such that the probability of θ being included in the interval (t_1, t_2) has a given value, say α, then, $P(t_1 \leq \theta \leq t_2) = \alpha$.

The two quantities t_1 and t_2 which denote the lower and upper limits of the interval are known as confidence limits, and the probability, α, with which the confidence interval will include the true value of the parameter is known as confidence coefficient of the interval.

The computation of confidence limits is based on the knowledge of sampling distribution of an appropriate statistic. Let us assume a random sample of size n from a normal population $N(\mu, \sigma^2)$, where the variance σ^2 is known. We want to find 95% confidence limits for the unknown parameter μ.

Since, sample mean $(\bar{y})$ follows normal distribution with mean μ and variance σ^2/n, therefore

$$z = (\bar{y} - \mu)/\left(\sigma/\sqrt{n}\right)$$

has a standard normal distribution.

Since 95% of the area under the standard normal curve lies between the ordinates at $z = \pm 1.96$, we get

$$P\left[-1.96 \leq (\bar{y} - \mu)/\left(\sigma/\sqrt{n}\right) \leq 1.96\right] = 0.95$$

which implies that in 95% of cases the following inequalities hold

$$-1.96 \leq (\bar{y} - \mu)/\left(\sigma/\sqrt{n}\right) \leq 1.96 \quad \Rightarrow \quad \bar{y} - \frac{1.96\sigma}{\sqrt{n}} \leq \mu \leq \bar{y} + \frac{1.96\sigma}{\sqrt{n}}$$

The interval $\left(\bar{y} - 1.96\sigma/\sqrt{n}, \bar{y} + 1.96\sigma/\sqrt{n}\right)$ is known as the 95% confidence interval for μ, and the 95% confidence limits for μ are $\bar{y} \pm 1.96\sigma/\sqrt{n}$.

Also, 99% of area under the standard normal curve lies between the ordinates at $z = \pm 2.58$, and 99.73% (*i.e.*, almost whole) of the area lies between the ordinates at $z = \pm 3$. Hence, the 99% confidence limits for μ are $\bar{y} \pm 2.58\sigma/\sqrt{n}$ and almost sure limits for μ are $\bar{y} \pm 3\sigma/\sqrt{n}$.

APPROXIMATE CONFIDENCE LIMITS—(FOR ANY DISTRIBUTION OF LARGE SAMPLES)

(a) for mean μ : 95% confidence limits $= \bar{y} \pm 1.96 \times S.E.$ of $\bar{y}$

99% confidence limits $= \bar{y} \pm 2.58 \times S.E.$ of $\bar{y}$

Almost sure limits $= \bar{y} \pm 3 \times S.E.$ of $\bar{y}$

(b) for proportion P : 95% confidence limits $= p \pm 1.96 \times S.E.$ of p

99% confidence limits $= p \pm 2.58 \times S.E.$ of p

Almost sure limits $= p \pm 3 \times S.E.$ of p

(c) for difference of means $(\mu_1 - \mu_2)$:

95% confidence limits $= (\bar{y}_1 - \bar{y}_2) \pm 1.96 \times S.E.$ of $(\bar{y}_1 - \bar{y}_2)$

99% confidence limits $= (\bar{y}_1 - \bar{y}_2) \pm 2.58 \times S.E.$ of $(\bar{y}_1 - \bar{y}_2)$

Almost sure limits $= (\bar{y}_1 - \bar{y}_2) \pm 3 \times S.E.$ of $(\bar{y}_1 - \bar{y}_2)$

(d) for difference of proportions $(P_1 - P_2)$:

95% confidence limits $= (p_1 - p_2) \pm 1.96 \times S.E.$ of $(p_1 - p_2)$

99% confidence limits $= (p_1 - p_2) \pm 2.58 \times S.E.$ of $(p_1 - p_2)$

Almost sure limits $= (p_1 - p_2) \pm 3 \times S.E.$ of $(p_1 - p_2)$

EXACT CONFIDENCE LIMITS FOR NORMAL DISTRIBUTION—(FOR SAMPLE OF ANY SIZE)

(a) for mean μ, (s.d. known): 95% confidence limits $= \bar{y} \pm 1.96 \times \sigma/\sqrt{n}$

99% confidence limits $= \bar{y} \pm 2.58 \times \sigma/\sqrt{n}$

(b) for difference of means $(\mu_1 - \mu_2)$, (s.d. known):

95% confidence limits $= (\bar{y}_1 - \bar{y}_2) \pm 1.96 \times \sqrt{\sigma_1^2/n_1 + \sigma_2^2/n_2}$

99% confidence limits $= (\bar{y}_1 - \bar{y}_2) \pm 2.58 \times \sqrt{\sigma_1^2/n_1 + \sigma_2^2/n_2}$

(c) for mean μ, (s.d. unknown): 95% confidence limits $= \bar{y} \pm t_{0.025}\left(S/\sqrt{n-1}\right)$

99% confidence limits $= \bar{y} \pm t_{0.005}\left(S/\sqrt{n-1}\right)$

Here S is the sample s.d. and $t_{0.025}$ denotes the upper 2.5% point of t distribution with $(n - 1)$ d.f.

(d) for difference of means $(\mu_1 - \mu_2)$, (common s.d. unknown):

95% confidence limits $= (\bar{y}_1 - \bar{y}_2) \pm t_{0.025} \times s\sqrt{1/n_1 + 1/n_2}$

99% confidence limits $= (\bar{y}_1 - \bar{y}_2) \pm t_{0.005} \times s\sqrt{1/n_1 + 1/n_2}$

where, $s = \sqrt{(n_1 S_1^2 + n_2 S_2^2)/(n_1 + n_2 - 2)}$

(e) for variance σ^2, (mean known):

95% confidence interval is: $\Sigma(y_i - \mu)^2/\chi^2_{0.025} \le \sigma^2 \le \Sigma(y_i - \mu)^2/\chi^2_{0.975}$

(f) for variance σ^2, (mean unknown):

95% confidence interval is: $nS^2/\chi^2_{0.025} \le \sigma^2 \le nS^2/\chi^2_{0.975}$

(g) for variance ratio σ_1^2/σ_2^2, (means unknown):

95% confidence interval is: $(1/F_{0.025})\cdot(s_1^2/s_2^2) \le \sigma_1^2/\sigma_2^2 \le (1/F_{0.975})\cdot(s_1^2/s_2^2)$

where, s_1^2 and s_2^2 denote unbiased estimators of σ_1^2, σ_2^2 respectively from the two samples.

METHOD OF MINIMUM VARIANCE

In this method, those estimators are chosen which are unbiased and have minimum variance, *i.e.*, if t_n is the estimator for θ, $E(t_n) = \theta$ and $\text{var}(t_n)$ should be minimum.

In other words, if $L(y_1, y_2, y_3, ..., y_n, \theta') \equiv L$,

$$\int L dy = 1.dy_1 dy_2 dy_3 ... dy_n = dy,$$

$$\int ... \int \int t_n L(y_1, y_2, y_3, ..., y_n, \theta') dy_1\, dy_2\, dy_3 ... dy_n = \theta \quad \Rightarrow \int t_n L dy = \theta$$

and $\int ... \int \int (t_n - \theta)^2 L(y_1, y_2, y_3, ..., y_n, \theta) dy_1 dy_2 dy_3 ... dy_n$

or, $\int (t_n - \theta)^2 L dy = \text{minimum}$

Thus, if there exists an estimator t_n satisfying the above conditions, we can express

$$\partial \log L/\partial\theta = (t_n - \theta)/\lambda$$

where λ may depend on θ, but not on $y_1, y_2, y_3, ..., y_n$, and λ is the variance of t_n.

METHOD OF MINIMUM CHI-SQUARE

In this method, we choose estimators for the population parameters in such a way as to minimise the distribution, $\chi^2 = \sum_i (o_i - e_i)^2/e_i$, where o_i's are the observed and e_i's are expected frequencies calculated from the population. In practice a modified version of this expression is used where we minimise $\chi'^2 = \sum_i (o_i - e_i)^2/o_i$.

METHOD OF INVERSE PROBABILITY

A given '*a priori*' density function $g(\theta)$ of the parameter θ, estimator θ is choosen such that '*a posteriori*' probability function for θ as obtained from the given random sample is a maximum. According to Bay's theorem, we have

A posteriori probability $P \propto$ likelihood function + *a priori* density function of θ

i.e., $$\partial \log P/\partial\theta \propto \partial \log L/\partial\theta + \partial \log g(\theta)/\partial\theta$$

On the R.H.S., the first factor has ordinarily a variance of order n, and the second term is independent of n. Thus, for large n, the inverse probability estimators will approach the maximum likelihood estimators, whatever be the density function $g(\theta)$.

METHOD OF LEAST SQUARES

When there are n equations of the form

$$\phi(y_i, \theta_0, \theta_1, \theta_2, ..., \theta_p) = 0; \qquad i = 1, 2, 3, ..., n$$

to solve for $(p + 1)$ parameters $\theta_0, \theta_1, \theta_2, ..., \theta_p$ and $n > p+1$, the method of least squares is applied and the parameters are chosen so as to minimise the sum of squares of residuals, *i.e.,* θ is chosen such as to satisfy the $(p + 1)$ equations:

$$\frac{\partial}{\partial\theta_j}\Sigma\{\phi(y_i, \theta_0, \theta_1, \theta_2, ..., \theta_p)\}^2 = 0, \qquad j = 0, 1, 2, 3, ..., p$$

When the residuals are distributed normally with variance σ^2, the logarithm of likelihood function is given by

$$\log L = \text{constant} - n \log\sigma - 1/2\sigma^2 - \sum_{i=1}^{p}\phi^2(y_i, \theta_0, \theta_1, \theta_2, ..., \theta_p)$$

This can be examined by minimising the sum of squares of residuals so that in this case the method of least squares will yield the same result as the method of maximum likelihood.

THEORY OF TEST OF SIGNIFICANCE

In practical problems, we have to take decisions about a statistical population on the basis of sample observations. For instance, given a random sample, we may be required to decide whether the population, from which the sample has been obtained, is a normal distribution with mean = 170 and s.d. = 7.8. In order to reach such conclusions, it is necessary to make certain assumptions or gusses about the characteristics of population, particularly about the probability distribution or the values of its parameters. Such an assumption or statement about the population is called a *Statistical Hypothesis*. The validity of a hypothesis is tested by analysing the sample. The procedure which enables us to decide whether a certain hypothesis is true or not, is called *Test of Significance* or *Test of Hypothesis*, which was introduced by J. Neyman and E.S. Pearson who used this technique to arrive at decisions in certain situations having an element of uncertainty on the basis of a sample whose size is known in advance.

In this method, a statistic is calculated from the sample drawn from a population and on the basis of this, it is verified whether the sample so drawn belongs to the parent population with certain specified characteristics. The calculated value of the statistic may differ from the hypothetical value of the parameter due to sampling fluctuations. In case of small difference, it is considered that the difference has arisen due to sampling fluctuations and hence the difference is accepted. In case of large difference, it is assumed that the difference has not arisen due to sampling fluctuations, but it is due to some other reasons. Hence, the difference is considered to be significant and the hypothesis is rejected. Thus, the test of hypothesis shows the fact whether the difference between the calculated statistic and the hypothetical parameter is significant or otherwise.

Statistical Hypothesis: It is some assumption or statement, which may or may not be true, about a population or equivalently about the probability distribution characterising the given population, which we want to test on the basis of the evidence from a random sample.

Simple Hypothesis: It is a statistical hypothesis which specifies the population completely, *i.e.*, the probability distribution and all parameters are known.

Composite Hypothesis: It is a statistical hypothesis which does not specify the population completely, *i.e.*, either the form of probability distribution or some parameters remain unknown.

Test of Hypothesis or Test of Significance: It is a procedure which specifies a set of 'rules for decision' whether to 'accept' or 'reject' the hypothesis under consideration, *i.e.*, the null hypothesis.

Null Hypothesis: It is a statistical hypothesis about the population characteristics which is set up and whose validity is tested for possible rejection on the basis of sample observations. It is denoted by H_0 and tested against alternatives. Tests of hypotheses deal with rejection or acceptance of null hypothesis only.

Alternative Hypothesis: It is a statistical hypothesis which differs from the null hypothesis and is denoted by H_1. It is not tested, but its acceptance (or rejection) depends on the rejection (or acceptance) of the null hypothesis. The choice of an appropriate critical region depends on the type of alternative hypothesis, *viz*., whether both-sided, one-sided (left or right) or specified alternative.

Test Statistic: It is a function of sample observations, *i.e*., statistic, whose calculated value determines the final decision regarding acceptance or rejection of H_0. The appropriate test statistic has to be chosen very carefully and a knowledge of its sampling distribution under H_0, *i.e*., when the null hypothesis is true, is essential in framing the decision rules. The H_0 is rejected if the value of the test statistic falls in the critical region.

Critical Region: The set of values of the test statistic which lead to rejection of the null hypothesis is called *critical region* of the test. The probability with which a true null hypothesis is rejected by the test is often referred to as *size* of the *critical region*.

Geometrically, a sample $y_1, y_2, y_3, ..., y_n$ of size n is looked upon as just a point y, called *sample point*, within the region of all possible samples, called the *sample space*. The critical region is then defined as a subset of those sample points which lead to the rejection of the null hypothesis.

Level of Significance: The maximum probability with which a true null hypothesis is rejected is known as *level of significance* of the test, and is denoted by α; and it is arbitrarily chosen in advance depending on the consequences of a statistical decision. Generally, 5% or 1% level of significance is taken, although other levels such as 2% or 1/2% is also used. The level of significance is used to indicate the upper limit of the probability of committing Type I error, *i.e*., the size of the critical region.

Errors in Hypothesis Testing: Generally, two types of errors are committed in hypothesis testing—*Type I Error* and *Type II Error*.

Type I Error is committed when we reject a correct or true null hypothesis and Type II Error is committed when we accept a wrong or incorrect hypothesis. Type I Error is denoted by α and Type II Error by β. Thus,

$$\alpha = \text{probability of Type I Error}$$
$$= \text{probability of rejecting } H_0 \text{ when } H_0 \text{ is true.}$$
$$\beta = \text{probability of Type II Error}$$
$$= \text{probability of accepting } H_0 \text{ when } H_0 \text{ is not true.}$$

If the difference between two mean is zero and if our test indicated rejection of the null hypothesis, we commit Type I Error. On the other hand, if the difference between two mean is not zero but our test suggests acceptance of null hypothesis we commit Type II Error.

Power of a Test: The probability of rejecting a false null hypothesis is called *power of the test*. Thus, if $\beta(\theta_1)$ represents the probability of Type II Error, when $\theta = \theta_r$ is true, the complementary probability $1 - \beta(\theta_1)$ is called power of the test against the specified alternative $H_1(\theta = \theta_1)$. Hence,

$$\text{Power} = 1 - \text{Probability of Type II Error}$$
$$= \text{Probability of rejecting } H_0 \text{ when } H_1 \text{ is true.}$$

Treated as a function of θ, the expression $P(\theta) = 1 - \beta(\theta)$ is called *Power Function* of the test for θ_0 against θ. The curve obtained by plotting $P(\theta)$ against all possible values of θ, is known as *Power Curve.*

The power of a test determines the extent to which a test can discriminate between true and false null hypotheses.

Steps for Hypothesis Testing:

(1) Set up the null hypothesis or hypothesis of no difference H_0 and the alternative hypothesis H_1 on the basis of the given problem. The null hypothesis usually specifies the values of some parameters involved in the population, *i.e.*, $H_0(\theta = \theta)$. The alternative hypothesis can take any one of the following types: $H_1(\theta \neq \theta_0)$, $H_1(\theta > \theta_0)$, and $H_1(\theta < \theta_0)$. Whether to use a two-tailed or one-tailed test (left or right tail) is determined by the type of alternative hypothesis.

(2) State the appropriate test statistic T and also its sampling distribution, when the null hypothesis is true. In case of large sample tests the statistic $k = (T - \theta_0)/S.E.(T)$, which approximately follows standard normal distribution, is often used. In case of small sample tests, the population is assumed to be normal and various test statistics are used which follow standard normal, chi-square, t or F distribution exactly.

(3) Select the level of significance α of the test, if it is not specified in the given problem. This represents the maximum probability of committing a Type I Error, *i.e.*, of making a wrong decision by the test procedure when in fact the null hypothesis is true. Generally, 5% or 1% level of significance is used.

(4) Find the critical region of the test at the chosen level of significance. This represents the set of values of the test statistic which lead to rejection of the null hypothesis and always appears in one or both tails of the distribution, depending on whether the alternative hypothesis is one-sided or both-sided. The area in the tails must be equal to the level of significance α, and for a one-tailed test α appears in one tail and for a two-tailed test $\alpha/2$ appears in each tail of the distribution. Thus, the critical region is

$$T \geq T_{\alpha/2} \text{ or } T \leq T_{1-\alpha/2} \qquad \text{when } H_1(\theta \neq \theta_0)$$

$$T \geq T_{\alpha} \qquad \text{when } H_1(\theta > \theta_0)$$

$$T \leq T_{1-\alpha} \qquad \text{when } H_1(\theta < \theta_0)$$

where T_α is the value of T such that the area to its right is α.

(5) Compute the value of the test statistic T on the basis of sample data and the null hypothesis.

(6) If the computed value of test statistic T lies in the critical region, H_0 is rejected; otherwise accepted.

(7) The last step in hypothesis testing is to draw conclusions about accepting or rejecting a hypothesis and writing the conclusion in plain non-technical language.

Two Tailed and One Tailed Test of Hypothesis: In a two tailed test, the region of rejection, which is generally 5%, is kept on both sides of the normal distribution, *i.e*, 2.5% of the critical region is on the extreme left of the normal curve and 2.5% on the extreme right. The middle 95% is the acceptance region. Whereas in a single tail test, the 5% area would be either on the extreme left of the normal curve or on the extreme right. The remaining 95% area would be the acceptance region. Diagrammatically:

Two Tailed Test Diagrams

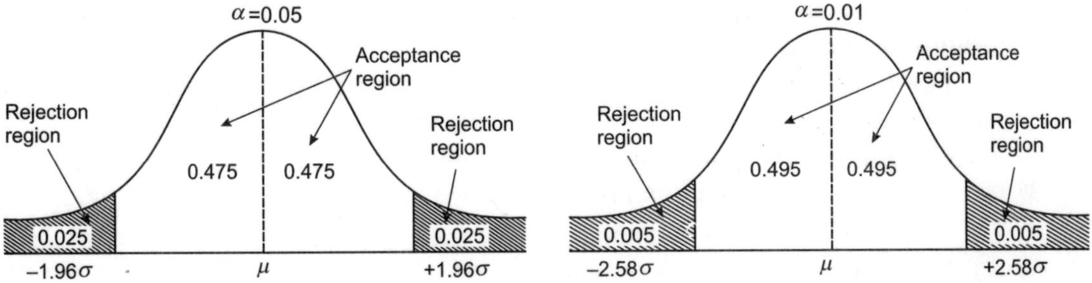

One Tailed Test Diagrams

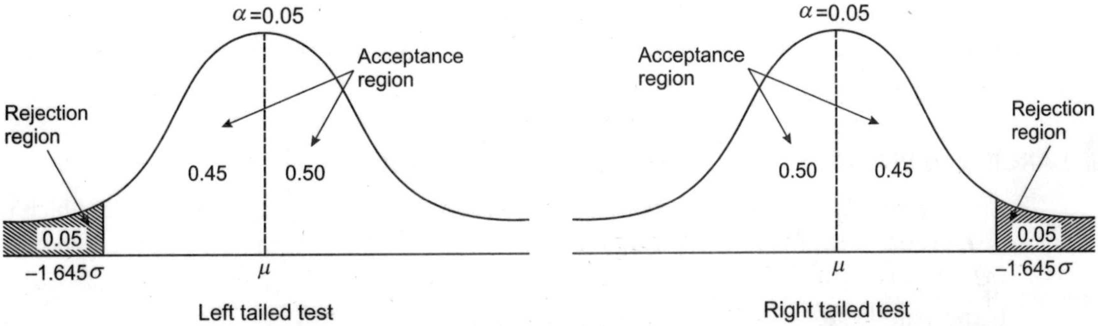

THE NEYMEN-PEARSON LEMMA

We want to test a simple hypothesis $H_0(\theta = \theta_0)$ against a simple alternative $H_1(\theta = \theta_1)$. Let $y_1, y_2, y_3, ..., y_n$ be a random sample from the population $f(y, \theta)$ under consideration.

Then, the likelihood function under H_0 and H_1 are: $L_0 = \prod_{i=1}^{n} f(y_i, \theta_0)$ and $L_1 = \prod_{i=1}^{n} f(y_i, \theta_1)$ respectively.

The Neymen-Pearson Lemma states that if there exists a critical region ω of the size α and a positive constant k such that

$$(L_1/L_0) > k \text{ inside } \omega \text{ and } (L_1/L_0) \leq k \text{ outside } \omega$$

then ω is the most powerful critical region for taking $H_0(\theta = \theta_0)$ against $H_1(\theta = \theta_1)$. When H_0 is true, $P(y \in \omega/H_0)$ is the probability that $y \in \omega$.

THE LIKELIHOOD RATIO TEST

This is the generalisation of Neymen-Pearson Lemma for testing simple null hypothesis H_0 against simple alternative hypothesis H_1.

Let us consider the problem of testing a composite hypothesis or a simple H_0 against a complete alternative hypothesis H_1. Let the maximum value of the likelihood function L under the null hypothesis H_0 be L_0, which is obtained by substituting the maximum likelihood estimates of the parameter in L under H_0. If H_0 specifies all the parameters in L then max $L_0 = L_0$ itself.

Let max L_0 be the maximum of L with respect to the parameter. Then the likelihood ratio statistic or the λ – criterion is given by $\lambda = \max L_n / \max L$.

According to the likelihood ratio test, a uniformly most powerful critical region C for testing H_0 (simple or composite) is obtained by the inequality: $\lambda \le k$ *inside* ω.

The constant k is usually determined by the inequality that the probability of Type I Error $\le \alpha$, does not always gives uniform powerful test. Sometimes it leads to a non-admissible test. Evidently $0 < \lambda \le 1$ and intuitively λ is a reasonable test statistic for H_0.

In the case of a composite hypothesis, we can formulate in general

$$H_0(\theta \in \omega_0) \text{ and } H_1(\theta \in \omega_0 \cap \upsilon)$$

where $\omega_0 \subset \upsilon$ and υ is the parameter space and generated by the possible values of the parameter.

LARGE SAMPLE TESTS—(A) USING NORMAL DISTRIBUTION

(1) *Test for a Specified Proportion*: Let us consider a random sample of size n in which the proportion of members possessing a certain attribute is p. We want to test the hypothesis that the proportion P in the population has a specified value P_0, *i.e.*, $H_0(P = P_0)$.

If the sample size n is large, the test statistic is

$$z = (p - P_0)/S.E. \text{ of } p$$

which follows standard normal distribution.

Rejection rules for $H_0(P = P_0)$:

Alternative hypothesis	*Critical Region*	
H_1	*5% level*	*1% level*
$P \ne P_0$	$\lvert z\rvert \ge 1.96$	$\lvert z\rvert \ge 2.58$
$P > P_0$	$z \ge 1.645$	$z \ge 2.33$
$P < P_0$	$z \le -1.645$	$z \le -2.33$

Otherwise, the null hypothesis H_0 is accepted.

Confidence limits for the population proportion P are:

$$95\% \text{ confidence limits} = p \pm 1.96 \times S.E. \text{ of } p$$

$$99\% \text{ confidence limits} = p \pm 2.58 \times S.E. \text{ of } p$$

where S.E. of $p = \sqrt{pq/n}$, if H_0 is rejected.

(2) *Test for Equality of Two Proportions*: Let p_1 and p_2 be the proportions in two large random samples of sizes n_1 and n_2 drawn respectively from two populations. We want to test whether the proportions in the two populations are equal or not, *i.e.*, whether the difference $(p_1 - p_2)$ as observed in the samples has arisen only due to fluctuations of sampling, *i.e.*, $H_0(P_1 = P_2)$ against $H_1(P_1 \neq P_2)$.

If the null hypothesis H_0 is true, *i.e.*, the two population proportions are equal, say P, the estimate of P is given by $P = (n_1p_1 + n_2p_2)/(n_1 + n_2)$.

The expected value of the difference $(p_1 - p_2)$ is then 0, and

$$S.E. \text{ of } (p_1 - p_2) = \sqrt{pq(1/n_1 + 1/n_2)}, \qquad \text{where } p + q = 1$$

If the sample sizes n_1 and n_2 are large, the test statistic is given by $z = (p_1 - p_2)/S.E.$, and it follows standard normal distribution. The critical regions are:

$$|z| \geq 1.96 \text{ at 5\% level of significance and } |z| \geq 2.58 \text{ at 1\% level of significance.}$$

Confidence limits for $(P_1 - P_2)$ are given by

$$95\% \text{ confidence limits} = (p_1 - p_2) \pm 1.96 \times S.E.$$

$$99\% \text{ confidence limits} = (p_1 - p_2) \pm 2.58 \times S.E.$$

where $S.E. = \sqrt{p_1q_1/n_1 + p_2q_2/n_2}$, if H_0 is rejected.

(3) *Test for a Specified Mean*: Let us consider a large random sample of size n and mean $\bar{y}$. We want to test the hypothesis that the population mean μ has a specified value μ_0, *i.e.*, $H_0(\mu = \mu_0)$.

Since, for large n, the sampling distribution of sample mean $(\bar{y})$ is approximately normal and therefore, when H_0 is true, the test statistic, $z = (\bar{y} - \mu_0)/S.E. \text{ of } \bar{y}$ approximately follows standard normal distribution.

Rejection rules for $H_0(\mu = \mu_0)$:

Alternative hypothesis	*Critical Region*	
H_1	*5% level*	*1% level*
$\mu \neq \mu_0$	$\lvert z\rvert \geq 1.96$	$\lvert z\rvert \geq 2.58$
$\mu > \mu_0$	$z \geq 1.645$	$z \geq 2.33$
$\mu < \mu_0$	$z \leq -1.645$	$z \leq -2.33$

Confidence limits for the population mean μ are:

$$95\% \text{ confidence limits } = \bar{y} \pm 1.96 \times S.E. \text{ of } \bar{y}$$

$$99\% \text{ confidence limits } = \bar{y} \pm 2.58 \times S.E. \text{ of } \bar{y}$$

where *S.E.* of $\bar{y} = \sigma/\sqrt{n}$, if H_0 is rejected. In case population *s.d.* σ is not known, it is replaced by the sample s.d. *S*.

(4) *Test for Equality of Two Means*: Let us consider two independent random samples of large size n_1 and n_2 from two populations with sample means $\bar{y}_1$ and $\bar{y}_2$ respectively. We want to test whether the two populations have the same mean.

The standard error of the difference of means $(\bar{y}_1 - \bar{y}_2)$ is given by

$$S.E. = \sqrt{\sigma_1^2/n_1 + \sigma_2^2/n_2}$$

where σ_1 and σ_2 are the standard deviations in the two populations.

To test the null hypothesis that the means are equal, *i.e.*, $H_0(\mu_1 = \mu_2)$, we use the test statistic $z = (\bar{y}_1 - \bar{y}_2)/S.E.$, which approximately follows standard normal distribution. For testing equality of means against the alternative hypothesis $H_1(\mu_1 \neq \mu_2)$ the critical regions are:

$|z| \geq 1.96$ at 5% level of significance and $|z| \geq 2.58$ at 1% level of significance.

Confidence limits for $(\mu_1 - \mu_2)$ are:

$$95\% \text{ confidence limits } = (\bar{y}_1 - \bar{y}_2) \pm 1.96 \times S.E. \text{ of } (\bar{y}_1 - \bar{y}_2)$$

$$99\% \text{ confidence limits } = (\bar{y}_1 - \bar{y}_2) \pm 2.58 \times S.E. \text{ of } (\bar{y}_1 - \bar{y}_2).$$

(B) USING CHI-SQUARE (χ^2) DISTRIBUTION

(1) *Test for Goodness of Fit (Pearsonian χ^2)*: This test is used to decide whether the observations are in good agreement with a hypothetical distribution. The observed frequencies (f_0) of different classes are compared with the expected frequencies (f_e) by the test statistic $\chi^2 = \Sigma\{(f_0 - f_e)^2/f_e\}$.

When, the null hypothesis, H_0 (Data are in agreement with the hypothetical population) is true, the statistic approximately follows chi-square distribution with (k – 1) d.f., where k is the number of classes. If the observed value of the statistic exceeds the tabulated value of χ^2 at a given level, the null hypothesis is rejected.

(2) *Test for Independence of Attributes*: Let the observations be classified simultaneously according to two attributes, and the frequencies (f_0) in the difference categories are shown in a two-way table, known as *Contingency Table*. On the basis of cell frequencies, we want to test whether the two attributes are associated or not. Under the null hypothesis, H_0 (Attributes are independent), the expected frequency (f_e) of any cell is given by

$$f_e = \frac{\textit{Row total} \times \textit{Column total}}{\textit{Total frequency}} = \frac{X_i \times Y_i}{N}$$

Then, the test statistic is given by $\chi^2 = \Sigma\{(f_0 - f_e)^2/f_e\}$, which approximately follows chi-square distribution with

$$\text{d.f.} = (\text{number of rows} - 1) \times (\text{number of columns} - 1)$$

If the computed value of χ^2 exceeds the tabulated value for the given d.f. at a specified level, the null hypothesis is rejected, and it is concluded that the attributes are not independent, but associated.

Let us consider a simple case of 2 × 2 contingency table, *i.e.*, 2 rows and 2 columns with the four cell frequencies *a*, *b*, *c*, *d* as shown below:

			Total
	a	*b*	*R*1
	c	*d*	*R*2
Total	C1	C2	N

$R_1 = a + b,\ R_2 = c + d,\ C_1 = a + c,\ C_2 = b + d,\ N = a + b + c + d = R_1 + R_2 = C_1 + C_2$

In this case the test statistic is

$$\chi^2 = \frac{N(ad - bc)^2}{R_1R_2C_1C_2} \quad \text{with d.f.} = (2 - 1) \times (2 - 1) = 1$$

Yates' Correction—For a 2 × 2 contingency table, there is only one d.f., *i.e.*, only one of the four cell frequencies can be arbitrarily given, if the row and column totals should remain fixed. Therefore, it is necessary to make a correction to formula so that its approximation to the continuous chi-square distribution can be improved.

If $ad > bc$, reduce *a* and *d* by 1/2, and increase *b* and *c* by 1/2.

If $ad < bc$, increase *a* and *d* by 1/2, and reduce *b* and *c* by 1/2.

This is known as Yates' correction for continuity. Hence,

$$\chi^2(corrected) = \frac{N\{|ad - bc| - N/2\}^2}{R_1R_2C_1C_2}$$

Note: Yates' correction is applied for only 1 d.f.

SMALL SAMPLE TESTS—(A) USING NORMAL DISTRIBUTION

(1) *Test for a Specified Population Mean (s.d. Known)*: Let us consider a random sample of size *n* from a normal population with known s.d. σ. We want to test the null hypothesis that the mean of population has a specified value, $H_0(\mu = \mu_0)$.

Since, the sampling distribution of sample mean $(\bar{y})$ is exactly normal with $S.E. = \sigma/\sqrt{n}$, therefore when the null hypothesis is true, the test statistic

$$z = (\bar{y} - \mu_0)/S.E. = (\bar{y} - \mu_0)/(\sigma/\sqrt{n})$$

follows standard normal distribution exactly.

Rejection rules for $H_0(\mu = \mu_0)$:

Alternative hypothesis	*Critical Region*	
H_1	*5% level*	*1% level*
$\mu \neq \mu_0$	$\lvert z\rvert \geq 1.96$	$\lvert z\rvert \geq 2.58$
$\mu > \mu_0$	$z \geq 1.645$	$z \geq 2.33$
$\mu < \mu_0$	$z \leq -1.645$	$z \leq -2.33$

Confidence limits for the population mean μ are:

$$95\% \text{ confidence limits } = \bar{y} \pm 1.96 \times (\sigma/\sqrt{n})$$

$$99\% \text{ confidence limits } = \bar{y} \pm 2.58 \times (\sigma/\sqrt{n})$$

(2) *Test for Equality of Two Means (s.d. Known)*: Let us consider two independent samples of sizes n_1 and n_2 from two normal population with unknown means, but known s.d. σ_1 and σ_2 respectively. We want to test the hypothesis that the population means are equal, *i.e.*, $H_0(\mu_1 = \mu_2)$.

The sampling distribution of the difference of sample means $(\bar{y}_1 - \bar{y}_2)$ is exactly normal with $S.E. = \sqrt{\sigma_1^2/n_1 + \sigma_2^2/n_2}$.

The test statistic is given by

$$z = \frac{\bar{y}_1 - \bar{y}_2}{S.E.} = \frac{\bar{y}_1 - \bar{y}_2}{\sqrt{\sigma_1^2/n_1 + \sigma_2^2/n_2}}$$

which follows standard normal distribution exactly. If the value of z falls in the critical region, the null hypothesis is rejected.

Rejection rules for $H_0(\mu_1 = \mu_2)$:

Alternative hypothesis	*Critical Region*	
H_1	*5% level*	*1% level*
$\mu_1 \neq \mu_2$	$\lvert z\rvert \geq 1.96$	$\lvert z\rvert \geq 2.58$
$\mu_1 > \mu_2$	$z \geq 1.645$	$z \geq 2.33$
$\mu_1 < \mu_2$	$z \leq -1.645$	$z \leq -2.33$

Confidence limits for the difference of population means $(\mu_1 - \mu_2)$ are:

$$95\% \text{ confidence limits } = (\bar{y}_1 - \bar{y}_2) \pm 1.96 \times S.E. \text{ of } (\bar{y}_1 - \bar{y}_2)$$
$$= (\bar{y}_1 - \bar{y}_2) \pm 1.96 \times \left(\sqrt{\sigma_1^2/n_1 + \sigma_2^2/n_2}\right)$$

$$99\% \text{ confidence limits } = (\bar{y}_1 - \bar{y}_2) \pm 2.58 \times S.E. \text{ of } (\bar{y}_1 - \bar{y}_2)$$
$$= (\bar{y}_1 - \bar{y}_2) \pm 2.58 \times \left(\sqrt{\sigma_1^2/n_1 + \sigma_2^2/n_2}\right).$$

(B) USING t DISTRIBUTION

(1) *Test for a Specified Mean (s.d. Unknown)*: For testing the null hypothesis $H_0(\mu = \mu_0)$, if the population s.d. (σ) is not known, formula $z = (\bar{y} - \mu_0)/(\sigma/\sqrt{n})$ cannot be used. The unbiased estimator of the population variance (σ^2) is $s^2 = nS^2/(n-1)$, where S^2 is the variance of a sample of size n. Thus, the estimate of σ is $s = \left(\sqrt{n/(n-1)}\right)\cdot S$. If this is substituted for σ in the formula for z, we get the statistic, known as Student's t as

$$t = (\bar{y} - \mu_0)/\left(s/\sqrt{n}\right) = (\bar{y} - \mu_0)/\left(S/\sqrt{n-1}\right)$$

which follows t distribution with $(n - 1)$ d.f.

If $t_{0.025}$ denotes the value of t distribution corresponding to the upper tail area, then the null hypothesis is rejected, if $|t| \geq t_{0.025}$.

Confidence limits for the population mean μ are:

$$95\% \text{ confidence limits } = \bar{y} \pm t_{0.025} \times s/\sqrt{n}, \text{ or, } \bar{y} \pm t_{0.025} \times S/\sqrt{n-1}$$

$$99\% \text{ confidence limits } = \bar{y} \pm t_{0.005} \times s/\sqrt{n}, \text{ or, } \bar{y} \pm t_{0.005} \times S/\sqrt{n-1}.$$

(2) *Test for Equality of Two Means (s.d. Unknown)*: Let us consider two independent random samples of sizes n_1 and n_2 from two normal populations with means μ_1 and μ_2. We want to test the hypothesis that the means are equal, *i.e.*, $H_0(\mu_1 = \mu_2)$.

Since, the two population standard deviations are assumed to be equal, an unbiased estimator of the common variance is given by

$$s^2 = \left(n_1 S_1^2 + n_2 S_2^2\right)/\left(n_1 + n_2 - 2\right)$$

where S_1 and S_2 are the sample standard deviations.

The test statistic is, $t = (\bar{y}_1 - \bar{y}_2)/s\sqrt{1/n_1 + 1/n_2}$ (known as Fisher's t) and it follows t distribution with $(n_1 + n_2 - 2)$ d.f.

Confidence limits for the difference of population means $(\mu_1 = \mu_2)$ are:

$$95\% \text{ confidence limits } = (\bar{y}_1 - \bar{y}_2) \pm t_{0.025} \times s\sqrt{1/n_1 + 1/n_2}$$

$$99\% \text{ confidence limits } = (\bar{y}_1 - \bar{y}_2) \pm t_{0.005} \times s\sqrt{1/n_1 + 1/n_2}.$$

(3) *Test for Equality of Two Means (Paired t)*: Let us consider a random sample of n pairs of observations (x_1, y_1), (x_2, y_2), (x_3, y_3), ..., (x_n, y_n) from a bivariate normal population. The values of x and y are correlated. We want to test the hypothesis whether the means of x and y in the population are equal, *i.e.*, $H_0(\mu_x = \mu_y)$.

Let $d_i = x_i - y_i$ denote the difference in the values of x and y for the i-th pair, where i = 1, 2, 3, ..., n.

The test statistic is given by $t = \bar{d}/\left(S/\sqrt{n-1}\right)$, where $\bar{d}$ and S denote respectively the mean and s.d. of the difference d_i, *i.e.*, $\bar{d} = \Sigma d_i/n$ and $S^2 = \Sigma(d_i - \bar{d})^2/n = \Sigma d_i^2/n - (\Sigma d_i/n)^2$

When null hypothesis H_0 is true, the test statistic t (known as paired t) follows t distribution with $(n - 1)$ d.f. The critical region is given by two tails or one tail (left or right tail) of t distribution depending on the alternative hypothesis: $H_1(\mu_x \neq \mu_y)$, $H_1(\mu_x > \mu_y)$, or $H_1(\mu_x < \mu_y)$.

(C) USING CHI-SQUARE (χ^2) DISTRIBUTION

(1) *Test for a Specified Value of s.d.*: Let us consider a random sample $y_1, y_2, y_3, \ldots, y_n$ from a normal population with mean μ and s.d. σ (both unknown). We want to test the hypothesis that the population s.d. σ has a specified value σ_0. For this we set a null hypothesis $H_0(\sigma = \sigma_0)$ against alternative hypothesis $H_1(\sigma > \sigma_0)$. Our test statistic is

$$\chi^2 = \Sigma(y_i - \bar{y})^2/\sigma_0^2 = nS^2/\sigma_0^2$$

where, S is the s.d. of the sample and when H_0 is true, it follows chi-square distribution with $(n - 1)$ d.f. If the calculated value of χ^2 exceeds the tabulated 5% value, it is concluded that the population s.d. is not equal to σ_0.

(D) USING F DISTRIBUTION

(1) *Test for Equality of Two s.ds. (Means Unknown)*: Let us consider two independent random samples of sizes n_1 and n_2 from two normal populations with unknown means. We want to test the hypothesis that the population s.ds. are equal, *i.e.*, $H_0(\sigma_1 = \sigma_2)$.

The estimates of variances σ_1^2 and σ_2^2 are obtained from the samples as:

$$s_1^2 = \{n_1/(n_1 - 1)\}S_1^2 \text{ and } s_2^2 = \{n_2/(n_2 - 1)\}S_2^2$$

where S_1 and S_2 are the s.ds. of the samples. If the null hypothesis H_0 is true, we consider the test statistic $F = s_1^2/s_2^2$, which follows F distribution with $(n_1 - 1, n_2 - 1)$ d.f. If the computed value of F is larger than the 5% tabulated value, the null hypothesis H_0 is rejected and it is concluded that the s.ds. in the two populations are not equal.

(E) TESTS FOR CORRELATION COEFFICIENT

(1) *Test for a Specified Correlation Coefficient (Very Large Sample)*: In the case of large sample size (n), say more than 500, the sample correlation coefficient (r) is assumed to follow roughly a normal distribution with mean ρ, *i.e.*, population correlation coefficient, and S.E. of $r = (1 - \rho^2)/\sqrt{n}$. In this case our test statistic is

$$z = (r - \rho)/\text{S.E. of } r = (r - \rho)/\left\{(1 - \rho^2)/\sqrt{n}\right\}$$

and it follows approximately a standard normal distribution. Therefore, this may be used to test the null hypothesis $H_0(\rho = \rho_0)$, *i.e.*, whether the sample may be assumed to have arisen from a population with correlation coefficient $\rho = \rho_0$, and also for finding approximate confidence limits for ρ.

The confidence limits for ρ are:

$$95\% \text{ confidence limits} = r \pm 1.98 \times S.E. \text{ of } r$$

$$99\% \text{ confidence limits} = r \pm 2.58 \times S.E. \text{ of } r$$

where $S.E. \text{ of } r = (1 - r^2)/\sqrt{n}$ approx.

(2) *Test for Significance of Observed Correlation Coefficient*: Let us consider a random sample of n pairs of observations $(x_1, y_1), (x_2, y_2), (x_3, y_3), ..., (x_n, y_n)$ from a bivariate normal population and let r be the observed correlation coefficient in the sample. We want to test if this sample correlation coefficient is significant of any correlation in the population. In other words, we want to test whether the value of the population correlation coefficient ρ is zero and the computed value of r has arisen due to fluctuations of sampling.

According to R.A. Fisher, when the null hypothesis $H_0(\rho = 0)$ is true, the test statistic, $t = r\sqrt{n-2}/\sqrt{1-r^2}$ is used and it follows Student's t distribution with $(n - 2)$ d.f. This test statistic is used for testing the significance of r.

Fisher's Z-transformation of Correlation Coefficient: If $\rho \neq 0$, the test statistic, $t = r\sqrt{n-2}/\sqrt{1-r^2}$ does not follow either normal or any other commonly used distribution even for large samples. In this case, the following transformation of r to a new variable has been suggested by Fisher.

$$Z = (1/2)\log_e\{(1 + r)/(1 - r)\}$$

Even for small sample sizes, Z approximately follows normal distribution with mean μ and s.d. σ, given by

$$\mu = (1/2)\log_e\{(1 + \rho)/(1 - \rho)\} \text{ and } \sigma = 1/\sqrt{n-3}$$

Hence, $U = (Z - \mu)\sqrt{n-3}$ is approximately a standard normal variate.

Z-transformation is used to—(a) test for a specified value of $\rho \neq 0$, and find confidence limits for ρ; and (b) test for equality of two correlation coefficients.

(3) *Test for a Specified Correlation Coefficient (Any Sample Size)*: Let us consider a random sample of n pairs of observations from a bivariate normal population with correlation coefficient r. We want to test the hypothesis that the population correlation coefficient ρ has a specified value, *i.e.*, $H_0(\rho = \rho_0)$.

When the null hypothesis H_0 is true, using Fisher's Z-transformation, the test statistic is given by $U = (Z - \mu_0)\sqrt{n-3}$, where μ_0 is the value of μ when $\rho = \rho_0$ and it follows standard normal distribution approximately.

(4) *Test for Equality of Two Correlation Coefficients*: Let us consider two independent random samples of n_1 and n_2 pairs of observations drawn from bivariate normal populations having correlation coefficients r_1 and r_2 respectively. We want to test the hypothesis that the correlation coefficients in the two populations are equal, *i.e.*, $H_0(\rho_1 = \rho_2)$.

When the null hypothesis H_0 is true, using Fisher's Z-transformation, the test statistics are

$$Z_1 = (1/2)\log_e\{(1 + r_1)/(1 - r_1)\} \text{ and } Z_2 = (1/2)\log_e\{(1 + r_2)/(1 - r_2)\}$$

which are approximately normally distributed and independent with the same mean, but variances $\sigma_1^2 = 1/(n_1 - 3)$ and $\sigma_2^2 = 1/(n_2 - 3)$ respectively. Hence, $(Z_1 - Z_2)$ is normally distributed with mean 0 and variance $1/(n_1 - 3) + 1/(n_2 - 3)$. Thus, the standard normal variate is given by

$$U = (Z_1 - Z_2)\Big/\sqrt{(1/(n_1 - 3) + 1/(n_2 - 3))}$$

If the alternative hypothesis is $H_1(\rho_1 \neq \rho_2)$, the calculated value of $|U|$ is compared with the tabulated value of 1.96 or 2.58 for significance at 5% or 1% level respectively. If the calculated value of U is significant, it is concluded that the two population correlation coefficients are not equal.

CHI-SQUARE TEST

In chapter 9 we have discussed various tests of significance which were based on the assumption that the samples had been taken from a normal population or the sampling distribution (whether of mean or of standard deviation etc.) adhered to the properties of the normal distribution. We have assumed that the sampling distributions give a normal probability curve, even if the population was not normally distributed. In all the test, Z-test, t-test, and F-test, we have to make assumptions about the population values or parameters, therefore such tests are called *Parametric Tests*.

Parametric Tests are not applicable in situations where it is not possible to make any dependable assumptions about the form of the parent distribution from which samples have been drawn. To overcome such situations various tests have been evolved which do not require any assumptions about the parameters. Such tests are distribution-free tests and are called *Non-parametric Tests*.

Chi-Square test is a non-parametric test, which was first developed by Hamlet and later Karl Pearson derived it independently and applied it as a *Test of Goodness of Fit*. It describes the magnitude of difference between observed frequencies and the frequencies expected under certain assumptions; and therefore, with the help of such test, it is possible to find out whether such differences are significant or are insignificant and could have arisen due to fluctuations of sampling.

The Chi-Square test involves following steps:

Step 1: Calculate the expected frequencies. Let us denote them by E.

Step 2: Find out the difference between observed frequencies (denoted by O) and the expected frequencies, *i.e.*, $(O - E)$

Step 3: Find out $(O - E)^2/E$.

Step 4: Total all the values of $(O - E)^2/E$ and this will be the value of χ^2, *i.e.*,

$$\chi^2 = \Sigma\{(O - E)^2/E\}.$$

Step 5: Compare the calculated value of χ^2 with the tabulated value of χ^2 for the given degrees of freedom and at the desired level of significance.

Step 6: If the calculated value of χ^2 is greater than the tabulated value, the difference between observed and expected values is significant. If the calculated value is less than the tabulated value, the differences between the observed and expected frequencies is not significant and could have arisen due to fluctuations of sampling.

Degrees of Freedom: The term degrees of freedom refers to the number of independent constraints in a set of data. In case of association table, the degrees of freedom are calculated by the formula: $v = (c - 1), (r - 1)$ where v stands for the degrees of freedom, c for the number of columns and r for the number of rows.

Form of χ^2 Distribution: For a large sample sizes, the sampling (probability) distribution of χ^2 can be closely approximated by a continuous curve known as the Chi-square distribution. A random variable y is said to follow Chi-square distribution if its p.d.f. is of the form

$$f(y) = K e^{-y/2}\, y^{(v/2-1)}; \qquad (0 < y < \infty)$$

where, K is a constant depending only on the degrees of freedom v.

The Chi-Square distribution has only one parameter, the degree of freedom, therefore it varies as the degrees of freedom change. If the degrees of freedom are low, the χ^2 distribution is highly skewed to the right. As the degrees of freedom increase, the degree of skewness decreases, and if the degrees of freedom are substantial, χ^2 assumes the shape of a normal distribution with a mean $\sqrt{(2v-1)}$ and standard deviation equal to unity. Constants of Chi-Square distribution are:

(a) Mean $= v$; Mode $= v - 2$; Variance $= 2v$

(b) Moments $= \mu_0 = 0$; $\mu_2 = 2v$; $\mu_3 = 8v$; $\mu_4 = 48v + 12v^2$

(c) $\beta_1 = \mu_3^2/\mu_2^3 = 64v^2/8v^3 = 8/v$ and hence $\gamma_1 = \sqrt{8/v}$, and

$\beta_2 = \mu_4/\mu_2^2 = \left(48v + 12v^2\right)/4v^2 = 12/v + 3$ and hence $\gamma_2 = \beta_2 - 3 = 12/v$

The Chi-Square distribution is used in both large sample and small sample tests. It is mainly used in—(a) test for goodness of fit, (b) test for independence of attributes, and (c) test for a specified standard deviation (small sample test).

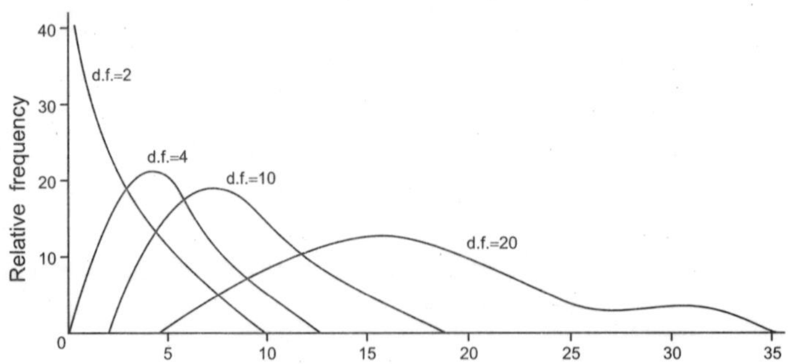

Probability curve of χ^2-distribution

Conditions for Validity of χ^2- test: It should be used only when:

(a) The total frequency, n, should be fairly large. It should be at least 50, *i.e.*, $n \geq 50$.

(b) No theoretical frequency should be small—say, less than 5. If it is less than 5, it should be pooled with the frequency of the neighbouring frequency. This reduces the value of degrees of freedom.

(c) All the individuals in a sample must be independent.
(d) The distribution should not be of proportions or percentage *etc.* It should be of original units.
(e) The number n, of classes or cells, should be neither too small nor too large. It is better if $5 \le n \le 20$. Sometimes the number of cells less than 5 can also be used provided the cell frequencies are not small.
(f) The constraints imposed on the cell frequencies must be linear, *i.e.*, the relation between cell frequencies should not be of more than one degree or exponential or logarithmic type *etc.* It can be of the following type: $b_1y_1 + b_2y_2 + \ldots + b_ny_n = b$.

TEST FOR GOODNESS OF FIT (PEARSONIAN χ^2)

Karl Pearson has devised this test to decide whether the observations are in good agreement with a hypothetical distribution, *i.e.*, whether the sample may be supposed to have arisen from a specified population. The test statistic for this test is given by

$$\chi^2 = \Sigma\{(f_0 - f_e)^2/f_e\}$$

where, f_o is observed frequencies and f_e is expected frequencies.

If the null hypothesis, H_0(Data are in agreement with the hypothetical population) is true, the statistic approximately follows Chi-Square distribution with $(n - 1)$ degrees of freedom, where n is the number of classes. If the calculated value of the statistic exceeds the tabulated value of χ^2 at a given level of significance, the null hypothesis is rejected.

CHI-SQUARE TEST FOR INDEPENDENCE OF ATTRIBUTES

Two attributes are said to be independent, if they are unrelated to each other. If there are m categories A_i for attribute A and n categories B_j for attribute B, then A and B will be independent if $(A_iB_j) = (A_i)(B_j)/N$, for all values of i and j. This implies that in the population each cell frequency will be equal to the product of the corresponding row and column totals divided by the total number of members in the population.

In particular, for a 2 × 2 contingency table we have: $(AB) = (A)(B)/N$.

Coefficient of Contingency—The coefficient of contingency is given by $C = \sqrt{\{\chi^2/(N + \chi^2)\}}$.

If the null hypothesis is correct, f_o and f_e will be equal for each cell and therefore $\chi^2 = 0$ and consequently, $C = 0$. Thus, C can never be unity and the maximum value of C depends on the number of cells.

When the number of rows and columns are equal then $C = \sqrt{(k-1)/k}$, where k is the number of rows or columns.

CHI-SQUARE TEST FOR ASSOCIATION OF ATTRIBUTES

The word association is used to indicate the degree of relationship between attributes (the corresponding word for variable is correlation). Let us consider the following (2 × 2) contingency table.

		Attribute A		
		A	α	*Total*
Attribute B	B	$(AB) = a$	$(\alpha B) = b$	$(B) = a + b$
	β	$(A\beta) = c$	$(\alpha\beta) = d$	$(\beta) = c+d$
	Total	$(A) = a+c$	$(\alpha) = b+d$	N

For a (2 × 2) contingency table, a measure of association is given by Yule's coefficient of association as:

$$Q = \frac{(AB)(\alpha\beta) - (A\beta)(B\alpha)}{(AB)(\alpha\beta) + (A\beta)(B\alpha)} = \frac{ad - bc}{ad + bc}$$

SIMPLIFIED FORMULA FOR (2 × 2) TABLE

Let us consider a contingency table having 2 rows and 2 columns with the four cell frequencies *a*, *b*, *c*, *d* as follows:

			Total
	a	b	R_1
	c	d	R_2
Total	C_1	C_2	N

$$R_1 = a + b, \quad R_2 = c + d, \quad C_1 = a + c, \quad C_2 = b + d$$

$$N = a + b + c + d = R_1 + R_2 = C_1 + C_2$$

In this case Chi-Square statistic is given by

$$\chi^2 = N(ad - bc)^2 / R_1 R_2 C_1 C_2$$

with degrees of freedom = (2 – 1) × (2 – 1) = 1.

YATES' CORRECTION

Since, there is only one degree of freedom for a (2 × 2) contingency table, *i.e.*, only one of the four cell frequencies can be arbitrarily given, the row and column totals should remain fixed. Therefore, it is necessary to make the following correction to the formula for Chi-Square statistic, so that its approximation to the continuous Chi-Square distribution can be improved.

If $ad > bc$, reduce *a* and *d* by 1/2, and increase *b* and *c* by 1/2.

If $ad < bc$, increase *a* and *d* by 1/2, and reduce *b* and *c* by 1/2.

Thus, Yates' correction formula is given by

$$\chi^2(corrected) = N\{|ad - bc| - N/2\}^2 / R_1R_2C_1C_2$$

and it is known as Yates' correction to continuity.

Note: Yates' correction can be applied for only 1 d.f. and it decreases slightly the value of χ^2.

DIRECT METHOD OF CALCULATING χ^2 IN 2 × *n* TABLE

We can calculate the value of χ^2 without finding out the expected frequencies directly as

b_1	b_2	b_3	b_4	b_5	...	b_n	T_b
c_1	c_2	c_3	c_4	c_5	...	c_n	T_c
T_1	T_2	T_3	T_4	T_5	...	T_n	T

$$\chi^2 = \frac{T^2}{T_b \times T_c} = \left[\frac{(b_1)^2}{T_1} + \frac{(b_2)^2}{T_2} + \frac{(b_3)^2}{T_3} + \ldots + \frac{(b_n)^2}{T_n} - \frac{(T_b)^2}{T}\right]$$

CHI-SQUARE TEST FOR POPULATION VARIANCE

When we want to ascertain whether the variance in the universe is a specified figure, we use the test statistic

$$\chi^2 = ns^2/\sigma^2 = \Sigma(y - \bar{y})^2/\sigma^2$$

with $(n - 1)$ degrees of freedom; where, y = random sample; n = size of the sample; s^2 = variance of the sample; and σ^2 = variance of the universe.

ADDITIVE PROPERTY OF χ^2

If the independent variates $y_1, y_2, y_3, \ldots, y_n$ all conform to χ^2 distribution with $\nu_1, \nu_2, \nu_3, \ldots, \nu_n$ degrees of freedom respectively, the sum of the variables $y_1 + y_2 + y_3 + \ldots + y_n$ is also distributed like χ^2 with $\nu_1 + \nu_2 + \nu_3 + \ldots + \nu_n$ degrees of freedom.

BARTLETT TEST

It is used for testing homogeneity of several independent estimates of population variances. Let n independent samples give unbiased estimates $s_1^2, s_2^2, s_3^2, \ldots, s_n^2$ of the population variance based on $\nu_1, \nu_2, \nu_3, \ldots, \nu_n$ d.f., then on the hypothesis that the samples are drawn from the same population:

$$s^2 = \Sigma\nu_i s_i^2 / \Sigma\nu_i = \Sigma\nu_i s_i^2/\nu; \quad \nu = \sum_{i=1}^{n}\nu_i \text{ and } s_i^2 = \frac{1}{\nu_i - 1}\sum_{i=1}^{n}(y_i - \bar{y})^2$$

is also an unbiased estimate of the population variance σ^2, since

$$E(s^2) = \frac{1}{\nu}\sum_{i=1}^{n} \nu_i [E(s_i^2)] = \frac{1}{\nu}\sum_{i=1}^{n} \nu_i \sigma^2 = \sigma^2$$

Bartlett has shown that on the hypothesis that the various samples come from the same population, the statistic

$$\sum_{i=1}^{n} \nu_i \log_e \left(s^2/s_i^2\right) \Big/ \left[1 + \frac{1}{3(n-1)}\left\{\sum_{i=1}^{n}(1/\nu_i) - \frac{1}{\nu}\right\}\right]$$

is distributed approximately as χ^2 with $(n - 1)$ d.f.

TESTING HOMOGENEITY OF SEVERAL INDEPENDENT ESTIMATES OF AN ORDINARY OR PARTIAL CORRELATION COEFFICIENT

Let $r_1, r_2, r_3, \ldots, r_k$ be k independent estimates of the correlation coefficient. We have to test whether these can be regarded as having come from the same bivariate population.

Let,

$$z_i = (1/2)\log_e \{(1 + r_i)/(1 - r_i)\} \text{ and } \bar{z} = \sum_{i=1}^{k}(n_i - 3)z_i \Big/ \sum_{i=1}^{k}(n_i - 3); \qquad i = 1, 2, 3, \ldots, k$$

Since, for large samples $z_1, z_2, z_3, \ldots, z_k$ are all distributed normally about the same mean, the sum $\sum_{i=1}^{k}(n_i - 3)(z_i - \bar{z})^2$ is distributed approximately as χ^2 with $(k - 1)$ d.f., one d.f. is lost as z is determined from the data and the test of independence can be carried out accordingly.

For partial coefficient of order k, n_i will be replaced by $n_i - k$.

TESTS OF SIGNIFICANCE BASED ON t, F AND z DISTRIBUTIONS

In case of large sample, the test of significance is based on the following asymptotic property of the sampling distribution of any statistic: The sampling distribution of any statistic in the standardised form is $N(0, 1)$ asymptotically, *i.e.*, as $n \to \infty$.

The asymptotically standard normal statistics are:

$$z = (y - np)/\sqrt{npq} = (y/n - p)/\sqrt{pq/n} \qquad \sim N(0,1) \text{ as } n \to \infty$$

or,

$$z = (y_1/n_1 - y_2/n_2)/\sqrt{\bar{p}\bar{q}(1/n_1 + 1/n_2)} \qquad \sim N(0,1) \text{ as } n \to \infty$$

where,

$$\bar{p} = (y_1 + y_2)/(n_1 + n_2), \quad \bar{q} = 1 - \bar{p}.$$

or,

$$z = (\bar{y} - \mu)/(s/\sqrt{n}) \qquad \sim N(0,1) \text{ as } n \to \infty$$

where,

$$\bar{y} = (1/n)\sum_{i=1}^{n} y_i; s^2 = (1/n)\sum_{i=1}^{n} (y_i - \bar{y})^2$$

or,

$$z = (\bar{y}_1 - \bar{y}_2)/s \times \sqrt{1/n_1 + 1/n_2} \qquad \sim N(0,1) \text{ as } n \to \infty$$

where,

$$s^2 = \left\{\sum_{i=1}^{n_1}(y_{1i} - \bar{y}_1)^2 + \sum_{i=1}^{n_2}(y_{2j} - \bar{y}_2)^2\right\} \Big/ (n_1 + n_2)$$

However, in case of small sample size, the distributions of the standardised statistics are far from normality and consequently normal test cannot be applied. Therefore, to deal with small samples, new techniques and tests of significance, which is known as *Exact Sample Tests*, have been developed. From the practical point of view, in most of the situations a sample is termed as small if $n \leq 30$. It should be noted that exact sample tests can be used even for large samples but large sample tests cannot be used for small samples.

The development of exact sample tests is attributed to an Irish brewery employee William S. Gosset who used to write under pen name of 'Student' in the beginning of 20th century and was later extended by Prof. R.A. Fisher. Gosset gave his statistic the name *Student's 't'* in 1908.

All the exact sample tests are based on two basic fundamental assumptions:

(a) The parent population from which the sample is drawn, should be normally distributed.

(b) The sample is random and independent of each other.

In the present chapter we shall discuss Gosset Student's t, Snedecor's F and Fisher's z distributions and tests of significance based on them.

STUDENT'S t

If $y_1, y_2, y_3, \ldots, y_n$ is a random sample of size n from a normal population with mean μ and variance σ^2, then Student's t statistic is defined by

$$t=(\bar{y}-\mu)\Big/\left(s/\sqrt{n}\right)=(\bar{y}-\mu)\Big/\sqrt{s^2/n}$$

where, $\bar{y}=\Sigma y_i/n$ is the sample mean and $s^2=\{1/(n-1)\}\,\Sigma(y_i-\bar{y})^2$ is an unbiased estimate of the population variance σ^2.

The statistic t follows Student's t distribution with $\nu=(n-1)$ degrees of freedom and probability density function (p.d.f.):

$$p(t)=\text{constant}\times\left(1+t^2/\nu\right)^{-(\nu+1)/2}; \qquad -\infty<t<\infty.$$

FISHER'S t

If $x \sim N(0,1)$ and y is an independent χ^2 variate with n d.f., then Fisher's t is defined by

$$t=x\Big/\sqrt{y/n}$$

and it follows Student's t distribution with n d.f. and p.d.f.

$$p(t)=\left\{1\Big/\beta(1/2,n/2)\sqrt{n}\right\}\left(1+t^2/n\right)^{-(n+1)/2}; \qquad -\infty<t<\infty$$

Fisher used this statistic to test the significance of—(a) difference between two sample means, and (b) observed sample correlation coefficient.

t-TEST FOR DIFFERENCE OF MEANS

(a) $x_1,x_2,x_3,\ldots,x_{n_1}$ and $y_1,y_2,y_3,\ldots,y_{n_2}$ are two independent random samples from the given normal populations. We set up the null hypothesis $H_0(\mu_x=\mu_y)$, *i.e.*, the samples have been drawn from the normal populations with the same means. In other words, the sample mean $\bar{x}$ and $\bar{y}$ do not differ significantly. Under the assumption that $\sigma_1^2=\sigma_2^2=\sigma^2$, *i.e.*, population variances are equal but unknown, the test statistic under H_0 is given by

$$t=(\bar{x}-\bar{y})\Big/s\sqrt{1/n_1+1/n_2} \qquad \sim t_{n_1+n_2-2}$$

where, $\bar{x}=(1/n_1)\Sigma x_i$, $\bar{y}=(1/n_2)\Sigma y_i$ and $s^2=\{1/(n_1+n_2-2)\}\left[\Sigma(x_i-\bar{x})^2+\Sigma(y_i-\bar{y})^2\right]$

s^2 is an unbiased estimate of the common population variance σ^2 based on both the samples. By comparing the calculated value of t with the tabulated value for (n_1+n_2-2) degrees of freedom and at desired level of significance, usually 5% or 1%, we reject or retain the null hypothesis H_0.

(b) Given two independent random samples from normal populations with the same variance, we have to test the hypothesis that the population means are μ_1 and μ_2 respectively. For this we use the test statistic:

$$t = \{(\bar{x} - \bar{y}) - (\mu_1 - \mu_2)\} / s\sqrt{1/n_1 + 1/n_2}$$

with $(n_1 + n_2 - 2)$ d.f.

BEHREN *d* TEST

For testing the significance of the difference between two sample means, an additional assumption is made that the variances of the two populations are the same. If the two variances are different, *t*-test is no longer valid and we apply Behren *d* test based on the fiducial intervals. In this case the test statistic is:

$$d = (\bar{x} - \bar{y}) / \sqrt{s_1^2 + s_2^2}$$

and we carry out the test of significance by using the tables of Sukhatme and Fisher for various values of n_1, n_2 and s_1^2/s_2^2, the values of *d* corresponding to various probability levels.

PAIRED t-TEST FOR DIFFERENCE OF MEANS

In case the sample sizes are equal, *i.e.*, $n_1 = n_2 = n$, the sample observations ($x_1, x_2, x_3, \ldots, x_n$) and ($y_1, y_2, y_3, \ldots, y_n$) are not completely independent but they are dependent in pairs, *i.e.*, the pairs of observations $(x_1, y_1), (x_2, y_2), (x_3, y_3), \ldots, (x_n, y_n)$ correspond to the 1st, 2nd, ..., *n*th unit respectively.

Let us denote the difference in the observations for the *i*th unit by d_i, *i.e.*, $d_i = x_i - y_i$ for $i = 1, 2, 3, \ldots, n$. In this case the test statistic is

$$t = \bar{d}/(s/\sqrt{n}) = \bar{d}/\sqrt{s^2/n}$$

where, $\bar{d} = \Sigma d_i/n$ and $s^2 = \Sigma(d_i - \bar{d})^2/(n-1) = \left[\Sigma d_i^2 - (\Sigma d_i)^2/n\right]/(n-1)$.

t-TEST FOR SIGNIFICANCE OF AN OBSERVED CORRELATION COEFFICIENT

Given a random sample (x_i, y_i), $i = 1, 2, 3, \ldots, n$ of size *n* from a bivariate normal distribution and *r* the observed sample correlation coefficient, the significance of *r* is tested using the statistic

$$t = r\sqrt{n-2}/\sqrt{1-r^2}$$

which follows Student's *t* distribution with $(n - 2)$ degrees of freedom.

If calculated value of $|t|$ exceeds $t_{0.05}$ for $(n - 2)$ d.f., the value of *r* is significant at 5% level of significance. If $|t| < t_{0.05}$ the sample is consistent with the hypothesis of an uncorrelated population.

t-TEST FOR SIGNIFICANCE OF AN OBSERVED PARTIAL CORRELATION COEFFICIENT

The sampling distribution of a partial correlation coefficient of order k is of the same form as that of correlation coefficient from a bivariate normal population with the sample size n reduced by k. In this case, the test statistic is

$$t = r\sqrt{n-k-2}\Big/\sqrt{1-r^2}$$

which follows Student's t distribution with $(n - k - 2)$ degrees of freedom.

t-TEST FOR SIGNIFICANCE OF A RANK CORRELATION COEFFICIENT

For a sample of moderate size, it can be shown that if r is the rank correlation coefficient from a sample of size n, and the hypothesis that the population rank correlation coefficient is zero is correct then the test statistic is

$$t = r\sqrt{(n-2)/(1-r^2)}$$

which follows Student's t distribution with $(n - 2)$ degrees of freedom.

t-TEST FOR SIGNIFICANCE OF AN OBSERVED REGRESSION COEFFICIENT

Given a random sample (x_1, y_1), (x_2, y_2), (x_3, y_3), ..., (x_n, y_n) from a bivariate normal population, the regression equation of y on x is given by

$$y - \bar{y} = b(x - \bar{x})$$

so that the estimated value of y corresponding to a given x_i is

$$\hat{y}_i = \bar{y} + b(x_i - \bar{x})$$

We have to test the hypothesis that the regression coefficient of y on x in the population is β. For this the test statistic is given by

$$t = (b - \beta)\left[(n-2)\Sigma(x_i - \bar{x})^2 / \Sigma(y_i - \hat{y}_i)^2\right]^{1/2}$$

which follows t distribution with $(n - 2)$ d.f.

F-STATISTIC

It is the ratio of two independent Chi-Square variates divided by their respective degrees of freedom, *i.e.*, if x is a χ^2 variate with n_1 d.f. and y is an independent χ^2 variate with n_2 d.f., then F-statistic is defined by

$$F = (x/n_1)/(y/n_2)$$

which follows G.W. Snedecor's F distribution with (n_1, n_2) degrees of freedom with probability density function given by

$$p(F) = y_0 \cdot F^{(n_1/2)-1} \Big/ \left(1 + \frac{n_1}{n_2}F\right)^{(n_1+n_2)/2} \quad ; \qquad 0 \le F < \infty$$

where y_0 is a constant which is so determined that total area under the probability curve is 1, *i.e.*, $\int_0^\infty p(F)\,dF = 1$.

This gives:

$$y_0 = (n_1/n_2)^{n_1/2} / \beta(n_1/2, n_2/2)$$

where $\beta(m,n)$ is defined as $\beta(m,n) = \Gamma m \Gamma n / \Gamma(m+n)$; $m>0, n>0$.

F-TEST FOR SIGNIFICANCE OF THE RATIO OF TWO INDEPENDENT ESTIMATES OF THE POPULATION VARIANCE

Given two independent random samples from normal population, for testing the hypothesis that the population variances are the same, it can be shown that the statistic s_1^2/s_2^2, where

$$s_1^2 = \sum_{i-1}^{n_1} (x_i - \bar{x})^2 \Big/ (n_1 - 1); \quad s_2^2 = \sum_{j=1}^{n_2} (y_j - \bar{y})^2 \Big/ (n_2 - 1)$$

and $s_1^2 > s_2^2$ confirms to the F distribution with $(n_1 - 1, n_2 - 1)$ degrees of freedom. Therefore, F statistic is given by $F = s_1^2/s_2^2$; $(s_1^2 > s_2^2)$.

If the calculated value exceeds the tabulated value of $F_{0.05}$ for $(n_1 - 1, n_2 - 1)$ degrees of freedom, we say that ratio is significant at 5% level of significance. However, if $F < F_{0.05}$ we say that the same could have come from two normal populations with the same variance.

F-TEST FOR SIGNIFICANCE OF CORRELATION RATIO

Let us consider a random sample (x_i, y_{ij}) $(i = 1, 2, 3, ..., h;\ j = 1, 2, 3, ..., n_i)$ of size $N(=\Sigma n_i)$ from a bivariate normal population and let η be the correlation ratio of y on x defined by

$$\sum_i \sum_j (y_{ij} - \bar{y}_i)^2 = (1 - \eta^2) \sum_i \sum_j (y_{ij} - \bar{y}_i)^2$$

To test the significance of η, the test statistic is

$$F = \{\eta^2/(1-\eta^2)\} \times \{(N-h)/(h-1)\}$$

which confirms to the F distribution with $\{(h - 1), (N - h)\}$ degrees of freedom.

F-TEST FOR THE SIGNIFICANCE OF A REGRESSION FUNCTION

For testing whether the given sample indicates any degree of association as represented by the regression equation, we use the test statistic

$$F = r^2(n-2)/(1-r^2)$$

which follows F distribution with $(1, n - 2)$ degrees of freedom.

F-TEST FOR NON-LINEARITY OF REGRESSION

For a sample of size N arranged in h arrays from a normal sample, the appropriate statistic for testing the hypothesis of linearity of regression is

$$F = \{(\eta^2 - r^2)/(1-\eta^2)\} \times \{(N-h)/(h-2)\}$$

with $(h - 2, N - h)$ degrees of freedom.

z-TRANSFORMATION

It is used for testing the significance of the coefficient of correlation in small samples. In this method, the coefficient of correlation (r) in the sample is transformed into z, which is calculated as:

$$z = (1/2)\log_e\{(1+r)/(1-r)\} = 1.1513\log_{10}\{(1+r)/(1-r)\}$$

The standard error of z is given by: $S.E._z = 1/(n-3)$.

According to Fisher, for the sample of a large size n, z is distributed asymptotically normally about mean ψ, where

$$\psi = (1/2)\log_e\{(1+\rho)/(1-\rho)\} = 1.1513\log_{10}\{(1+\rho)/(1-\rho)\}$$

with variance $1/(n-3)$.

To test the significance of a coefficient of correlation, the difference between z and ψ is calculated and the relationship of the difference and the standard error of z is then found out. If the difference is more than thrice or twice the standard error, it is supposed to be significant.

z-TRANSFORMATION TO TEST WHETHER r DIFFERS SIGNIFICANTLY FROM ρ

Let us consider a random sample (x_1, y_1), (x_2, y_2), ..., (x_n, y_n) from a bivariate normal population. We want to test the hypothesis that the population correlation coefficient is ρ. For this, we calculate z and ψ by applying Fisher's transformation and then calculate the asymptotically standard normal variate

$$\omega = (z - \psi)/\left(1/\sqrt{n-3}\right)$$

If the absolute value of ω exceeds 1.96, the difference is significant at 5% level of significance.

z-TRANSFORMATION TO TEST THE SIGNIFICANCE OF THE DIFFERENCE BETWEEN TWO INDEPENDENT CORRELATION COEFFICIENT

Suppose we want to test if two independent random samples of sizes n_1 and n_2 and correlation coefficients r_1 and r_2 respectively have been drawn from same bivariate normal population or from different normal populations with same correlation coefficient. If the hypothesis is correct, the statistic

$$z = (z_1 - z_2)/\sqrt{1/(n_1 - 3) + 1/(n_2 - 3)}$$

where, $z_1 = 1.1513\log_{10}\{(1+r_1)/(1-r_1)\}$, $z_2 = 1.1513\log_{10}\{(1+r_2)/(1-r_2)\}$

is asymptotically a standard normal variate. If the absolute value of z is greater than 1.96, the difference will be significant at 5% level of significance.

z-TRANSFORMATION TO TEST THE SIGNIFICANCE OF COMBINATION OF ESTIMATES OF A CORRELATION COEFFICIENT

Correlation coefficients $r_1, r_2, r_3, \ldots, r_k$ are obtained from samples of sizes $n_1, n_2, n_3, \ldots, n_k$ from the same bivariate normal population with correlation coefficient ρ, then $z_1, z_2, z_3, \ldots, z_k$ are independent estimates of the same quantity $\psi = 1.1513\log_{10}\{(1+\rho)/(1-\rho)\}$. To get a pooled estimate for ψ, we take weighted average of $z_1, z_2, z_3, \ldots, z_k$ and to get an estimate with the minimum variance we use weights inversely proportional to the variance.

Then, the estimate is accordingly given by $\bar{z} = \Sigma(n_i - 3)z_i/\Sigma(n_i - 3)$ and from this we can get an estimate for ρ from the relation $\rho = \tanh\bar{z}$.

Since, the variate z_i are approximately normally distributed about ψ with variance $1/(n_i - 3)$, the sum $\Sigma(n_i - 3)(z_i - \bar{z})^2$ is distributed approximately as χ^2 with $(k - 1)$ degrees of freedom. Then the significance can be ascertained from the table of χ^2 and it shows that the estimate r_i of correlation in the population may be regarded as homogeneous.

TIME SERIES ANALYSIS

A series of observations recorded in accordance with the time of occurrence is called time series. In other words, a chronologically arranged values of a variable over successive time periods is known as time series and the analysis of such values of the variable is called the time series analysis.

According to Wessel and Wellet, when quantitative data are arranged in the order of their occurrence, the resulting statistical series is called a time series.

According to Kenny and Keeping, a set of data depending on the time is called time series.

According to Moris Hamburg, a time series is a set of statistical observations arranged in chronological order.

Croxton and Cowden defined the time series as the chronologically arranged data.

According to Patterson, time series consists of statistical data which are collected, recorded or observed over successive increments.

According to Ya Lun Chou, a time series may be defined as a collection of readings belonging to different time periods, of some economic variable or composite of variables such as production of steel, per capita income, gross national product, price of tobacco or index of industrial production.

Cecil H. Meyers defined a time series as a sequence of repeated measurements of a variable made periodically through time.

According to Spiegel, a time series is a set of observations taken at specified times, usually at 'equal intervals'. Mathematically, a time series is defined by the values $y_1, y_2, ...$ of a variable y (temperature, closing price of a share, *etc.*) at times $t_1, t_2, ...$ Thus, y is a function of t, symbolised by $y = F(t)$.

Time series data are of particular interest in economics, business and commerce, where the values of a variable are observed chronologically by days, weeks, months, quarters, or years. Time series analysis is done primarily for the purpose of evaluating past performances and making forecasts for the future.

Time series analysis is used—(a) in the analysis of past behaviour of a variable, (b) in forecasting, (c) in evaluation of current achievement, (d) in making comparative studies, *etc.*

Components of a Time Series: Since, a time series is the result of the combined effects of different categories of forces, it has following four components:

(a) **Secular trends or Long Term Movements** (*T*)—They refer to long-term tendency of a phenomenon and represent the general tendency of the time series data to increase or to decrease or to remain segregated during a long period of time. The trend is that part which the series would have exhibited, had there been no other factors affecting the values.

(b) **Seasonal Variations** (*S*)—They refer to such movements in a time series which are due to forces which are rhythmic in nature and which repeat themselves periodically every season and where the period is not longer than one year. They are the result of such factors which uniformly and regularly rise and fall in magnitude and attributed to the causes—(i) resulting from natural forces, and (ii) resulting from local customs and traditions.

(c) **Cyclical Fluctuations** (C)—They are another types of periodic movements, where the period is more than a year. Such movements are fairly regular and oscillatory in nature. One complete period is called a cycle. Though cyclical movements, by and large are regular, they are not necessarily uniformly periodic. A cycle may not always complete itself with a fixed duration of time.

(d) **Irregular or Random Variations** (*I*)—They are such variations which are caused by factors of an erratic nature. These are completely unpredictable or caused by such unforeseen events as war, flood, earth-quake, strike and lockout, *etc.* and may sometimes be the result of many small forces, each of which has a negligible effect, but their combined effect is not negligible. They are generally mixed up with seasonal and cyclical variations and do not reveal any pattern of the repetitive tendency and may be considered as residual variation.

Classical or traditional approach assumes a multiplicative relationship between the four components, *i.e.*, any particular observation is considered to be the product of the effects of four components:

$$y_t = T \times S \times C \times I \qquad (\textit{Multiplicative model})$$

Another approach assumes an additive relationship between the four components:

$$y_t = T + S + C + I \qquad (\textit{Additive model})$$

The additive model facilitates easier calculations, but it has been found inappropriate in many practical situations, and therefore it is not generally used.

Adjustments to Time Series Data: Before analysing time series data, it is often necessary to adjust them for calendar variation, population changes and price changes, in order to make them homogeneous and comparable as far as possible.

Adjustment for calendar variation is done by dividing each monthly total by the number of days in months. And the daily average is then multiplied by 365/12 which is the average number of days in a month. In case of leap year, the multiplication factor would be 366/12.

When we are studying national income figures or figures of output, adjustment for population change become necessary because national income may be rising due to change in the number of working population or output may be going down due to the decline in the number of workers. In such cases, adjustment is done by dividing the income or output by the number of population concerned and thus we have per capita income figures or figures of output per worker.

When a current value series is given and we are interested in real value changes, adjustment for price changes is necessary. In such case, the current values should be deflated by the ratio of current prices to base year prices. In other words, we have to find out: *Current value* × (*Base Price Index*/*Current Price Index*). Such adjustments are also necessary when with a value series at our disposal, we are interested in studying quantity changes.

MEASUREMENT OF TREND

There are four methods by which trend values can be determined.

(a) *Free-hand or Graphic Method*—The given data are plotted as points on a graph paper against time. The time series data (y_t) are shown along the vertical axis and time (t) along the horizontal axis. Then, a free-hand smoothed curve is drawn through these points in such a way that the curve appears to represent their pattern of movement over time. The distance of this line, known as trend line, gives the trend value for each time period. In the process of drawing a smooth curve, the seasonal, cyclical and irregular variations are eliminated.

(b) *Semi-average Method*—In this method, the data is divided into two parts, and then an average for each part is calculated. These averages are plotted as points on a graph paper against the mid-point of the time interval covered by each part. The straight line joining these two points gives the trend line and the distances of trend line from the horizontal axis give the trend values.

(c) *Moving Average Method*—This method is used for the isolation of trend and smoothing out fluctuations in time series and obtaining trend values with a fair degree of accuracy. In this method, we calculate from the given data, a series of arithmetic means of successive observations, known as moving averages which are used as trend values. Moving averages of period n are a series of arithmetic means of groups of successive n observations, and are shown against the mid-points of time intervals covered by the respective groups.

If the period of moving average is odd, the trend values correspond to the given time. If the period of moving average is even, a two-point moving average of the moving averages so obtained, has to be found out for centering them. Sometimes trend values are determined by using weighted moving average. In this method, the moving totals are multiplied by the weights assigned to them and the weighted moving average is obtained by dividing this product by the sum of the weights.

(d) *Fitting Mathematical Curves (Method of Least Squares)*—In this method, an appropriate type of mathematical equations is selected for trend, and the constants appearing in the trend equation are determined on the basis of the given time series data. The line obtained through this process is called the Line of the Best Fit, and it is the line from which the sum of the deviations of various points on either side is equal to zero. The choice of the appropriate type of equation is facilitated by a graphical representation of the data. For example,

(i) A straight line of the form, $y = a + bx$ is used if the plotted data show approximately a straight line tendency on an ordinary graph paper.

(ii) An exponential curve of the form, $\log y = a + bx$ is used if the plotted data show a straight line on a semi-logarithmic graph paper.

(iii) Sometimes a parabola or higher order polynomial of the form, $y = a + bx + cx^2$ can also be used.

(iv) In certain cases, special types of curves are used such as

Modified exponential curve $\rightarrow$ $y = a + bc^x$

Logistic curve $\rightarrow$ $1/y = a + bc^x$

Gompertz curve $\rightarrow$ $\log y = a + bc^x$

MONTHLY TREND FROM ANNUAL DATA

In time series, annual data are usually available in two different forms: (a) monthly averages for each year and (b) annual totals.

***Case* I**: Monthly averages for odd number of years are given—If the data for the odd number of years are available, middlemost year is taken as origin, *i.e.*, x unit = 1 year, and the trend equation is fitted. Let the linear trend fitted to monthly averages be, $y = a + bx$ (let origin 1990; x unit = 1 year). The origin is precisely at the mid-point of the year 1990, *i.e.*, 30th June, 1990. The slope b represents the increase in monthly average for unit increase in x, *i.e.*, for 12 months. Therefore, the monthly trend increment is $b/12$, and the monthly trend equation is

$$y = a + \frac{b}{12}\cdot x \qquad \text{(origin: 30th June, 1990; } x \text{ unit = 1 month)}$$

For using the trend equation, some specified month as origin is taken, *i.e.*, precisely the mid-point of a month. If July, 1990 is used as origin, *i.e.*, the origin is shifted half a unit later from 30th June, 1990 to 15th July, 1990, we have to replace x by $(x + 1/2)$. Thus, the trend equation for monthly values is given by

$$y = a + \frac{b}{12}\left(x + \frac{1}{2}\right) \qquad \text{(origin: 15th July, 1990; } x \text{ unit = 1 month)}$$

If the origin is shifted half a unit earlier to 30th June, 1990; x is to be replaced by $(x - 1/2)$ and the trend equation will be

$$y = a + \frac{b}{12}\left(x - \frac{1}{2}\right) \qquad \text{(origin: 15th June, 1990; } x \text{ unit = 1 month).}$$

***Case* II**: Monthly averages for even number of years are given— If data for an even number of years are available, the mid-point of the two middlemost years are taken as origin and x unit = 6 months. For instance, suppose that monthly averages are given for 8 years from 1996 to 2003. Let the trend equation, fitted to monthly averages be

$$y = a + bx \qquad \text{(origin: 1999-2000; } x \text{ unit} = 6 \text{ months)}$$

The origin is at the point midway between 1999 and 2000, *i.e.*, 31st December, 1999. The slope b represents the increase in monthly average for unit increase in x, *i.e.*, for 6 months; hence the monthly trend increment is $b/6$. Thus, the monthly trend equation is given by

$$y = a + \frac{b}{6}\cdot x \qquad \text{(origin: 31st December, 1999; } x \text{ unit} = 1 \text{ month)}$$

For using the monthly trend equation, some specified month is chosen as origin; *i.e.*, precisely the mid-point of a month. For instance, if the origin is shifted half a unit later from 31st December, 1999 to 15th January, 2000, x is to be replaced by $(x + 1/2)$, then the monthly trend equation becomes

$$y = a + \frac{b}{6}\left(x + \frac{1}{2}\right) \qquad \text{(origin: 15th January, 2000; } x \text{ unit} = 1 \text{ month)}$$

***Case* III**: Annual totals are given—Let for a number of years (odd or even), the trend equation fitted to annual totals is given by $y = A + Bx$.

If monthly averages were first computed from the given annual totals, then a straight line trend fitted to these monthly averages would be

$$y = \frac{A}{12} + \frac{B}{12}\cdot x = a + bx \qquad \text{(say)}$$

where, $a = A/12$ and $b = B/12$.

The monthly trend equation may now be obtained by applying the methods discussed in Case I and Case II, depending on whether the number of years given is odd or even.

For odd number of years the trend equation will be

$$y = \frac{A}{12} + \frac{B}{144}\left(x + \frac{1}{2}\right) \qquad \text{(origin: 15th July, 1990; } x \text{ unit} = 1 \text{ month)}$$

and for even number of years it would be

$$y = \frac{A}{12} + \frac{B}{72}\left(x + \frac{1}{2}\right) \qquad \text{(origin: 15th January, 2000; } x \text{ unit} = 1 \text{ month).}$$

MEASUREMENT OF SEASONAL VARIATION

There are four methods of measuring seasonal fluctuations:

(a) *Method of Simple Averages*—It is applied when the time series data do not have trend or cyclical variations to any appreciable extent. From the quarterly data, we calculate the totals for each quarter Q_1, Q_2, Q_3, Q_4, and the averages A_1, A_2, A_3, A_4 for the four quarters. Thereafter, we calculate the Grand Average as

$$G = \frac{1}{4}(A_1 + A_2 + A_3 + A_4)$$

In additive model, the deviations of quarterly averages from the grand average give the seasonal variation as

$$S_1 = A_1 - G,\ S_2 = A_2 - G,\ S_3 = A_3 - G,\ S_4 = A_4 - G$$

In multiplicative model, each quarterly average is expressed as a percentage of the grand average giving seasonal indices as

$$S_1 = \frac{A_1}{G} \times 100,\ S_2 = \frac{A_2}{G} \times 100,\ S_3 = \frac{A_3}{G} \times 100,\ S_4 = \frac{A_4}{G} \times 100$$

If we have given monthly figures, we find the 12 averages $A_1, A_2, ..., A_{12}$ for the 12 months and then the seasonal index for each month is obtained proceeding the same way as above.

In additive model, the total (or average) seasonal variation is 0, and in the multiplicative model the average seasonal index is 100.

(b) *Method of Moving Averages*—From the quarterly figures, we estimate the trend by taking four-quarter moving averages and then from the original data effect of trend is eliminated.

In case of additive model, the moving average trend values are subtracted from the original data to give us deviations from trend. Since these deviations do not contain any effect of trend, the method of quarterly averages is applied to these deviations, using additive model.

In case of multiplicative model, we find the ratios to moving averages expressed as percentages, *i.e.*, the original values are expressed as percentages of the corresponding moving average values. These percentages are then arranged by quarters and the average for each quarter, P_1, P_2, P_3, P_4 are calculated. After that these are adjusted to a total of 400, multiplying each by $100/P$, where P is the Grand Average as

$$P = \frac{1}{4}(P_1 + P_2 + P_3 + P_4)$$

The seasonal indices are calculated as

$$S_1 = \frac{P_1}{P} \times 100,\ S_2 = \frac{P_2}{P} \times 100,\ S_3 = \frac{P_3}{P} \times 100,\ S_4 = \frac{P_4}{P} \times 100$$

respectively for the quarters Q_1, Q_2, Q_3, and Q_4.

(c) *Method of Trend-Ratio*—In this method, the multiplicative model is used. By fitting a mathematical curve, trend values are obtained and the original data are expressed as percentages of the corresponding trend. Then, similar to the moving average method, these percentages are arranged by quarters and the average trend-ratio for each quarter, P_1, P_2, P_3, P_4 are calculated. Each of these is now multiplied by

$100/P$, where $P = (1/4)(P_1 + P_2 + P_3 + P_4)$, to give the seasonal indices as

$$S_1 = \frac{P_1}{P} \times 100,\ S_2 = \frac{P_2}{P} \times 100,\ S_3 = \frac{P_3}{P} \times 100,\ S_4 = \frac{P_4}{P} \times 100$$

corresponding to the quarters Q_1, Q_2, Q_3, Q_4 respectively.

The total of the four seasonal indices is 400.

(d) *Method of Link Relative*—If the quarterly data are given, each value is expressed as a percentage of the value for the immediately preceding period. These are known as Link Relative (L.R.). Steps involved in the calculation of seasonal indices are:

(i) Calculate the link relative as

$$\textit{Link relative} = \frac{\textit{Current season's figure}}{\textit{Previous season's figure}} \times 100$$

(ii) Calculate the average of L.R. for each season or arithmetic mean can be used for this purpose.

(iii) Convert the L.R. into chain relatives (C.R.) on the basis of the first season as

$$\textit{Chain relative} = \frac{\text{L.R. } \textit{of the season} \times \text{C.R. } \textit{of previous season}}{100}$$

The chain relative for the first season should be taken as 100. Thus, if the data are given on a monthly basis, the chain relative for January would be 100 and

$$\text{C.R. } \textit{for February} = \frac{\text{L.R. } \textit{Feb.} \times \text{C.R. } \textit{of Jan.}}{100} = \frac{\text{L.R. } \textit{Feb.} \times 100}{100} = \text{L.R. } \textit{Feb.}$$

$$\text{C.R. } \textit{for March} = \frac{\text{L.R. } \textit{of March} \times \text{C.R. } \textit{of Feb.}}{100} \text{ and}$$

$$\text{C.R. } \textit{for December} = \frac{\text{L.R. } \textit{of Dec.} \times \text{C.R. } \textit{of Nov.}}{100}$$

(iv) New chain relative for January on the basis of C.R. of December is calculated as

$$\textit{New } \text{C.R. } \textit{for January} = \frac{\text{L.R.} \textit{of Jan.} \times \text{C.R. } \textit{of Dec.}}{100}$$

(v) Correction factor is calculated as

$$\textit{Correction factor (c.f.)} = \frac{\textit{New } \text{C.R. } \textit{for Jan.} - \textit{Old } \text{C.R. } \textit{of Jan.}}{12}$$

If the figures are given quarterly, the correction factor is given by

$$\textit{c.f.} = \frac{\textit{New } \text{C.R. } \textit{for 1st quarter} - \textit{Old } \text{C.R. } \textit{for 1st quarter}}{4}$$

(vi) The indices of seasonal variation are obtained by adjusting these chain relatives to a total of 1200 for monthly data and to a total of 400 for quarterly data. It implies that each corrected chain relative would be multiplied by 1200/*Sum of corrected C.Rs.* for monthly data and by 400/*Sum of corrected C.Rs.* for quarterly data.

DESEASONALISATIONS OF DATA

In case of additive model where $y_t = T + S + C + I$, deseasonalisation can be done by subtracting the seasonal variations from the value of Y. Thus, a deseasonalised series would be $y_t - S = T + C + I$.

In case of multiplicative model where $y_t = T \times S \times C \times I$ the deseasonalised values are obtained by dividing the given values by corresponding indices of seasonal variations. Thus,

$$Deseasonalised\ values = \frac{y_t}{S} = \frac{T \times S \times C \times I}{S} = T \times C \times I$$

MEASUREMENT OF CYCLICAL VARIATIONS

According to Croxton and Cowden, there are four methods of measuring cyclical variations.

(a) *Residual Method*—This method consists in removing the three components, *viz.*, trend, seasonal and irregular, from the original data, in order to isolate the cyclical component. At first, trend and seasonal effects are measured by suitable methods and are removed from the data.

In case of the multiplicative model, in order to get $C \times I$, we remove T and S from y_t by division. We may either divide y_t by T first, and then by S; or divide y_t by S first and then by T; or divide y_t by the product $T \times S$, called normal value. Thus, we have

$$\frac{y_t}{Normal\ value} = \frac{T \times S \times C \times I}{T \times S} = C \times I$$

In case of additive model, trend and seasonal components are removed by subtraction as

$$y_t - T - S = (T + S + C + I) - T - S = C + I$$

Thus, the residual consists of a combination of cyclical and irregular components, either $C \times I$ or $C + I$. At the next stage, the irregular component (I) is removed from residuals by smoothing, using moving averages. The appropriate period of the moving average depends on the average duration of irregular movements. Thus, now we are left with only the cyclical variations (C).

(b) *Reference Cycle Analysis*—When several time series are being studied, it is possible to compare the cyclical movements of one series with the cyclical movements of the other series. Following steps are taken:

(1) The data of peaks and troughs of a business cycle are selected as reference data.

(2) A cyclical pattern for each series for the period between each two successive troughs is obtained. To make the inter-series comparison possible, each period is chosen to be same for all series.

(3) Then, the data are adjusted for seasonal variations to obtain a cyclical pattern and thereafter they are divided into reference cycle segments, for each segment, the monthly values are expressed as percentages of the average of all values in the segment. These are called reference cycle relatives and this step converts the series into percentages and eliminates inter-cycle trend. However, this step does not eliminate intra-cycle trend.

(4) Each reference-cycle segment is divided into nine stages to correspond to the same nine stages in the business cycle and the reference cycle relatives are averaged for each of the nine stages. The nine stages are: (i) the three months centred on the initial trough; (ii) the first third of the expansion period; (iii) the second third of the expansion period; (iv) the last third of the expansion period; (v) the three months centred on the peak; (vi) the first third of the contraction period; (vii) the second third of the contraction period; (viii) the last third of the contraction period; and (ix) the three months centred on the terminal trough.

The erratic movements in a series are reduced by the nine stage averages for each reference-cycle segment. In this way, they give a reference cycle pattern for the particular series under consideration.

(c) *Harmonic Analysis*—In case, the cyclical movements in a series have same duration and amplitude, a sine-cosine or a curve of a similar type is fitted. Such a curve is fitted to the *CI* data or the data after the irregular movements have been smoothed.

(d) *Direct Analysis*—In this method, seasonal variation and trend are eliminated. The values of each month are expressed as percentages of the values of the same month in the previous year.

EXPONENTIAL SMOOTHING

It is a technique for "forecasting" by smoothing out fluctuations in time series data. The forecasts are obtained by weighted moving averages of all observations upto the current period. Thus, all past data are involved in computing the forecast. More weight is given to more recent observations and less to those in the past.

In exponential smoothing, the weights used are: α for the current observation, $\alpha(1-\alpha)$ for the immediately preceding observation, $\alpha(1-\alpha)^2$ for the still preceding observation, and so on, where α is a given constant, called the "*smoothing coefficient*", and has a value lying between 0 and 1. It may be noted that the weights of observations from the current period backwards are diminishing with a common ratio $(1-\alpha)$; *i.e.*, the weights form an infinite G.P. series as

$$\alpha,\ \alpha(1-\alpha),\ \alpha(1-\alpha)^2,\ \alpha(1-\alpha)^3,\ \ldots$$

and the sum of these weights is 1.

$$\text{Sum of weights} = \alpha + \alpha(1-\alpha) + \alpha(1-\alpha)^2 + \ldots = \frac{\alpha}{1-(1-\alpha)} = 1.$$

Weights used in Exponential Smoothing

Time	*Observation*	*Weight*	*Forecast*
⋮	⋮	⋮	⋮
$t-2$	y_{t-2}	$w_{t-2} = \alpha(1-\alpha)^2$	u_{t-2}
$t-1$	y_{t-1}	$w_{t-1} = \alpha(1-\alpha)$	u_{t-1}
t	y_t	$w_t = \alpha$	u_t

The "*exponentially smoothed average*" for time t is the weighted average of observations y_t, y_{t-1}, y_{t-2} ... corresponding to time t and backwards with weights w_1, w_{t-1}, w_{t-2}, ..., and is thus given by

$$u_t = \frac{w_t y_t + w_{t-1} y_{t-1} + w_{t-2} y_{t-2} + \ldots}{w_t + w_{t-1} + w_{t-2} + \ldots}$$

$$= w_t y_t + w_{t-1} y_{t-1} + w_{t-2} y_{t-2} + \ldots$$

$$= \alpha y_t + \alpha(1-\alpha) y_{t-1} + \alpha(1-\alpha)^2 y_{t-2} + \ldots$$

The exponentially smoothed average, which is used as a "forecast" for time t, is thus given by

$$u_t = \alpha y_t + \alpha(1-\alpha) y_{t-1} + \alpha(1-\alpha)^2 y_{t-2} + \ldots$$

This may also be written as a recurrence relation

$$u_t = \alpha y_t + (1-\alpha) u_{t-1}$$

where u_t and u_{t-1} are the forecasts at time t and $t-1$ respectively, y_t is the observation for time t, and α is the smoothing coefficient.

We can also write,

$$u_t = u_{t-1} + \alpha(y_t - u_{t-1}), \text{ i.e., } u_t = u_{t-1} + \alpha e_t$$

where, $e_t = (y_t - u_{t-1})$ is the "error" or discrepancy of the latest observation from the forecast in the previous period.

For numerical computations, the following steps are followed:

(a) Find the error or discrepancy in the latest observation from the previous forecast: $e_t = y_y - u_{t-1}$.

(b) Multiply the error by the smoothing coefficient to obtain the correction (αe_t).

(c) The correction (αe_t) is added to the previous forecast (u_{t-1}) to obtain the forecast (u_t) for the current period t.

In exponential smoothing, the first forecast has to be obtained by some subjective method, or by taking the average of the first few time periods. Subsequent forecasts are then obtained by repeatedly using the relation, $u_t = u_{t-1} + \alpha.e_t$.

When $\alpha = 0$, no adjustment is necessary to the previous forecast. When $\alpha = 1$, the new forecast will always be the latest observation. Between 0 and 1, the smoothing coefficient α is selected on the basis of experiments with different values. For practical purposes, a value of α between 0.1 and 0.4 is ordinarily chosen.

ANALYSIS OF VARIANCE (ANOVA)

The statistical methodology for comparing several means is called analysis of variance, or in short ANOVA. It is defined as the statistical technique for the separation of variation due to a group of causes from the variation due to other groups.

This technique was developed by R.A. Fisher for separation of the experimentally observed variance into a number of components traceable to specific sources. It is based upon an unusual result that the equality of several population means can be tested by comparing the sample variances using F distribution. Essentially, it provides a test of the homogeneity of a set of data. In an experiment generally there are several factors at work, each one of which may cause a certain amount of variability in the observation made. The aim of ANOVA is to find how much of the total variability is due to each factor and by comparing these contributory amounts of variation, we can test the homogeneity of the observations.

The problem in this method is to decide whether the differences among sample means is only due to chance or whether the differences occur because the means of the populations, from which the samples have been drawn, are actually different. Therefore, this method takes into consideration the problem of testing the hypothesis that the means of several populations are equal. Rejection of the hypothesis will lead to the conclusion that at least one of the population means is different from others.

The technique of ANOVA uses F-test to determine whether the sample means differ from one another (variance between groups) to a great extent than the test scores differ from their own sample mean (variance within groups) using the ratio: F = var *between groups* / var *within groups*.

In ANOVA following necessary assumptions are made:

(a) The samples are independently drawn.

(b) The populations are normally distributed, with a common variance.

(c) The effects of various components are additive.

The ANOVA is extremely useful in all types of researches where the effects of one or two variables on a problem under study had to be determined on the basis of a number of experiments conducted simultaneously. It can tell us whether different sample data classified on the basis of a single variable are meaningful.

Variance within and Between Classes: The ANOVA involves the separation of the total variation in the whole number of observations into two parts—the variance between the classes, or the variance attributable to the different conditions, and the variance which arises from individual difference within the classes. The first part is due to assignable causes whereas the the second part is due to various chance causes.

Techniques of ANOVA: The techniques of ANOVA has been divided into two—(a) one-way classification, and (b) two-way classification. In both the cases, a comparison is made between the variance of sample mean with the residual variance. However, in the first case, the total variance is divided in two parts only—variance between the samples and variance within the samples. Whereas in the second case, the total variance is divided in three parts—(a) variance due to variable number one, (b) variance due to variable number two, and (c) residual variance.

TECHNIQUE IN ONE-WAY CLASSIFIED DATA

In this method, we examine the relationship between one independent and one dependent variable. Let us consider k independent random samples or groups of observations. We take one group from each of k normal populations with means μ_1, μ_2, ..., μ_k and a common variance σ^2 (unknown). On the basis of the data, it is required to test the null hypothesis that the population means are equal, *i.e.*, H_0 ($\mu_1 = \mu_2 = ... = \mu_k$) against the alternative hypothesis H_1 (all μ_i are not equal).

The mathematical model can be written as

$$y_{ij} = \mu_i + e_{ij}$$

where, $y_{ij} \rightarrow$ j-th observation in the i-th group; $\mu_i \rightarrow$ mean of the i-th population; and $e_{ij} \rightarrow$ error due to many unspecified causes.

It is assumed that e_{ij} are independently and identically distributed normal variates with mean 0 and variance σ^2. This model may also be written in the form

$$y_{ij} = \mu + \alpha_i + e_{ij}$$

where, $\mu \rightarrow$ the general effect; $\alpha_i \rightarrow$ effect special to the i-th population; and $e_{ij} \rightarrow$ error component.

Then, the null hypothesis is equivalent to stating that there is no special effect due to any population, *i.e.*, H_0 ($\alpha_1 = \alpha_2 = ... = \alpha_k = 0$) against the alternative hypothesis H_1 (all α_i are not zero).

We introduce the following notations:

$k \rightarrow$ number of independent samples or groups,

$n_i \rightarrow$ number of observations in the i-th sample,

$N \rightarrow$ total number of observations in all the samples $= \Sigma n_i$,

$y_{ij} \rightarrow$ the j-th observation in the i-th group,

$T_i \rightarrow$ total of the i-th sample $= \sum_j y_{ij}$,

$T \rightarrow$ grand total of all observations $= \sum T_i = \sum \sum y_{ij}$,

$\bar{y}_i \rightarrow$ mean of the i-th sample $= T_i / n_i$,

$\bar{y} \rightarrow$ grand mean of all observations $= T/N$.

It can be shown algebraically that

$$\sum_i \sum_j (y_{ij} - \bar{y})^2 = \sum_i \sum_j (y_{ij} - \bar{y}_i)^2 + \sum_i n_i (\bar{y}_i - \bar{y})^2$$

i.e., $TSS = SSW + SSB$

where, $TSS \rightarrow$ total sum of squares; $SSW \rightarrow$ sum of squares within groups, or sum of squares due to error (SSE); and SSB $\rightarrow$ sum of squares between groups.

The degrees of freedom can similarly be split up. Thus, we have

Sum of squares : $TSS = SSW + SSB$

Degrees of freedom : $N - 1 = (N - k) + (k - 1)$

The mean squares are now calculated on dividing the sum of squares by the corresponding degrees of freedom.

Mean square within groups (MSW) $=$ $SSW \div (N - k)$

Mean square between groups (MSB) $=$ $SSB \div (k - 1)$

Assuming normal distribution in the population with a common variance (σ^2), it can be shown that the expected value of MSW is σ^2, *i.e.*, $E(MSW) = \sigma^2$ irrespective of whether the population means are equal or not. But $E(MSB) \geq \sigma^2$, where the sign of equality holds only when the population means are equal.

Thus, when the null hypothesis H_0 is true, both the mean squares MSB and MSW provide independent unbiased estimates of the same population variance (σ^2). Hence, we have the test statistic $F = MSB/MSW$, which under H_0 follows F distribution with degrees of freedom $(k - 1, N - k)$.

If the observed value of the statistic equals or exceeds the theoretical F value from Statistical Tables at a specified level of significance, we reject the null hypothesis and conclude that all the population means are not equal. On the other hand, if the observed value of the statistic is less than the theoretical value all the population means may be considered to be equal.

The various sums of squares, degrees of freedom, mean squares corresponding to the different sources of variation, and also the F values (both observed and theoretical) are shown in the form of a table known as "*Analysis of Variance Table*", as follows:

Source of variation	*Sum of square*	*Degree of freemdom*	*Mean square*	*Variance ratio (F)*
Between samples	SSB	$k - 1$	$MSB = SSB/(k - 1)$	
Within samples	SSW	$N - k$	$MSW = SSW/(N - k)$	MSB/MSW
Total	TSS	$N - 1$		

STEPS IN COMPUTATION (ONE-WAY CLASSIFIED DATA)

(1) By subtracting a suitable constant reduce the sample observations.

(2) From these reduced figures, obtain: (a) totals (T_i) for each group; (b) grand total ($T = \Sigma T_i$), *i.e.*, sum of group totals T_i; (c) total of the squares of all figure ($\Sigma\Sigma y_{ij}^2$).

(3) Calculate: (a) correction factor (CF) = T^2/N; (b) $TSS = \Sigma\Sigma y_{ij}^2 - CF$; (c) $SSB = \Sigma(T_i^2/n_i) - CF$; (d) $SSW = TSS - SSB$.

(4) Find the degrees of freedom (D.F.): (a) D.F. for $SSB = (k - 1)$; (b) D.F. for $SSW = (N - k)$.

(5) Calculate the mean squares: (a) $MSB = SSB \div (k - 1)$; (b) $MSW = SSW \div (N - k)$.

(6) Obtain the observed value of F on dividing MSB by MSW: $F = MSB \div MSW$.

(7) Consult the F-tables and obtain the theoretical values of F at (say) 5% level corresponding to the degrees of freedom ($k - 1$, $N - k$).

(8) If the observed value of F at (6) equals or exceeds the theoretical value of F at (7), the null hypothesis is rejected, and we conclude that all the population means are not equal. Otherwise, they may be taken to be equal.

LOCATING UNEQUAL PAIRS OF MEANS

If the number of observations in each sample is the same, say n, then the appropriate statistic for testing equality of means μ_i and μ_j is given by

$$t = \frac{\bar{y}_i - \bar{y}_j}{s\sqrt{(1/n + 1/n)}} = \frac{\bar{y}_i - \bar{y}_j}{s\sqrt{2/n}}$$

where, $s^2 \rightarrow$ the unbiased estimator of the common (but unknown) population variance σ^2.

In the analysis of variance tests, σ^2 is estimated by the mean square error (MSE), *i.e.*, MSW in one-way classified data. Therefore, $s^2 = MSE$.

The population means μ_i and μ_j will be equal at 5% (say) level of significance, if $|t| \le t_{0.025}$, where $t_{0.025}$ denotes the upper 2.5% value of t distribution with $N - k$ degrees of freedom. Thus, we have

$$\left|\frac{\bar{y}_i - \bar{y}_j}{s\sqrt{2/n}}\right| \le t_{0.025} \quad i.e., \quad |\bar{y}_i - \bar{y}_j| \le s\sqrt{2/n}\cdot t_{0.025}$$

Multiplying both sides by n, we get

$$|T_i - T_j| \le s\sqrt{2n}\cdot t_{0.025}$$

where, T_i and T_j denotes the totals of the i-th and j-th samples respectively.

Thus, in case when all the samples are of the same size (n), critical difference between sample totals is given as

$$\text{Critical difference (C.D.)} = s\sqrt{2n} \cdot t_{0.025}$$

If the difference between any particular pair, say 1 and 3, *viz.*, $|T_1 - T_3|$, exceeds the C.D., then the population means μ_1 and μ_3 are considered to be unequal.

TECHNIQUE IN TWO-WAY CLASSIFIED DATA

Let us consider a random sample having hk observations classified according to two factors—into h classes according to factor A, and into k classes according to factor B. In each of the hk cells, there is one observation corresponding to a class of factor A and simultaneously a class of factor B. Therefore, hk observations can be arranged in the form of two-way table with h rows and k columns.

The mathematical model can be written as

$$y_{ij} = \mu + \alpha_i + \beta_j + e_{ij}$$

where, $\mu \rightarrow$ the general effect,

$\alpha_i \rightarrow$ the effect special to the i-th class according to factor A,

$\beta_j \rightarrow$ the effect special to the j-*th* class according to factor B, and

$e_{ij} \rightarrow$ the error component.

Let us also assume that e_{ij} are independently distributed normal variables each with mean 0 and variance σ^2 (unknown).

This technique can be used to decide on two types of problems simultaneously, *viz.*, (a) whether there is any differential effect due to classification by factor A, represented by null hypothesis

$$H_{01}\ (\alpha_1 = \alpha_2 = \ldots = \alpha_h = 0)$$

and/or (b) whether there is any differential effect due to classification by factor B, represented by null hypothesis

$$H_{02}\ (\beta_1 = \beta_2 = \ldots = \beta_k = 0)$$

We introduce the following notations:

$h \rightarrow$ number of classes by factor A (representing number of rows),

$k \rightarrow$ number of classes by factor B (representing number of columns),

$N = hk \rightarrow$ total number of observations,

$T_i \rightarrow$ total of the i-th row, (i = 1, 2, ... h),

$T_j \rightarrow$ total of the j-th column, (j = 1, 2, ... k),

$T \rightarrow$ grand total of all observations $= \Sigma T_i = \Sigma T_j = \Sigma\Sigma y_{ij}$,

$\bar{y}_i = T_i/k \rightarrow$ mean of the i-th row,

$\bar{y}_j = T_j/h \rightarrow$ mean of the j-th column,

$\bar{y} = T/N \rightarrow$ grand mean of all observations.

It can be shown algebraically that

$$\Sigma\,\Sigma(y_{ij} - \bar{y})^2 = k\,\Sigma(\bar{y}_i - \bar{y})^2 + h\,\Sigma(\bar{y}_j - \bar{y})^2 + \Sigma\,\Sigma(y_{ij} - \bar{y}_i - \bar{y}_j + \bar{y})^2$$

i.e., $TSS = SSA + SSB + SSE$

where, $TSS \rightarrow$ total sum of squares; $SSA \rightarrow$ sum of squares between factor A; $SSB \rightarrow$ sum of squares between factor B; and $SSE \rightarrow$ sum of squares due to error.

The degrees of freedom can similarly be split up. Thus, we have

Sum of squares : $TSS = SSA + SSB + SSE$

Degrees of freedom : $hk - 1 = (h - 1) + (k - 1) + (h - 1)(k - 1)$

The mean squares are now calculated on dividing the sum of squares by the corresponding degrees of freedom.

Mean square between classes by factor A, $MSA = SSA \div (h - 1)$

Mean square between classes by factor B, $MSB = SSB \div (k - 1)$

Mean square error, $MSE = SSE \div (h - 1)(k - 1)$

Assuming normal distribution in the population with a constant variance σ^2, it can be shown that the expected value of MSE is σ^2, *i.e.*, $E(MSE) = \sigma^2$, irrespective of whether the null hypotheses H_{01} and H_{02} are true or not. In general, however,

$$E(MSA) \geq \sigma^2 \text{ and } E(MSB) \geq \sigma^2$$

But, $E(MSA) = \sigma^2$ only if H_{01} is true; and $E(MSB) = \sigma^2$ only if H_{02} is true. Thus, when H_{01} is true the statistic $F_1 = MSA/MSE$ follows F distribution with the appropriate degrees of freedom. Similarly, when H_{02} is true, the statistic $F_2 = MSB/MSE$ follows F distribution with the appropriate degrees of freedom.

If all the data are drawn from the same population with a variance σ^2, the analysis of variance table for two-way classification is as follows:

Source of variation	*Sum of square*	*Degree of freedom*	*Mean square*	*Variance ratio* (F)
Between samples	SSA	$h - 1$	$MSA = SSA \div (h - 1)$	$F_1 = MSA/MSE$
Between B classes	SSB	$k - 1$	$MSB = SSB \div (k - 1)$	$F_2 = MSB/MSE$
Residual	SSE	$(k - 1)(h - 1)$	$MSE = SSE \div (k - 1)(h - 1)$	
Total	TSS	$hk - 1$		

STEPS IN COMPUTATION (TWO-WAY CLASSIFIED DATA)

1. By subtracting a constant from each, reduce the sample observations.
2. From the reduced sample observations, obtain: (a) totals (T_i) for each group classified according to factor A, *i.e.*, for each row; (b) totals (T_j) for each group classified according to factor B, *i.e.*, for each column; (c) grand total $(T = \Sigma T_i = \Sigma T_j)$ for all figures; (d) total of the squares of all figures $(\Sigma\Sigma y_{ij}^2)$.
3. Calculate: (a) correction factor, $CF = T^2/N$, (b) $TSS = \Sigma\Sigma y_{ij}^2 - CF$, (c) $SSA = \Sigma(T_i^2/k) - CF$, (d) $SSB = \Sigma(T_j^2/h) - CF$, (e) $SSE = TSS - SSA - SSB$.

4. Find the degrees of freedom (D.F.): (a) D.F. for $TSS = hk - 1$, (b) D.F. for $SSA = h - 1$, (c) D.F. for $SSB = k - 1$ (d) D.F. for SSE = (D.F. for TSS) minus D.Fs. for SSA and SSB $= hk - h - k + 1$.
5. Calculate the mean squares: (a) $MSA = SSA \div (h - 1)$; (b) $MSB = SSB \div (k - 1)$; (c) $MSE = SSE \div (hk - h - k + 1)$.
6. Calculate the observed values of F on dividing MSA and MSB by MSE, *i.e.*, (a) $F_1 = MSA \div MSE$ (b) $F_2 = MSB \div MSE$.
7. From the F-tables obtain the theoretical values of F at 5% (say) level for the appropriate degrees of freedom.
8. (a) In case F_1 exceeds the theoretical value we conclude that classification according to factor A has a differential effect on the value of the variable, *i.e.*, the means of classes by factor A are significantly different, otherwise, there is no differential effect.

   (b) In case F_2 exceeds the theoretical value we conclude that classification according to factor B has a differential effect on the values, if not, then there is no differential effect.

DIRECT METHOD FOR ONE-WAY CLASSIFICATION

(a) Calculate: $\bar{Y}_1, \bar{Y}_2, \bar{Y}_3, \ldots, \bar{Y}_n$.

(b) Calculate grand mean:

$$\bar{Y} = \frac{\bar{Y}_1 + \bar{Y}_2 + \bar{Y}_3 + \ldots + \bar{Y}_n}{n_1 + n_2 + n_3 + \ldots + n_n}$$

If samples are of different size:

$$\bar{Y} = \frac{\bar{Y}_1 n_1 + \bar{Y}_2 n_2 + \bar{Y}_3 n_3 + \ldots + \bar{Y}_n n_n}{n_1 + n_2 + n_3 + \ldots + n_n}$$

(c) Calculate sum of squares between samples:

$$SSB = n_1(\bar{Y}_1 - \bar{Y})^2 + n_2(\bar{Y}_2 - \bar{Y})^2 + \ldots + n_n(\bar{Y}_n - \bar{Y})^2$$

(d) Calculate sum of squares within samples:

$$SSW = \Sigma(Y_1 - \bar{Y}_1)^2 + \Sigma(Y_2 - \bar{Y}_2)^2 + \ldots + \Sigma(Y_n - \bar{Y}_n)^2$$

(e) Calculate total sum of squares: $TSS = SSB + SSW$.

(f) Calculate the mean squares: (a) $MSB = SSB \div (k - 1)$; (b) $MSW = SSW \div (N - k)$.

(g) Calculate F-ratio: $F = MSB/MSW$.

SHORT CUT METHOD FOR ONE-WAY CLASSIFICATION

(a) Calculate the correction factor:

$$C.F. = \frac{T^2}{N} = \frac{\Sigma Y_1 + \Sigma Y_2 + \ldots + \Sigma Y_n}{N}.$$

(b) Calculate total sum of squares:

$$TSS = \left[\Sigma Y_1^2 + \Sigma Y_2^2 + \Sigma Y_3^2 + \ldots + \Sigma Y_n^2\right] - \frac{T^2}{N}.$$

(c) Calculate sum of squares between samples:

$$SSB = \left[\frac{(\Sigma Y_1)^2}{n_1} + \frac{(\Sigma Y_2)^2}{n_2} + \frac{(\Sigma Y_3)^2}{n_3} + \ldots + \frac{(\Sigma Y_n)^2}{n_n}\right] - \frac{T^2}{N}.$$

(d) Calculate sum of squares within samples: $SSW = TSS - SSB$.

(e) Calculate the mean squares: (a) $MSB = SSB \div (k - 1)$; (b) $MSW = SSW \div (N - k)$.

(f) Calculate *F*-ratio: $F = MSB/MSW$.

INDEX NUMBERS

Index numbers are statistical devices which express the relative change in magnitudes of a phenomenon with respect to time, geographical location or some other characteristic. They represent the continuous upward or downward changes in the value of the variable (such as price of a certain commodity, quantitative production of certain goods, the cost of living, *etc.*) and are known as economic barometers or economic indicators, since they help in understanding the changes in economic conditions of the society.

> *Edgeworth*: Index numbers are numbers which show by their variation, the change in magnitude which is not susceptible either of accurate measurement in itself or of direct valuation in practice.
>
> *Maslow*: Index numbers are numerical values characterising the changes in complex economic phenomena over a period of time or space.
>
> *Croxton* and *Cowden*: Index numbers are devices for measuring differences in the magnitude of a group of related variables.
>
> *Spiegal*: An index number is a statistical measure designed to show changes in variable or a group of related variables with respect to time, geographical location or other characteristics.
>
> *John Riffin*: Index numbers are used to measure changes over time in magnitudes which are not capable of direct measurement.
>
> *Tuttle*: Index number is a single ratio (usually in percentages) which measures the combined (*i.e.*, averaged) change of several variables between two different times, places or situations.
>
> *Wheldon*: Index number is a statistical device for indicating the relative movements of the data where measurement of actual movements is difficult or incapable of being made.

Carli, an Italian, constructed the first index number in 1764 to compare the changes in the price for the year 1750 with the price level of the year 1500.

The index number is called a univariate index if the variation in the level of a single item is being considered, *e.g.*, the index numbers of rice production in India during the last 20 years. On the other hand, if the changes in average level of a number of items is being studied together, the index number is called a composite index number, *e.g.*, the index number of agricultural production in India.

Index numbers are classified on the basis of the phenomena whose changes they measure and on this basis they can be classified into four main categories: (a) *Price Index Numbers*—measure the general changes in the prices. They are further sub-divided into *Wholesale Price Index Numbers* reflecting the changes in the general price level of a country, and *Retail Price Index Numbers* reflecting the general changes in the retail prices of various commodities such as consumption goods, stocks and shares, bank deposits, government bonds, *etc.* (b) *Quantity Index Numbers*—measure the changes in the volume of goods produced, consumed or distributed, like the indices of agricultural production, industrial production, imports and exports, *etc.* (c) *Value Index Numbers*—measure the changes in the total value (price multiplied by quantity) of production such as indices of retail sales or profits or inventories. (d) *Special Purpose Index Numbers.*

Methods of Construction of Index Numbers:

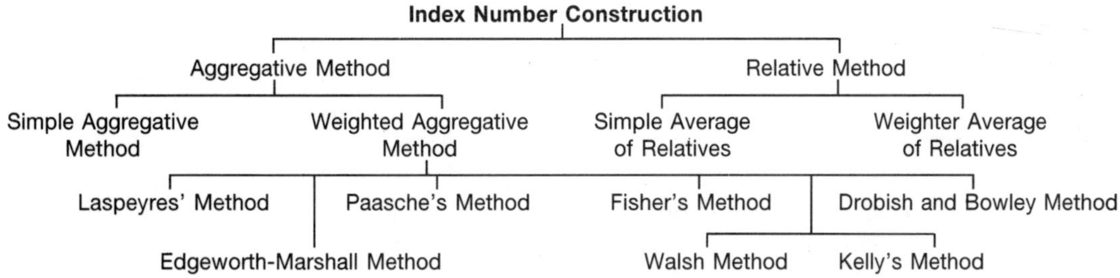

Tests of Adequacy of Index Numbers: Irving Fisher put forward certain tests to judge the efficiency of an index number formula as a measure of the level of a phenomenon from one period to another. The four most important tests of index numbers are:

(a) *Unit Test*—The formula used for the construction of index numbers should be such which is not affected by the unit in which prices or quantities have been measured.

(b) *Time Reversal Test*—A good index number formula should work both ways, forward and backward, with respect to time. In other words, the formula for calculating an index number should be such that it will give the same ratio between one point of comparison and the other no matter which of the two is taken as base.

(c) *Factor Reversal Test*—A good index number formula should be such that the price ratio multiplied by the quantity ratio between two points of time gives the ratio of total values. In other words, each formula for index number should permit interchange of two items without giving inconsistent result, so it ought to permit interchanging the price and quantities without giving inconsistent result, *i.e.*, the two results multiplied together should give the true value ratio.

(d) *Circular Test*—It is an extension of time reversal test. An index number formula is said to satisfy the circular test, if the time reversal test is satisfied through a number of intermediate periods.

Base Shifting: This is the method of changing the given base period of a series of index numbers and recasting them to form a new series with reference to a new base period. The reasons why base shifting is necessary are:

(a) When the base period is too old and is too far away from the current period, it is unsuitable of meaningful comparisons and therefore it is necessary to shift the base to a more recent period in order that data are more meaningful.

(b) For comparing two or more series of index number with different base periods, base shifting is very essential, because if all the series are expressed with a common base, the comparison is easier and quicker.

$$\text{Index number for any year with new base} = \frac{\text{Old index number for the year}}{\text{Old index number for the new base year}} \times 100$$

Splicing: By splicing of index numbers, we mean combining two or more overlapping series of index numbers with different base periods to obtain a single continuous series of index numbers with a common base period. This is done by the same technique as used in base shifting. This can be done only if the index numbers are constructed with the same items, and have an overlapping year.

Deflation: It refers to the method of adjusting the values of a series after making allowance for changes in price level. When prices rise, the purchasing power of money declines. If the money incomes of people remain constant and prices are doubled, the purchasing power of money income is reduced to half. Consequently, a raise in the money income will not really mean the same percentage increase in the standard of living, *i.e.*, the real income will be less than the money income.

Thus, for a true comparison, it becomes necessary to adjust the money income for changes in the cost of living index numbers. The real income is determined after deflating the money income by cost of living index as:

$$\text{Real income} = (\text{Money income}/\text{Cost of living index}) \times 100$$

Errors in Index Numbers: Index numbers are generally affected by three types of errors.

(a) *Formula error*—each formula for index numbers in common use has its own defect, and consequently some error is inherent in each index numbers.

(b) *Sampling error*—since all the items cannot be taken into consideration for making index number, the calculated index number can never represent the changes in the phenomenon accurately. Thus, the error introduced by selecting a sample of items is known as the sampling error.

(c) *Homogeneity error*—arises due to the fact that index numbers are constructed from such items which are marketed approximately in the same quality both in the base and the current periods. With the passage of time, new items replace many of the old items and hence the homogeneity in composition of the items cannot be strictly maintained. And hence, the error increases as the gap between the base and current periods increases.

SIMPLE AGGREGATIVE INDEX (I_{on}) $= (\Sigma p_n/\Sigma p_0) \times 100$

where, p_n = price in the current year, and p_0 = price in the base year.

WEIGHTED AGGREGATIVE INDEX (I_{on}) $= (\Sigma p_n w / \Sigma p_0 w) \times 100$

where w represents the weight. It should be noted that the same set of weights must be used both for base year as well as for current year.

LASPEYRES' INDEX (I_{on}) $= (\Sigma p_n q_0 / \Sigma p_0 q_0) \times 100$

where, q_0 = base year quantity.

PAASCHE'S INDEX (I_{on}) $= (\Sigma p_n q_n / \Sigma p_0 q_n) \times 100$

where, q_n = current year quantity.

EDGEWORTH-MARSHALL'S INDEX (I_{on}) $= \{\Sigma p_n (q_0 + q_n) / \Sigma p_0 (q_0 + q_n)\} \times 100$

FISHER'S IDEAL INDEX (I_{on}) $= \sqrt{Laspeyres'\ Index \times Paasche's\ Index}$

$$= \sqrt{(\Sigma p_n q_0 / \Sigma p_0 q_0) \times (\Sigma p_n q_n / p_0 q_n)} \times 100$$

BOWLEY'S INDEX (I_{on}) $= (1/2)(Laspeyres'\ Index + Paasche's\ Index)$

$$= (1/2)\left[(\Sigma p_n q_0 / \Sigma p_0 q_0) + (\Sigma p_n q_n / \Sigma p_0 q_n)\right] \times 100$$

WALSH'S INDEX (I_{on}) $= \left(\Sigma p_n \sqrt{q_0 q_n} / \Sigma p_0 \sqrt{q_0 q_n}\right) \times 100$

KELLY'S INDEX (I_{on}) $= (\Sigma p_n q / \Sigma p_0 q) \times 100$

where, q is the fixed weights for all periods, without any reference to base or current period. This formula is also known as *aggregative index with fixed weights*.

RELATIVE METHOD

In this technique, the price of each item in the current year is expressed as a percentage of the price in the base year. This is known as *Price Relatives* and is given by

$$Price\ Relative = \frac{Price\ in\ the\ given\ year}{Price\ in\ the\ base\ year} \times 100 = \frac{p_n}{p_0} \times 100$$

The average of price relatives, which shows the average percentage change for the whole group of items, gives the index number. Thus,

$$Price\ Index = Average\ of\ Price\ Relatives$$

SIMPLE ARITHMETIC MEAN OF RELATIVE INDEX (I_{on}) $= \Sigma(\textit{Price Relatives}) \div k$

where, k is the number of items included.

SIMPLE GEOMETRIC MEAN OF RELATIVE INDEX (I_{on}) $= \sqrt[k]{\textit{Product of price relatives}}$

WEIGHTED ARITHMETIC MEAN OF RELATIVE INDEX (I_{on}) $= \Sigma(\textit{Price relatives}) \times w/\Sigma w$

where, weights, w, are the values of the items, *i.e.*, w = price × quantity.

AGGREGATIVE FORMULAE BY RELATIVE METHOD

The weighted average of relatives leads to several index number formulae of the aggregative type, depending on the nature of weights used.

(i) The A.M. of relatives formula weighted by base year values (p_0q_0) gives exactly the same formula as Laspeyres':

$$I_{0n} = \frac{\Sigma\{(p_n/p_0)\times 100\}\,p_0q_0}{\Sigma p_0q_0} = \frac{\Sigma p_nq_0}{\Sigma p_0q_0} \times 100 = \textit{Laspeyres' Index}$$

(ii) The A.M. of relatives formula weighted by values of current year quantities at base year prices (p_0q_n) gives Paasche's formula:

$$I_{0n} = \frac{\Sigma\{(p_n/p_0)\times 100\}\,p_0q_n}{\Sigma p_0q_n} = \frac{\Sigma p_nq_n}{\Sigma p_0q_n} \times 100 = \textit{Paasche's Index}$$

(iii) The H.M. of relatives formula weighted by current year values (p_nq_n) gives the same formula as Paasche's:

$$I_{0n} = \frac{\Sigma p_nq_n}{\Sigma p_nq_n/\{(p_n/p_0)\times 100\}} = \frac{\Sigma p_nq_n}{\Sigma p_0q_n} \times 100 = \textit{Paasche's Index}$$

GENERAL INDEX $= \Sigma IW/\Sigma W$

where, I is the *Group Index* and W is the *Group Weight*.

QUANTITY INDEX NUMBERS

Measure and permit comparison of the physical quantity of goods produced or consumed or marketed or distributed.

Simple Aggregative Quantity Index $= (\Sigma q_n/\Sigma q_0) \times 100$

Laspeyres' Quantity Index $= (\Sigma q_np_0/\Sigma q_0p_0) \times 100$

Paasche's Quantity Index $= (\Sigma q_np_n/\Sigma q_0p_n) \times 100$

Edgeworth-Marshall's Quantity Index $= \{\Sigma q_n(p_0 + p_n)/\Sigma q_0(p_0 + p_n)\} \times 100$

Fisher's Ideal Quantity Index $= \sqrt{(\Sigma q_n p_0/\Sigma q_0 p_0) \times (\Sigma q_n p_n/\Sigma q_0 p_n)} \times 100$

Quantity Relative $= (q_n/q_0) \times 100$

Simple A.M. of Quantity Relatives Index $= \Sigma(\textit{quantity relatives}) \div k$

Weighted A.M. of Quantity Relatives Index $= \Sigma(\textit{quantity relative} \times \textit{weight})/\Sigma(\textit{weight})$

CHAIN BASE METHOD

For the construction of index numbers by the chain base method, using a suitable index number formula, we first compute index numbers for all the years, always using the preceding year as base year. These are known as *Link Index*.

If $I_{01}, I_{12}, I_{23}, I_{34}$, ... are the link indices, then we multiply these link indices successively (called *chaining process*) in order to relate them to a common base. The progressive products, expressed as percentages, give the required index numbers by the chain base method. Thus, a chain index number is the product of several index numbers, each calculated with preceding period as base. Hence,

$$\textit{Chain index for current year} = (\textit{Link relative of current year} \times \textit{Chain index of previous year})/100$$

CONVERSION OF CHAIN BASE INDEX INTO FIXED BASE INDEX

A chain base index number can be converted into a fixed base index number using following conversion formula:

$$\textit{Current years FBI} = (\textit{Current years CBI} \times \textit{Previous years FBI})/100$$

where, *FBI* = fixed base index and *CBI* = chain base index.

CONVERSION OF FIXED BASE INDEX INTO CHAIN BASE INDEX

A fixed base index can be converted into a chain base index using following conversion formula:

$$\textit{Chain base index} = (\textit{Current year's FBI} \times 100)/\textit{Previous year's FBI}$$

For the first year, the value would be 100.

TESTS OF ADEQUACY OF INDEX NUMBERS

(a) *Time reversal test*: $I_{0n} \times I_{n0} = 1$

(b) *Factor reversal test*: $P_{0n} \times Q_{0n} = \Sigma p_n q_n/\Sigma p_0 q_0$

(c) *Circular test*: $I_{01} \times I_{12} \times I_{23} \times \ldots \times I_{(n-1),n} \times I_{n0} = 1$

COST OF LIVING INDEX

At first a separate index number is computed for each group of items, using Laspeyres' formula in the form of weighted average of price relatives.

$$Group\ Index\,(I) = \Sigma w\{(p_n/p_0)\times 100\}/100$$

where, $w = (p_0q_0/\Sigma p_0q_0) \times 100$

$$Cost\ of\ Living\ Index = \Sigma IW/\Sigma W$$

where, weight (W) of a group index is the percentage of total expenditure of an average family spent on that group.

ASSOCIATION OF ATTRIBUTES

In statistics, there are two types of data—quantitative and qualitative. Quantitative data represent the actual magnitude of some variable and are known as *statistics of variables*. Measures of central tendency, measures of dispersion, skewness, and correlation are some of the important statistical methods used in the analysis of such data.

There may be some data whose magnitude cannot be measured. In such cases, we can only observe the presence or absence of a particular quality in the data. Such data in which the quantitative measurement of the magnitude is not possible and in which only the presence or absence of an attribute can be studied are called *statistics of attributes*.

If there exists a relation between two or more attributes, they are said to be associated and this is known as association of attributes. This association may be positive, negative, or zero (*i.e.*, the attributes may be independent).

Difference between Correlation and Association: Correlation is used to measure the degree of relationship between two such phenomena as are capable of direct quantitative measurement. Whereas the method of association of attributes is employed to measure the degree of relationship between two phenomena whose size we cannot measure and where we can only determine the presence or absence of a particular attribute.

Classification of Attributes: The data are classified in two ways according to attributes.

(a) *Classification by Dichotomy*—Under this data are classified into two groups according to the presence or absence of a certain attribute. These groups are mutually exclusive. For example, the data are classified into employed and unemployed under employment, into uneducated and educated under education and so on.

(b) *Manifold Classification*—Under this data are classified according to more than two attributes and sub-attributes. For example, the result of students is classified into three groups, *viz.*, pass, supplementary and fail.

Notation and Terminology: We use certain symbols to represent different classes and their frequencies. Normally, A, B and C *etc.*, are used to denote the presence of attributes and α, β and γ *etc.*, are used to denote absence of these attributes respectively. The number of units possessing a particular attribute represented by A would be termed belonging to *Class* A, and similarly, those in whom this attributes is absent would be termed belonging to *Class* α. If two attributes are being studied, their combination can be represented by the combination of the letters representing the two attributes, such as, AB, $A\alpha$, $B\beta$, $\alpha\beta$, *etc.*

Class Frequencies: The number of units in different classes are called class frequencies, which are denoted by enclosing the class symbols by brackets. Thus, (AB) represents the frequency of the class AB.

Number of Classes: If there is one attribute represented by A, the total number of classes would be 3 (if the total or 'N' is also taken as a class)—A, α and N. If the number of attributes is two, represented by A and B, the total number of classes (including N) would be 9—N, A, B, AB, α, β, $A\alpha$, $A\beta$ and $\alpha\beta$. If the number of attributes is three, the total number of classes (including N) would be 27. The total number of classes is always equal to 3^n where n is the number of attributes.

Order of Classes: The various classes and their frequencies are demarcated on the basis of following order:

$$N \rightarrow \text{is a class of zero order,}$$

$$\left.\begin{matrix} A & B \\ \alpha & \beta \end{matrix}\right\} \rightarrow \text{are classes of the first order, and}$$

$$\left.\begin{matrix} AB & \alpha\beta \\ A\beta & \alpha B \end{matrix}\right\} \rightarrow \text{are classes of the second order.}$$

Also, the frequencies of these classes are known as frequencies of the zero, first or second order.

Ultimate Class Frequencies: Every class frequency can be expressed in terms of the frequencies of the highest order, *i.e.*, of order n. Any frequency can be analysed into highest frequencies, and the process need stop only when we have reached the frequencies of the highest order.

The classes specified by n attributes, *i.e.*, those of the highest order, are termed as ultimate class frequencies. A given data can be completely specified if only the ultimate class frequencies are given. The number of classes of the ultimate order is equal to 2^n, where n stands for the number of attributes.

Positive and Negative Classes: The classes which represent the presence of an attribute or attributes are called positive classes. The classes which represent the absence of an attribute or attributes are called negative classes. The classes in which one attribute is present and the other absent are called *pairs of contraries*. Thus,

$$N, A, B \text{ and } AB \rightarrow \text{positive classes,}$$

$$\alpha \text{ and } \beta \rightarrow \text{negative classes, and}$$

$$A\beta \text{ and } \alpha B \rightarrow \text{pairs of contraries.}$$

Relationship Between Classes of Various Orders: In classifying statistical data according to attributes, the following rule should be followed—Any class frequency can always be expressed in terms of class frequencies of higher order.

Thus, the frequencies of first order can be expressed in terms of the frequencies of the second order which in turn can be expressed in terms of frequencies of the third order and so on.

For example, if there is one attribute only, represented by A, the frequency of the universe or N can be divided into two classes (A) and (α). Thus, $N = (A) + (\alpha)$.

Now, if one more attribute B is taken into account the first order classes, *i.e.*, A and α can each be divided into two classes—one in which attribute B is present and the other in which it is not present. Thus, $(A) = (AB) + (A\beta)$ and $(\alpha) = (\alpha B) + (\alpha\beta)$. Similarly, $(B) = (AB) + (\alpha B)$ and $(\beta) = (A\beta) + (\alpha\beta)$.

Consistency of Data: The consistency of the given data can be judged by applying a very simple test. The test is to find out whether any one or more of the ultimate class-frequencies is negative or not. If none of the class-frequencies is negative we can assume that the given data are consistent, *i.e.*, the frequencies do not conflict in any way with each other. On the other hand, if any of the ultimate class frequencies comes out to be negative the given data are inconsistent. Thus, the necessary and sufficient condition for the consistency of a set of independent class-frequencies is that no ultimate class frequency is negative.

Association and Disassociation: There is perfect positive association between the attributes A and B if all A's are present in B's and all B's are present in A's, *i.e.*, $(A) = (B)$. If (A) and (B) are not equal in number, A and B are associated if either all A's are B's, or all B's are A's depending on whether the number of A's is less than the number of B's or *vice versa*.

In the same way, there is complete disassociation (or perfect negative association) between A and B, if none of the A's is B's and none of the α's is β's, *i.e.*, $(AB) = 0$ and $(\alpha\beta) = 0$.

There may be complete negative association between the two attributes even if the value of (AB) may not be 0. Since, (AB) cannot be less than $(A) + (B) - N$, if $(A) + (B) - N$ is more than 0, (AB) would also be more than 0; and if (AB) is just equal to $(A) + (B) - N$, there would be disassociation between A and B. In such cases, the value of $(\alpha\beta)$ only would be zero.

Methods of Studying Association: The study of association can be done by any of the following methods—(a) comparison of actual and observed frequencies, (b) comparison of various proportions and products, (c) calculation of Yule's coefficient of association, (d) calculation of coefficient of collignation, and (e) calculation of coefficient of contingency.

Partial Association: There may not be direct association between A and B, however, the association between attributes A and B may be due to their association with a third attribute, say, C. Thus, if A is positively associated with C and B is associated with C, A may be found to be positively associated with B. This type of association between A and B is not direct, it is the effect of their association with a third attribute C. To find out their association with C, it is necessary to study the association of A and B in the sub-populations of C and γ. If A and B are associated in both the sub-populations of C and γ, they are really associated with each other. Such type of association between A and B in sub-populations are called *partial associations*, to distinguish them from the total associations between A and B in the universe as a whole.

CONSISTENCE OF DATA

If any class frequency is negative, the data are said to be inconsistent.

If there is only one attribute represented by A:

$$(A) \nless 0, \text{ otherwise (A) will be negative.}$$

$$(A) \ngtr N, \text{ otherwise } (\alpha) \text{ will be negative, since, } N = (A) + (\alpha).$$

If there are two attributes represented by A and B:

$$(AB) \nless 0, \text{ otherwise } (AB) \text{ will be negative.}$$

$$(AB) \nless (A)+(B) - N \text{ otherwise } (\alpha\beta) \text{ will be negative.}$$

$$(AB) \ngtr (A), \text{ otherwise } (A\beta) \text{ will be negative, since, } (A) = (AB) + (A\beta).$$

$$(AB) \ngtr (B), \text{ otherwise } (\alpha B) \text{ will be negative, since, } (B) = (AB) + (\alpha B).$$

COMPARISON OF ACTUAL AND OBSERVED FREQUENCIES

If two attributes A and B are studied in a universe and if the frequency of A is represented by (A) and of B by (B), then

the probability of $(A) = (A)/N$; and the probability of $(B) = (B)/N$.

The combined probability of two independent events is equal to the product of their individual probabilities. Thus, the combined probability of (A) and (B) would be $(A)/N \times (B)/N$ and the expectation of (A) and (B) would be $(A)/N \times (B)/N \times N = (A) \times (B)/N$.

Therefore, if attributes A and B are independent, the expected frequency of (AB) would be equal to $(A) \times (B)/N$.

Criterion of Independence: Two attributes A and B are said to be independent if the observed frequency of (AB) is equal to its expected frequency, *i.e.*, $(A) \times (B)/N$.

Limitation: The main limitation of studying association by a comparison of actual and expected values of AB is that it only determines the nature of association between A and B, whether the association is positive or negative. It does not tell us about the degree of association, whether it is high or low.

COMPARISON OF VARIOUS PROPORTIONS AND PRODUCTS

Two attributes A and B are:

(a) independent if $(AB)/(B) = (A\beta)/(\beta)$

(b) positively associated if $(AB)/(B) > (A\beta)/(\beta)$ and

(c) negatively associated if $(AB)/(B) < (A\beta)/(\beta)$.

If the relation (a) holds good, the corresponding relationship

$$(\alpha B)/(B) = (\alpha\beta)/(\beta); \quad (AB)/(A) = (\alpha B)/(\alpha); \text{and } (A\beta)/(A) = (\alpha\beta)/(\alpha)$$

would also hold good.

A and B would be independent if:

(a) $(AB)/(A) = (B)/(N)$; $(AB)/(B) = (A)/(N)$ and $(AB)/(N) = (A)/(N) \times (B)/(N)$

(b) $(AB)(\alpha\beta) = (A\beta)(\alpha B)$

YULE'S COEFFICIENT OF ASSOCIATION

It gives some idea about the extent of association between two attributes. The value of coefficient of association is 0 when the two attributes are independent, +1 when they are perfectly associated and –1 when they are perfectly disassociated. Yule's coefficient of association is given by

$$Q = \{(AB)(\alpha\beta) - (A\beta)(\alpha B)\}/\{(AB)(\alpha\beta) + (A\beta)(\alpha\beta)\}$$

When two attributes A and B are independent, the value of $(AB)(\alpha\beta) = (A\beta)(\alpha B)$ and therefore the value of the numerator in the above formula would be 0 and the value of the coefficient of association would also be 0. If there is perfect association between the two attributes A and B the value of $(A\beta)(\alpha B)$ would be 0 and since it will be so both in the numerator and the denominator, the value of the coefficient of association would be +1. Similarly, if there is perfect disassociation between two attributes A and B, the value of $(AB)(\alpha\beta)$ would be 0 and since it will be so both in the numerator and denominator, the coefficient of association would be –1.

COEFFICIENT OF COLLIGNATION

It is independent of the relative proportions of A's and α's, like the Yule's coefficient of association. Coefficient of collignation is given by

$$\gamma = \left\{1 - \sqrt{(A\beta)(\alpha B)/(AB)(\alpha\beta)}\right\} / \left\{1 + \sqrt{(A\beta)(\alpha B)/(AB)(\alpha\beta)}\right\}$$

This coefficient was given by Yule.

PARTIAL ASSOCIATION

The association of A and B in the sub-populations of C and γ are called partial association and the association of A and B in the universe as a whole, total association. A and B are positively associated in the sub-population of C if: $(ABC) > (AC)(BC)/(C)$ and negatively associated if: $(ABC) < (AC)(BC)/(C)$.

Similarly, A and B are positively associated in the sub-population of γ if: $(AB\gamma) > (A\gamma)(B\gamma)/(\gamma)$ and negatively associated if: $(AB\gamma) < (A\gamma)(B\gamma)/(\gamma)$.

NUMBER OF PARTIAL ASSOCIATIONS

The total number of association in the case of n attributes $= 3n(n-1)/2$. Out of these $n(n-1)/2$ are total associations and the remainder are partial associations.

ASSOCIATION OF THREE ATTRIBUTES

The total number of classes = $3^n = 3^3 = 27$. Out of these 27 classes, the number of classes of the ultimate order = $2^n = 2^3 = 8$.

The number of classes of the various order in case of three attributes are:

			Nature	*Number*	*Order*
N			positive	1	zero order
(A)	(B)	(C)	positive	6	first order
(α)	(β)	(γ)	negative		
(AB)	(AC)	(BC)	positive		
$(A\beta)$	$(A\gamma)$	$(B\gamma)$	pairs of	12	second order
(αB)	(αC)	(βC)	contraries		
$(\alpha\beta)$	$(\alpha\gamma)$	$(\beta\gamma)$	negative		
(ABC)			positive		
$(AB\gamma)$	$(A\beta C)$	(αBC)	pairs of	8	third order
$(A\beta\gamma)$	$(\alpha B\gamma)$	$(\alpha\beta C)$	contraries		
$(\alpha\beta\gamma)$			negative		

In case of three attributes also, any class frequency can be expressed in terms of the frequencies of a higher order as:

$$(AB) = (ABC) + (AB\gamma);\quad (AC) = (ABC) + (A\beta C);\quad (BC) = (ABC) + (\alpha BC) \text{ and so on.}$$

A higher order frequency can be expressed in terms of lower order frequencies and some higher order frequencies as:

$$(ABC) = (AB) - (A\beta C);\ (A\beta C) = (AC) - (ABC) \text{ and } (\alpha\beta C) = (\alpha C) - (\alpha BC)$$

All frequencies can be expressed in terms of the frequencies of the highest order or ultimate order as:

$$(A) = (AB) + (A\beta) = (ABC) + (A\beta\gamma) + (A\beta C) + (A\beta\gamma) \text{ and so on.}$$

The following chart describe the inter-relationship between frequencies of the different orders:

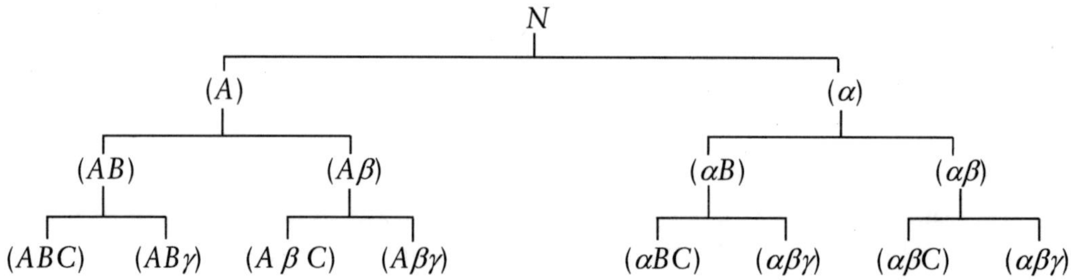

Similarly,

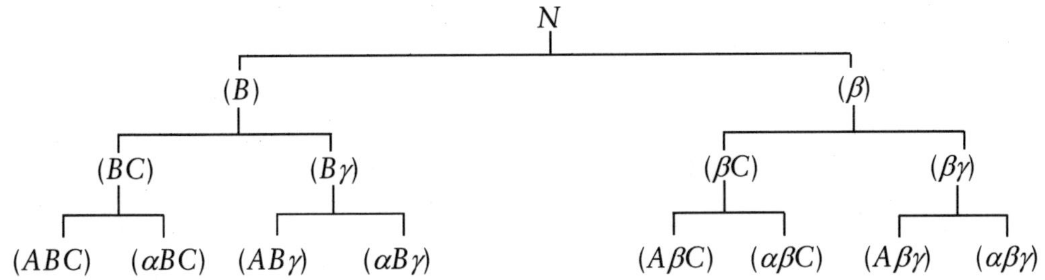

and

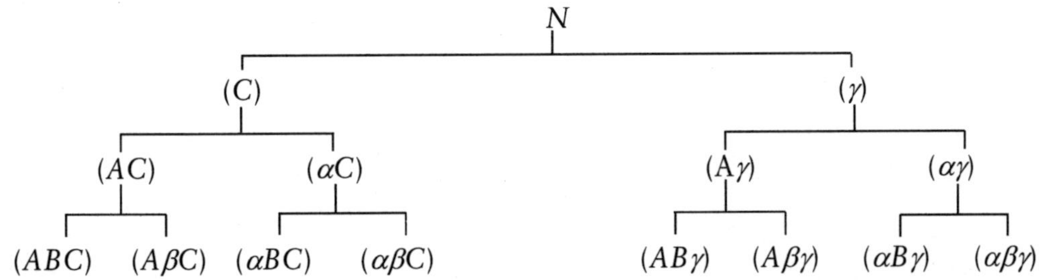

CONSISTENCY OF DATA IN CASE OF THREE ATTRIBUTES

The given data will be inconsistent if any class-frequency is negative. The following rules of consistency are applicable in case of three attributes:

(a) $(ABC) \nless 0$ otherwise (ABC) will be negative

(b) $(ABC) \nless (AB) + (AC) - (A)$ otherwise $(A\beta\gamma)$ will be negative

(c) $(ABC) \nless (AB) + (BC) - (B)$ otherwise $(\alpha\beta\gamma)$ will be negative

(d) $(ABC) \nless (AC) + (BC) - (C)$ otherwise $(\alpha\beta C)$ will be negative

(e) $(ABC) \ngtr (AB)$ otherwise $(AB\gamma)$ will be negative

(f) $(ABC) \ngtr (AC)$ otherwise $(A\beta C)$ will be negative

(g) $(ABC) \ngtr (BC)$ otherwise (αBC) will be negative

(h) $(ABC) \ngtr (AB) + (AC) + (BC) - (A) - (B) - (C) + N$ otherwise $(\alpha\beta\gamma)$ will be negative

From the above eight rules, we can derive four additional rules by combining upper and lower limits of ABC. Thus, $(ABC) \nless 0$. This is the minimum value of (ABC). Now, $(ABC) \ngtr (AB) + (AC) + (BC) - (A) - (B) - (C) + N$. This is the maximum value of ABC. This maximum value cannot be less than the minimum value and we get a new rule:

(i) $(AB) + (AC) + (BC) - (A) - (B) - (C) + N \nless 0 \Rightarrow (AB) + (AC) + (BC) \nless (A) + (B) + (C) - N$

Similarly, by combining rule (b) and rule (g), rule (c) and rule (f); and rule (d) and (e), we get the following rules:

(j) $(AB) + (AC) - (BC) \ngtr (A)$

(k) $(AB) - (AC) + (BC) \ngtr (B)$

(l) $(AC) + (BC) - (AB) \ngtr (C)$ or, $(AB) + (AC) - (A) \ngtr (BC)$

or, $(AB) + (BC) - (B) \ngtr (AC)$ or, $(AC) + (BC) - (C) \ngtr (AB)$

One more rule of consistency applicable to any number of attributes is that for n attributes $(A, B, C, D, E, ..., M)$.

(m) $(ABCD...M) \nless [(A) + (B) + (C) + (D) + ... + (M)] - (n-1)N$

where N is the total frequency and n is the number of attributes.

Thus, $(ABCD) \nless [(A) + (B) + (C) + (D)] - (4-1)N \nless (A) + (B) + (C) + (D) - 3N$.

COEFFICIENT OF CONTINGENCY

The attribute A can be divided into a number of groups A_1, A_2, A_3, ... and the attribute B can be sub-divided as B_1, B_2, B_3, ... It is observed that each one of the classes A_1, A_2, A_3,... of the first attribute can be divided into a number of heads like B_1, B_2, B_3, ..., when a second attribute B is taken into consideration. Such a classification is called *Manifold Classification.*

In the case of two attributes, if attribute A is sub-divided in s classes and attribute B in t classes, we get a table of the following type:

			Attribute A			
Attribute	A_1	A_2	A_3	...	A_s	*Total*
B_1	(A_1B_1)	(A_2B_1)	(A_3B_1)	...	(A_sB_1)	(B_1)
B_2	(A_1B_2)	(A_2B_2)	(A_3B_2)	...	(A_sB_2)	(B_2)
B_3	(A_1B_3)	(A_2B_3)	(A_3B_3)	...	(A_sB_3)	(B_3)
...	...	...	...	...	...	...
...	...	...	...	...	...	...
...	...	...	...	...	...	...
B_t	(A_1B_t)	(A_2B_t)	(A_3B_t)	...	(A_sB_t)	(B_t)
Total	(A_1)	(A_2)	(A_3)	...	(A_s)	N

In the above table, the totals of various columns A_1, A_2, A_3, ..., A_s and the totals of various rows B_1, B_2, B_3, ..., B_t would give the first order frequencies and the frequencies in various cells would give second order frequencies. The total of either A_1, A_2, A_3, ... or B_1, B_2, B_3, ... would give the grand total N. Such a table is called *Contingency Table*.

If A and B are completely independent of each other in the universe at large, then the actual values A_1B_1, A_2B_2, A_3B_3, ... must be equal to their expected values which are, in turn, equal to $(A_1)(B_1)/N$, $(A_2)(B_2)/N$, $(A_3)(B_3)/N$, ... respectively.

Thus, if the observed frequency in each of the cells of a contingency table is equal to the expected frequency of that cell, A and B would be completely independent of each other. If these values are not equal in all the cells, it is an indication of association between the attributes A and B. To test the intensity of association, the difference between the actual and expected frequencies of various cells is calculated. With these difference, the value of Chi-square is computed as:

$$\chi^2 = \Sigma\left[(\textit{Differences of actual and expected frequencies})^2 / \textit{Expected frequencies}\right]$$

If O stands for actual or observed frequency of a class and E for expected frequency, the value of χ^2 is given by

$$\chi^2 = \Sigma\left[(O - E)^2/E\right]$$

This value is called "Square Contingency" and if the mean of the square contingency is calculated, it is called "Mean Square Contingency".

Thus, square contingency = χ^2

and mean square contingency, $\phi^2 = \chi^2/N$

χ^2 can also be calculated as: $\chi^2 = \Sigma\left[(O)^2/E\right] - N$

If the actual and the expected values are equal in all cases, the values of χ^2 and ϕ^2 would be 0.

Karl Pearson's *coefficient of mean square contingency* is given by

$$C = \sqrt{\chi^2/(N+\chi^2)} = \sqrt{\phi^2/(1+\chi^2)}$$

If χ^2 is calculated by the formula: $\chi^2 = \Sigma\left[O^2/E\right] - N$

and if $\Sigma\left[O^2/E\right]$ is represented by S, the coefficient of mean square contingency is given by

$$C = \sqrt{(S - N)/(N + S - N)} = \sqrt{(S - N)/S}$$

The drawback of this coefficient is that it never reaches the limit of 1. The limit of 1 is reached by it only if the number of classes is infinite. The maximum value of the coefficient depends on the values of s and t, *i.e.*, the number of sub-division of the two attributes A and B.

TSCHUPROW'S COEFFICIENT

Tschuprow's coefficient is given by

$$T = \sqrt{C^2/(1 - C^2)\sqrt{(s - 1)(t - 1)}}$$

where C stands for the coefficient of contingency, s stands for the sub-divisions of attribute A and t stands for sub-divisions of attribute B.

TABLES

CUMULATIVE BINOMIAL PROBABILITIES

Cumulative binomial probabilities are given by

$$\sum_{x=r}^{n} \binom{n}{x} p^{x}(1-p)^{n-x}$$

where, p = probability of success in a single trial and n = number of trials. The following table gives the probability of obtaining *r or more* successes in *n* independent trials.

	p =	0.01	0.02	0.03	0.04	0.05	0.06	0.07	0.08	0.09
n = 2	r = 0	1.0000	1.0000	1.0000	1.0000	1.0000	1.0000	1.0000	1.0000	1.0000
	1	.0199	.0396	.0591	.0784	.0975	.1164	.1351	.1536	.1719
	2	.0001	.0004	.0009	.0016	.0025	.0036	.0049	.0064	.0081
n = 5	r = 0	1.0000	1.0000	1.0000	1.0000	1.0000	1.0000	1.0000	1.0000	1.0000
	1	.0490	.0961	.1413	.1846	.2262	.2661	.3043	.3409	.3760
	2	.0010	.0038	.0085	.0148	.0226	.0319	.0425	.0544	.0674
	3		.0001	.0003	.0006	.0012	.0020	.0031	.0045	.0063
	4						.0001	.0001	.0002	.0003
n = 10	r = 0	1.0000	1.0000	1.0000	1.0000	1.0000	1.0000	1.0000	1.0000	1.0000
	1	.0956	.1829	.2626	.3352	.4013	.4614	.5160	.5656	.6106
	2	.0043	.0162	.0345	.0582	.0861	.1176	.1517	.1879	.2254
	3	.0001	.0009	.0028	.0062	.0115	.0188	.0283	.0401	.0540
	4			.0001	.0004	.0010	.0020	.0036	.0058	.0088
	5					.0001	.0002	.0003	.0006	.0010
										.0001
n = 20	r = 0	1.0000	1.0000	1.0000	1.0000	1.0000	1.0000	1.0000	1.0000	1.0000
	1	.1821	.3324	.4562	.5580	.6415	.7099	.7658	.8113	.8484
	2	.0169	.0599	.1198	.1897	.2642	.3395	.4131	.4831	.5484
	3	.0010	.0071	.0210	.0439	.0755	.1150	.1610	.2121	.2666
	4		.0006	.0027	.0074	.0159	.0290	.0471	.0706	.0993
	5			.0003	.0010	.0026	.0056	.0107	.0183	.0290
	6				.0001	.0003	.0009	.0019	.0038	.0068
	7						.0001	.0003	.0006	.0013
	8								.0001	.0002
n = 50	r = 0	1.0000	1.0000	1.0000	1.0000	1.0000	1.0000	1.0000	1.0000	1.0000
	1	.3950	.6358	.7819	.8701	.9231	.9547	.9734	.9845	.9910
	2	.0894	.2642	.4447	.5995	.7206	.8100	.8735	.9173	.9468
	3	.0138	.0784	.1892	.3233	.4595	.5838	.6892	.7740	.8395
	4	.0016	.0178	.0628	.1391	.2396	.3527	.4673	.5747	.6697
	5	.0001	.0032	.0168	.0490	.1036	.1794	.2710	.3710	.4723
	6		.0005	.0037	.0144	.0378	.0776	.1350	.2081	.2928
	7		.0001	.0007	.0036	.0118	.0289	.0583	.1019	.1596
	8			.0001	.0008	.0032	.0094	.0220	.0438	.0768
	9				.0001	.0008	.0027	.0073	.0167	.0328
	10					.0002	.0007	.0022	.0056	.0125
	11						.0002	.0006	.0017	.0043
	12							.0001	.0005	.0013
	13								.0001	.0004
	14									.0001

Cumulative Binomial Probabilities (*Contd.*)

	p =	0.01	0.02	0.03	0.04	0.05	0.06	0.07	0.08	0.09
n = 100	r = 0	1.0000	1.0000	1.0000	1.0000	1.0000	1.0000	1.0000	1.0000	1.0000
	1	.6340	.8674	.9524	.9831	.9941	.9979	.9993	.9998	.9999
	2	.2642	.5967	.8054	.9128	.9629	.9848	.9940	.9977	.9991
	3	.0794	.3233	.5802	.7679	.8817	.9434	.9742	.9887	.9952
	4	.0184	.1410	.3528	.5705	.7422	.8570	.9256	.9633	.9827
	5	.0034	.0508	.1821	.3711	.5640	.7232	.8368	.9097	.9526
	6	.0005	.0155	.0808	.2116	.3840	.5593	.7086	.8201	.8955
	7	.0001	.0041	.0312	.1064	.2340	.3936	.5557	.6968	.8060
	8		.0009	.0106	.0475	.1280	.2517	.4012	.5529	.6872
	9		.0002	.0032	.0190	.0631	.1463	.2660	.4074	.5506
	10			.0009	.0068	.0282	.0775	.1620	.2780	.4125
	11			.0002	.0022	.0115	.0376	.0908	.1757	.2882
	12				.0007	.0043	.0168	.0469	.1028	.1876
	13				.0002	.0015	.0069	.0224	.0559	.1138
	14					.0005	.0026	.0099	.0282	.0645
	15					.0001	.0009	.0041	.0133	.0341
	16						.0003	.0016	.0058	.0169
	17						.0001	.0006	.0024	.0078
	18							.0002	.0009	.0034
	19							.0001	.0003	.0014
	20								.0001	.0005
	21									.0002
	22									.0001

	p =	0.10	0.15	0.20	0.25	0.30	0.35	0.40	0.45	0.50
n = 2	r = 0	1.0000	1.0000	1.0000	1.0000	1.0000	1.0000	1.0000	1.0000	1.0000
	1	.1900	.2775	.3600	.4375	.5100	.5775	.6400	.6975	.7500
	2	.0100	.0225	.0400	.0625	.0900	.1225	.1600	.2025	.2500
n = 5	r = 0	1.0000	1.0000	1.0000	1.0000	1.0000	1.0000	1.0000	1.0000	1.0000
	1	.4095	.5563	.6723	.7627	.8319	.8840	.9222	.9497	.9688
	2	.0815	.1648	.2627	.3672	.4718	.5716	.6630	.7438	.8125
	3	.0086	.0266	.0579	.1035	.1631	.2352	.3174	.4069	.5000
	4	.0005	.0022	.0067	.0156	.0308	.0540	.0870	.1312	.1875
	5		.0001	.0003	.0010	.0024	.0053	.0102	.0185	.0313
n = 10	r = 0	1.0000	1.0000	1.0000	1.0000	1.0000	1.0000	1.0000	1.0000	1.0000
	1	.6513	.8031	.8926	.9437	.9718	.9865	.9940	.9975	.9990
	2	.2639	.4557	.6242	.7560	.8507	.9140	.9536	.9767	.9893
	3	.0702	.1798	.3222	.4744	.6172	.7384	.8327	.9004	.9453
	4	.0128	.0500	.1209	.2241	.3504	.4862	.6177	.7430	.8281
	5	.0016	.0099	.0328	.0781	.1503	.2485	.3669	.4956	.6230
	6	.0001	.0014	.0064	.0197	.0473	.0949	.1662	.2616	.3770
	7		.0001	.0009	.0035	.0106	.0260	.0548	.1020	.1719
	8			.0001	.0004	.0016	.0048	.0123	.0274	.0547
	9					.0001	.0005	.0017	.0045	.0107
	10							.0001	.0003	.0010
n = 20	r = 0	1.0000	1.0000	1.0000	1.0000	1.0000	1.0000	1.0000	1.0000	1.0000
	1	.8784	.9612	.9885	.9968	.9992	.9998	1.0000	1.0000	1.0000
	2	.6083	.8244	.9308	.9757	.9924	.9979	.9995	.9999	1.0000
	3	.3231	.5951	.7939	.9087	.9645	.9879	.9964	.9991	.9998
	4	.1330	.3523	.5886	.7748	.8929	.9556	.9840	.9951	.9987

Cumulative Binomial Probabilities (*Contd.*)

	p =	0.10	0.15	0.20	0.25	0.30	0.35	0.40	0.45	0.50
	5	.0432	.1702	.3704	.5852	.7625	.8818	.9490	.9811	.9941
	6	.0113	.0673	.1958	.3828	.5836	.7546	.8744	.9447	.9793
	7	.0024	.0219	.0867	.2142	.3920	.5834	.7500	.8701	.9423
	8	.0004	.0059	.0321	.1018	.2277	.3990	.5841	.7480	.8684
	9	.0001	.0013	.0100	.0409	.1133	.2376	.4044	.5857	.7483
	10		.0002	.0026	.0139	.0480	.1218	.2447	.4086	.5881
	11			.0006	.0039	.0171	.0532	.1275	.2493	.4119
	12			.0001	.0009	.0051	.0196	.0565	.1308	.2517
	13				.0002	.0013	.0060	.0210	.0580	.1316
	14					.0003	.0015	.0065	.0214	.0577
	15						.0003	.0016	.0064	.0207
	16							.0003	.0015	.0059
	17								.0003	.0013
	18									.0002
n = 50	r = 0	1.0000	1.0000	1.0000	1.0000	1.0000	1.0000	1.0000	1.0000	1.0000
	1	.9948	.9997	1.0000	1.0000	1.0000	1.0000	1.0000	1.0000	1.0000
	2	.9662	.9971	.9998	1.0000	1.0000	1.0000	1.0000	1.0000	1.0000
	3	.8883	.9858	.9987	.9999	1.0000	1.0000	1.0000	1.0000	1.0000
	4	.7497	.9540	.9943	.9995	1.0000	1.0000	1.0000	1.0000	1.0000
	5	.5688	.8879	.9815	.9979	.9998	1.0000	1.0000	1.0000	1.0000
	6	.3839	.7806	.9520	.9930	.9993	.9999	1.0000	1.0000	1.0000
	7	.2298	.6387	.8966	.9806	.9975	.9998	1.0000	1.0000	1.0000
	8	.1221	.4812	.8096	.9547	.9927	.9992	.9999	1.0000	1.0000
	9	.0579	.3319	.6927	.9084	.9817	.9975	.9998	1.0000	1.0000
	10	.0245	.2089	.5563	.8363	.9598	.9933	.9992	.9999	1.0000
	11	.0094	.1199	.4164	.7378	.9211	.9840	.9978	.9998	1.0000
	12	.0032	.0628	.2893	.6184	.8610	.9658	.9943	.9994	1.0000
	13	.0010	.0301	.1861	.4890	.7771	.9339	.9867	.9982	.9998
	14	.0003	.0132	.1106	.3630	.6721	.8837	.9720	.9955	.9995
	15	.0001	.0053	.0607	.2519	.5532	.8122	.9460	.9896	.9987
	16		.0019	.0308	.1631	.4308	.7199	.9045	.9780	.9967
	17		.0007	.0144	.0983	.3161	.6111	.8439	.9573	.9923
	18		.0002	.0063	.0551	.2178	.4940	.7631	.9235	.9836
	19		.0001	.0025	.0287	.1406	.3784	.6644	.8727	.9675
	20			.0009	.0139	.0848	.2736	.5535	.8026	.9405
	21			.0003	.0063	.0478	.1861	.4390	.7138	.8987
	22			.0001	.0026	.0251	.1187	.3299	.6100	.8389
	23				.0010	.0123	.0710	.2340	.4981	.7601
	24				.0004	.0056	.0396	.1562	.3866	.6641
	25				.0001	.0024	.0207	.0978	.2840	.5561
	26					.0009	.0100	.0573	.1966	.4439
	27					.0003	.0045	.0314	.1279	.3359
	28					.0001	.0019	.0160	.0780	.2399
	29						.0007	.0076	.0444	.1611
	30						.0003	.0034	.0235	.1013
	31						.0001	.0014	.0116	.0595
	32							.0005	.0053	.0325
	33							.0002	.0022	.0164
	34							.0001	.0009	.0077
	35								.0003	.0033
	36								.0001	.0013
	37									.0005
	38									.0002

Cumulative Binomial Probabilities (*Contd.*)

	p =	0.10	0.15	0.20	0.25	0.30	0.35	0.40	0.45	0.50
n = 100	r = 0	1.0000	1.0000	1.0000	1.0000	1.0000	1.0000	1.0000	1.0000	1.0000
	1	1.0000	1.0000	1.0000	1.0000	1.0000	1.0000	1.0000	1.0000	1.0000
	2	.9997	1.0000	1.0000	1.0000	1.0000	1.0000	1.0000	1.0000	1.0000
	3	.9981	1.0000	1.0000	1.0000	1.0000	1.0000	1.0000	1.0000	1.0000
	4	.9922	.9999	1.0000	1.0000	1.0000	1.0000	1.0000	1.0000	1.0000
	5	.9763	.9996	1.0000	1.0000	1.0000	1.0000	1.0000	1.0000	1.0000
	6	.9424	.9984	1.0000	1.0000	1.0000	1.0000	1.0000	1.0000	1.0000
	7	.8828	.9953	.9999	1.0000	1.0000	1.0000	1.0000	1.0000	1.0000
	8	.7939	.9878	.9997	1.0000	1.0000	1.0000	1.0000	1.0000	1.0000
	9	.6791	.9725	.9991	1.0000	1.0000	1.0000	1.0000	1.0000	1.0000
	10	.5487	.9449	.9977	1.0000	1.0000	1.0000	1.0000	1.0000	1.0000
	11	.4168	.9006	.9943	.9999	1.0000	1.0000	1.0000	1.0000	1.0000
	12	.2970	.8365	.9874	.9996	1.0000	1.0000	1.0000	1.0000	1.0000
	13	.1982	.7527	.9747	.9990	1.0000	1.0000	1.0000	1.0000	1.0000
	14	.1239	.6526	.9531	.9975	.9999	1.0000	1.0000	1.0000	1.0000
	15	.0726	.5428	.9196	.9946	.9998	1.0000	1.0000	1.0000	1.0000
	16	.0399	.4317	.8715	.9889	.9996	1.0000	1.0000	1.0000	1.0000
	17	.0206	.3275	.8077	.9789	.9990	1.0000	1.0000	1.0000	1.0000
	18	.0100	.2367	.7288	.9624	.9978	.9999	1.0000	1.0000	1.0000
	19	.0046	.1628	.6379	.9370	.9955	.9999	1.0000	1.0000	1.0000
	20	.0020	.1065	.5398	.9005	.9911	.9997	1.0000	1.0000	1.0000
	21	.0008	.0663	.4405	.8512	.9835	.9992	1.0000	1.0000	1.0000
	22	.0003	.0393	.3460	.7886	.9712	.9983	1.0000	1.0000	1.0000
	23	.0001	.0221	.2611	.7136	.9521	.9966	.9999	1.0000	1.0000
	24		.0119	.1891	.6289	.9245	.9934	.9997	1.0000	1.0000
	25		.0061	.1314	.5383	.8864	.9879	.9994	1.0000	1.0000
	26		.0030	.0875	.4465	.8369	.9789	.9988	1.0000	1.0000
	27		.0014	.0558	.3583	.7756	.9649	.9976	.9999	1.0000
	28		.0006	.0342	.2776	.7036	.9442	.9954	.9998	1.0000
	29		.0003	.0200	.2075	.6232	.9152	.9916	.9996	1.0000
	30		.0001	.0112	.1495	.5377	.8764	.9852	.9992	1.0000
	31			.0061	.1038	.4509	.8270	.9752	.9985	1.0000
	32			.0031	.0693	.3669	.7669	.9602	.9970	.9999
	33			.0016	.0446	.2893	.6971	.9385	.9945	.9998
	34			.0007	.0276	.2207	.6197	.9087	.9902	.9996
	35			.0003	.0164	.1629	.5376	.8697	.9834	.9991
	36			.0001	.0094	.1161	.4542	.8205	.9728	.9982
	37			.0001	.0052	.0799	.3731	.7614	.9571	.9967
	38				.0027	.0530	.2976	.6932	.9349	.9940
	39				.0014	.0340	.2301	.6178	.9049	.9895
	40				.0007	.0210	.1724	.5379	.8657	.9824
	41				.0003	.0125	.1250	.4567	.8169	.9716
	42				.0001	.0072	.0877	.3775	.7585	.9557
	43				.0001	.0040	.0594	.3033	.6913	.9334
	44					.0021	.0389	.2365	.6172	.9033
	45					.0011	.0246	.1789	.5387	.8644
	46					.0005	.0150	.1311	.4587	.8159
	47					.0003	.0088	.0930	.3804	.7579
	48					.0001	.0050	.0638	.3069	.6914
	49					.0001	.0027	.0423	.2404	.6178

Cumulative Binomial Probabilities (*Contd.*)

p =	0.10	0.15	0.20	0.25	0.30	0.35	0.40	0.45	0.50
50						.0015	.0271	.1827	.5398
51						.0007	.0168	.1346	.4602
52						.0004	.0100	.0960	.3822
53						.0002	.0058	.0662	.3086
54						.0001	.0032	.0441	.2421
55							.0017	.0284	.1841
56							.0009	.0176	.1356
57							.0004	.0106	.0967
58							.0002	.0061	.0666
59							.0001	.0034	.0443
60								.0018	.0284
61								.0009	.0176
62								.0005	.0105
63								.0002	.0060
64								.0001	.0033
65									.0018
66									.0009
67									.0004
68									.0002
69									.0001

The above table gives binomial probabilities only for a limited range of values of n and p since, in practice, either the more compact tabulation of the Poisson distribution or that of the Normal distribution can usually be used to give an adequate approximation.

Working rule:

(i) Use the Poisson approximation if $p < 0.1$, putting $m = np$.

(ii) Use the Normal approximation if $0.1 \leq p \leq 0.9$ and $np > 5$, putting $\mu = np$ and $\sigma = \sqrt{np(1-p)}$.

(iii) Use the Poisson approximation if $p > 0.9$, putting $m = n(1 - p)$ and working in terms of 'failures'.

Note: For values of $p > 0.5$, work in terms of 'failures' which will have probability q (= $1 - p$).

BINOMIAL DISTRIBUTION FUNCTION

$$B(x; n, p) = \sum_{k=0}^{x} {}^{n}C_{k}\, P^{k} (1-p)^{n-k}$$

							p					
	x	0.05	0.10	0.20	0.30	0.40	0.50	0.60	0.70	0.80	0.90	0.95
n = 1	0	0.950	0.900	0.800	0.700	0.600	0.500	0.400	0.300	0.200	0.100	0.050
	1	1.000	1.000	1.000	1.000	1.000	1.000	1.000	1.000	1.000	1.000	1.000
n = 2	0	0.903	0.810	0.640	0.490	0.360	0.250	0.160	0.090	0.040	0.010	0.003
	1	0.998	0.990	0.960	0.910	0.840	0.750	0.640	0.510	0.360	0.190	0.098
	2	1.000	1.000	1.000	1.000	1.000	1.000	1.000	1.000	1.000	1.000	1.000
n = 3	0	0.857	0.729	0.512	0.343	0.216	0.125	0.064	0.027	0.008	0.001	0.000
	1	0.993	0.972	0.896	0.784	0.648	0.500	0.352	0.216	0.104	0.028	0.007
	2	1.000	0.999	0.992	0.973	0.936	0.875	0.784	0.657	0.488	0.271	0.143
	3	1.000	1.000	1.000	1.000	1.000	1.000	1.000	1.000	1.000	1.000	1.000
n = 4	0	0.815	0.656	0.410	0.240	0.130	0.063	0.026	0.008	0.002	0.000	0.000
	1	0.986	0.948	0.819	0.652	0.475	0.313	0.179	0.084	0.027	0.004	0.000
	2	1.000	0.996	0.973	0.916	0.821	0.688	0.525	0.348	0.181	0.052	0.014
	3	1.000	1.000	0.998	0.992	0.974	0.938	0.870	0.760	0.590	0.344	0.185
	4	1.000	1.000	1.000	1.000	1.000	1.000	1.000	1.000	1.000	1.000	1.000
n = 5	0	0.774	0.590	0.328	0.168	0.078	0.031	0.010	0.002	0.000	0.000	0.000
	1	0.977	0.919	0.737	0.528	0.337	0.188	0.087	0.031	0.007	0.000	0.000
	2	0.999	0.991	0.942	0.837	0.683	0.500	0.317	0.163	0.058	0.009	0.001
	3	1.000	1.000	0.993	0.969	0.913	0.813	0.663	0.472	0.263	0.081	0.023
	4	1.000	1.000	1.000	0.998	0.990	0.969	0.922	0.832	0.672	0.410	0.226
	5	1.000	1.000	1.000	1.000	1.000	1.000	1.000	1.000	1.000	1.000	1.000
n = 6	0	0.735	0.531	0.262	0.118	0.047	0.016	0.004	0.001	0.000	0.000	0.000
	1	0.967	0.886	0.655	0.420	0.233	0.109	0.041	0.011	0.002	0.000	0.000
	2	0.998	0.984	0.901	0.744	0.544	0.344	0.179	0.070	0.017	0.001	0.000
	3	1.000	0.999	0.983	0.930	0.821	0.656	0.456	0.256	0.099	0.016	0.002
	4	1.000	1.000	0.998	0.989	0.959	0.891	0.767	0.580	0.345	0.114	0.033
	5	1.000	1.000	1.000	0.999	0.996	0.984	0.953	0.882	0.738	0.469	0.265
	6	1.000	1.000	1.000	1.000	1.000	1.000	1.000	1.000	1.000	1.000	1.000
n = 7	0	0.698	0.478	0.210	0.082	0.028	0.008	0.002	0.000	0.000	0.000	0.000
	1	0.956	0.850	0.577	0.329	0.159	0.063	0.019	0.004	0.000	0.000	0.000
	2	0.996	0.974	0.852	0.647	0.420	0.227	0.096	0.029	0.005	0.000	0.000
	3	1.000	0.997	0.967	0.874	0.710	0.500	0.290	0.126	0.033	0.003	0.000
	4	1.000	1.000	0.995	0.971	0.904	0.773	0.580	0.353	0.148	0.026	0.004
	5	1.000	1.000	1.000	0.996	0.981	0.938	0.841	0.671	0.423	0.150	0.044
	6	1.000	1.000	1.000	1.000	0.998	0.992	0.972	0.918	0.790	0.522	0.302
	7	1.000	1.000	1.000	1.000	1.000	1.000	1.000	1.000	1.000	1.000	1.000
n = 8	0	0.663	0.430	0.168	0.058	0.017	0.004	0.001	0.000	0.000	0.000	0.000
	1	0.943	0.813	0.503	0.255	0.106	0.035	0.009	0.001	0.000	0.000	0.000
	2	0.994	0.962	0.797	0.552	0.315	0.145	0.050	0.011	0.001	0.000	0.000
	3	1.000	0.995	0.944	0.806	0.594	0.363	0.174	0.058	0.010	0.000	0.000
	4	1.000	1.000	0.990	0.942	0.826	0.637	0.406	0.194	0.056	0.005	0.000
	5	1.000	1.000	0.999	0.989	0.950	0.855	0.685	0.448	0.203	0.038	0.006
	6	1.000	1.000	1.000	0.999	0.991	0.965	0.894	0.745	0.497	0.187	0.057
	7	1.000	1.000	1.000	1.000	0.999	0.996	0.983	0.942	0.832	0.570	0.337
	8	1.000	1.000	1.000	1.000	1.000	1.000	1.000	1.000	1.000	1.000	1.000
n = 9	0	0.630	0.387	0.134	0.040	0.010	0.002	0.000	0.000	0.000	0.000	0.000
	1	0.929	0.775	0.436	0.196	0.071	0.020	0.004	0.000	0.000	0.000	0.000
	2	0.992	0.947	0.738	0.463	0.232	0.090	0.025	0.004	0.000	0.000	0.000
	3	0.999	0.992	0.914	0.730	0.483	0.254	0.099	0.025	0.003	0.000	0.000

Binomial Distribution Function (*Contd.*)

	x	0.05	0.10	0.20	0.30	0.40	p 0.50	0.60	0.70	0.80	0.90	0.95
	4	1.000	0.999	0.980	0.901	0.733	0.500	0.267	0.099	0.020	0.001	0.000
	5	1.000	1.000	0.997	0.975	0.901	0.746	0.517	0.270	0.086	0.008	0.001
	6	1.000	1.000	1.000	0.996	0.975	0.910	0.768	0.537	0.262	0.053	0.008
	7	1.000	1.000	1.000	1.000	0.996	0.980	0.929	0.804	0.564	0.225	0.071
	8	1.000	1.000	1.000	1.000	1.000	0.998	0.990	0.960	0.866	0.613	0.370
	9	1.000	1.000	1.000	1.000	1.000	1.000	1.000	1.000	1.000	1.000	1.000
n = 10	0	0.599	0.349	0.107	0.028	0.006	0.001	0.000	0.000	0.000	0.000	0.000
	1	0.914	0.736	0.376	0.149	0.046	0.011	0.002	0.000	0.000	0.000	0.000
	2	0.988	0.930	0.678	0.383	0.167	0.055	0.012	0.002	0.000	0.000	0.000
	3	0.999	0.987	0.879	0.650	0.382	0.172	0.055	0.011	0.001	0.000	0.000
	4	1.000	0.998	0.967	0.850	0.633	0.377	0.166	0.047	0.006	0.000	0.000
	5	1.000	1.000	0.994	0.953	0.834	0.623	0.367	0.150	0.033	0.002	0.000
	6	1.000	1.000	0.999	0.989	0.945	0.828	0.618	0.350	0.121	0.013	0.001
	7	1.000	1.000	1.000	0.998	0.988	0.945	0.833	0.617	0.322	0.070	0.012
	8	1.000	1.000	1.000	1.000	0.998	0.989	0.954	0.851	0.624	0.264	0.086
	9	1.000	1.000	1.000	1.000	1.000	0.999	0.994	0.972	0.893	0.651	0.401
	10	1.000	1.000	1.000	1.000	1.000	1.000	1.000	1.000	1.000	1.000	1.000
n = 11	0	0.569	0.314	0.086	0.020	0.004	0.000	0.000	0.000	0.000	0.000	0.000
	1	0.898	0.697	0.322	0.113	0.030	0.006	0.001	0.000	0.000	0.000	0.000
	2	0.985	0.910	0.617	0.313	0.119	0.033	0.006	0.001	0.000	0.000	0.000
	3	0.998	0.981	0.839	0.570	0.296	0.113	0.029	0.004	0.000	0.000	0.000
	4	1.000	0.997	0.950	0.790	0.533	0.274	0.099	0.022	0.002	0.000	0.000
	5	1.000	1.000	0.988	0.922	0.753	0.500	0.247	0.078	0.012	0.000	0.000
	6	1.000	1.000	0.998	0.978	0.901	0.726	0.467	0.210	0.050	0.003	0.000
	7	1.000	1.000	1.000	0.996	0.971	0.887	0.704	0.430	0.161	0.019	0.002
	8	1.000	1.000	1.000	0.999	0.994	0.967	0.881	0.687	0.383	0.090	0.015
	9	1.000	1.000	1.000	1.000	0.999	0.994	0.970	0.887	0.678	0.303	0.102
	10	1.000	1.000	1.000	1.000	1.000	1.000	0.996	0.980	0.914	0.686	0.431
	11	1.000	1.000	1.000	1.000	1.000	1.000	1.000	1.000	1.000	1.000	1.000
n = 12	0	0.540	0.282	0.069	0.014	0.002	0.000	0.000	0.000	0.000	0.000	0.000
	1	0.882	0.659	0.275	0.085	0.020	0.003	0.000	0.000	0.000	0.000	0.000
	2	0.980	0.889	0.558	0.253	0.083	0.019	0.003	0.000	0.000	0.000	0.000
	3	0.998	0.974	0.795	0.493	0.225	0.073	0.015	0.002	0.000	0.000	0.000
	4	1.000	0.996	0.927	0.724	0.438	0.194	0.057	0.009	0.001	0.000	0.000
	5	1.000	0.999	0.981	0.882	0.665	0.387	0.158	0.039	0.004	0.000	0.000
	6	1.000	1.000	0.996	0.961	0.842	0.613	0.335	0.118	0.019	0.001	0.000
	7	1.000	1.000	0.999	0.991	0.943	0.806	0.562	0.276	0.073	0.004	0.000
	8	1.000	1.000	1.000	0.998	0.985	0.927	0.775	0.507	0.205	0.026	0.002
	9	1.000	1.000	1.000	1.000	0.997	0.981	0.917	0.747	0.442	0.111	0.020
	10	1.000	1.000	1.000	1.000	1.000	0.997	0.980	0.915	0.725	0.341	0.118
	11	1.000	1.000	1.000	1.000	1.000	1.000	0.998	0.986	0.931	0.718	0.460
	12	1.000	1.000	1.000	1.000	1.000	1.000	1.000	1.000	1.000	1.000	1.000
n = 13	0	0.513	0.254	0.055	0.010	0.001	0.000	0.000	0.000	0.000	0.000	0.000
	1	0.865	0.621	0.234	0.064	0.013	0.002	0.000	0.000	0.000	0.000	0.000
	2	0.975	0.866	0.502	0.202	0.058	0.011	0.001	0.000	0.000	0.000	0.000
	3	0.997	0.966	0.747	0.421	0.169	0.046	0.008	0.001	0.000	0.000	0.000
	4	1.000	0.994	0.901	0.654	0.353	0.133	0.032	0.004	0.000	0.000	0.000
	5	1.000	0.999	0.970	0.835	0.574	0.291	0.098	0.018	0.001	0.000	0.000
	6	1.000	1.000	0.993	0.938	0.771	0.500	0.229	0.062	0.007	0.000	0.000
	7	1.000	1.000	0.999	0.982	0.902	0.709	0.426	0.165	0.030	0.001	0.000
	8	1.000	1.000	1.000	0.996	0.968	0.867	0.647	0.346	0.099	0.006	0.000
	9	1.000	1.000	1.000	0.999	0.992	0.954	0.831	0.579	0.253	0.034	0.003

Binomial Distribution Function (*Contd.*)

	x	0.05	0.10	0.20	0.30	0.40	p 0.50	0.60	0.70	0.80	0.90	0.95
	10	1.000	1.000	1.000	1.000	0.999	0.989	0.942	0.798	0.498	0.134	0.025
	11	1.000	1.000	1.000	1.000	1.000	0.998	0.987	0.936	0.766	0.379	0.135
	12	1.000	1.000	1.000	1.000	1.000	1.000	0.999	0.990	0.945	0.746	0.487
	13	1.000	1.000	1.000	1.000	1.000	1.000	1.000	1.000	1.000	1.000	1.000
n = 14	0	0.488	0.229	0.044	0.007	0.001	0.000	0.000	0.000	0.000	0.000	0.000
	1	0.847	0.585	0.198	0.047	0.008	0.001	0.000	0.000	0.000	0.000	0.000
	2	0.970	0.842	0.448	0.161	0.040	0.006	0.001	0.000	0.000	0.000	0.000
	3	0.996	0.956	0.698	0.355	0.124	0.029	0.004	0.000	0.000	0.000	0.000
	4	1.000	0.991	0.870	0.584	0.279	0.090	0.018	0.002	0.000	0.000	0.000
	5	1.000	0.999	0.956	0.781	0.486	0.212	0.058	0.008	0.000	0.000	0.000
	6	1.000	1.000	0.988	0.907	0.692	0.395	0.150	0.031	0.002	0.000	0.000
	7	1.000	1.000	0.998	0.969	0.850	0.605	0.308	0.093	0.012	0.000	0.000
	8	1.000	1.000	1.000	0.992	0.942	0.788	0.514	0.219	0.044	0.001	0.000
	9	1.000	1.000	1.000	0.998	0.982	0.910	0.721	0.416	0.130	0.009	0.000
	10	1.000	1.000	1.000	1.000	0.996	0.971	0.876	0.645	0.302	0.044	0.004
	11	1.000	1.000	1.000	1.000	0.999	0.994	0.960	0.839	0.552	0.158	0.030
	12	1.000	1.000	1.000	1.000	1.000	0.999	0.992	0.953	0.802	0.415	0.153
	13	1.000	1.000	1.000	1.000	1.000	1.000	0.999	0.993	0.956	0.771	0.512
	14	1.000	1.000	1.000	1.000	1.000	1.000	1.000	1.000	1.000	1.000	1.000
n = 15	0	0.463	0.206	0.035	0.005	0.000	0.000	0.000	0.000	0.000	0.000	0.000
	1	0.829	0.549	0.167	0.035	0.005	0.000	0.000	0.000	0.000	0.000	0.000
	2	0.964	0.816	0.398	0.127	0.027	0.004	0.000	0.000	0.000	0.000	0.000
	3	0.995	0.944	0.648	0.297	0.091	0.018	0.002	0.000	0.000	0.000	0.000
	4	0.999	0.987	0.836	0.515	0.217	0.059	0.009	0.001	0.000	0.000	0.000
	5	1.000	0.998	0.939	0.722	0.403	0.151	0.034	0.004	0.000	0.000	0.000
	6	1.000	1.000	0.982	0.869	0.610	0.304	0.095	0.015	0.001	0.000	0.000
	7	1.000	1.000	0.996	0.950	0.787	0.500	0.213	0.050	0.004	0.000	0.000
	8	1.000	1.000	0.999	0.985	0.905	0.696	0.390	0.131	0.018	0.000	0.000
	9	1.000	1.000	1.000	0.996	0.966	0.849	0.597	0.278	0.061	0.002	0.000
	10	1.000	1.000	1.000	0.999	0.991	0.941	0.783	0.485	0.164	0.013	0.001
	11	1.000	1.000	1.000	1.000	0.998	0.982	0.909	0.703	0.352	0.056	0.005
	12	1.000	1.000	1.000	1.000	1.000	0.996	0.973	0.873	0.602	0.184	0.036
	13	1.000	1.000	1.000	1.000	1.000	1.000	0.995	0.965	0.833	0.451	0.171
	14	1.000	1.000	1.000	1.000	1.000	1.000	1.000	0.995	0.965	0.794	0.537
	15	1.000	1.000	1.000	1.000	1.000	1.000	1.000	1.000	1.000	1.000	1.000
n = 16	0	0.440	0.185	0.028	0.003	0.000	0.000	0.000	0.000	0.000	0.000	0.000
	1	0.811	0.515	0.141	0.026	0.003	0.000	0.000	0.000	0.000	0.000	0.000
	2	0.957	0.789	0.352	0.099	0.018	0.002	0.000	0.000	0.000	0.000	0.000
	3	0.993	0.932	0.598	0.246	0.065	0.011	0.001	0.000	0.000	0.000	0.000
	4	0.999	0.983	0.798	0.450	0.167	0.038	0.005	0.000	0.000	0.000	0.000
	5	1.000	0.997	0.918	0.660	0.329	0.105	0.019	0.002	0.000	0.000	0.000
	6	1.000	0.999	0.973	0.825	0.527	0.227	0.058	0.007	0.000	0.000	0.000
	7	1.000	1.000	0.993	0.926	0.716	0.402	0.142	0.026	0.001	0.000	0.000
	8	1.000	1.000	0.999	0.974	0.858	0.598	0.284	0.074	0.007	0.000	0.000
	9	1.000	1.000	1.000	0.993	0.942	0.773	0.473	0.175	0.027	0.001	0.000
	10	1.000	1.000	1.000	0.998	0.981	0.895	0.671	0.340	0.082	0.003	0.000
	11	1.000	1.000	1.000	1.000	0.995	0.962	0.833	0.550	0.202	0.017	0.001
	12	1.000	1.000	1.000	1.000	0.999	0.989	0.935	0.754	0.402	0.068	0.007
	13	1.000	1.000	1.000	1.000	1.000	0.998	0.982	0.901	0.648	0.211	0.043
	14	1.000	1.000	1.000	1.000	1.000	1.000	0.997	0.974	0.859	0.485	0.189
	15	1.000	1.000	1.000	1.000	1.000	1.000	1.000	0.997	0.972	0.815	0.560
	16	1.000	1.000	1.000	1.000	1.000	1.000	1.000	1.000	1.000	1.000	1.000

Binomial Distribution Function (*Contd.*)

						p						
	x	0.05	0.10	0.20	0.30	0.40	0.50	0.60	0.70	0.80	0.90	0.95
n = 17	0	0.418	0.167	0.023	0.002	0.000	0.000	0.000	0.000	0.000	0.000	0.000
	1	0.792	0.482	0.118	0.019	0.002	0.000	0.000	0.000	0.000	0.000	0.000
	2	0.950	0.762	0.310	0.077	0.012	0.001	0.000	0.000	0.000	0.000	0.000
	3	0.991	0.917	0.549	0.202	0.046	0.006	0.000	0.000	0.000	0.000	0.000
	4	0.999	0.978	0.758	0.389	0.126	0.025	0.003	0.000	0.000	0.000	0.000
	5	1.000	0.995	0.894	0.597	0.264	0.072	0.011	0.001	0.000	0.000	0.000
	6	1.000	0.999	0.962	0.775	0.448	0.166	0.035	0.003	0.000	0.000	0.000
	7	1.000	1.000	0.989	0.895	0.641	0.315	0.092	0.013	0.000	0.000	0.000
	8	1.000	1.000	0.997	0.960	0.801	0.500	0.199	0.040	0.003	0.000	0.000
	9	1.000	1.000	1.000	0.987	0.908	0.685	0.359	0.105	0.011	0.000	0.000
	10	1.000	1.000	1.000	0.997	0.965	0.834	0.552	0.225	0.038	0.001	0.000
	11	1.000	1.000	1.000	0.999	0.989	0.928	0.736	0.403	0.106	0.005	0.000
	12	1.000	1.000	1.000	1.000	0.997	0.975	0.874	0.611	0.242	0.022	0.001
	13	1.000	1.000	1.000	1.000	1.000	0.994	0.954	0.798	0.451	0.083	0.009
	14	1.000	1.000	1.000	1.000	1.000	0.999	0.988	0.923	0.690	0.238	0.050
	15	1.000	1.000	1.000	1.000	1.000	1.000	0.998	0.981	0.882	0.518	0.208
	16	1.000	1.000	1.000	1.000	1.000	1.000	1.000	0.998	0.977	0.833	0.582
	17	1.000	1.000	1.000	1.000	1.000	1.000	1.000	1.000	1.000	1.000	1.000
n = 18	0	0.397	0.150	0.018	0.002	0.000	0.000	0.000	0.000	0.000	0.000	0.000
	1	0.774	0.450	0.099	0.014	0.001	0.000	0.000	0.000	0.000	0.000	0.000
	2	0.942	0.734	0.271	0.060	0.008	0.001	0.000	0.000	0.000	0.000	0.000
	3	0.989	0.902	0.501	0.165	0.033	0.004	0.000	0.000	0.000	0.000	0.000
	4	0.998	0.972	0.716	0.333	0.094	0.015	0.001	0.000	0.000	0.000	0.000
	5	1.000	0.994	0.867	0.534	0.209	0.048	0.006	0.000	0.000	0.000	0.000
	6	1.000	0.999	0.949	0.722	0.374	0.119	0.020	0.001	0.000	0.000	0.000
	7	1.000	1.000	0.984	0.859	0.563	0.240	0.058	0.006	0.000	0.000	0.000
	8	1.000	1.000	0.996	0.940	0.737	0.407	0.135	0.021	0.001	0.000	0.000
	9	1.000	1.000	0.999	0.979	0.865	0.593	0.263	0.060	0.004	0.000	0.000
	10	1.000	1.000	1.000	0.994	0.942	0.760	0.437	0.141	0.016	0.000	0.000
	11	1.000	1.000	1.000	0.999	0.980	0.881	0.626	0.278	0.051	0.001	0.000
	12	1.000	1.000	1.000	1.000	0.994	0.952	0.791	0.466	0.133	0.006	0.000
	13	1.000	1.000	1.000	1.000	0.999	0.985	0.906	0.667	0.284	0.028	0.002
	14	1.000	1.000	1.000	1.000	1.000	0.996	0.967	0.835	0.499	0.098	0.011
	15	1.000	1.000	1.000	1.000	1.000	0.999	0.992	0.940	0.729	0.266	0.058
	16	1.000	1.000	1.000	1.000	1.000	1.000	0.999	0.986	0.901	0.550	0.226
	17	1.000	1.000	1.000	1.000	1.000	1.000	1.000	0.998	0.982	0.850	0.603
	18	1.000	1.000	1.000	1.000	1.000	1.000	1.000	1.000	1.000	1.000	1.000
n = 19	0	0.377	0.135	0.014	0.001	0.000	0.000	0.000	0.000	0.000	0.000	0.000
	1	0.755	0.420	0.083	0.010	0.001	0.000	0.000	0.000	0.000	0.000	0.000
	2	0.933	0.705	0.237	0.046	0.005	0.000	0.000	0.000	0.000	0.000	0.000
	3	0.987	0.885	0.455	0.133	0.023	0.002	0.000	0.000	0.000	0.000	0.000
	4	0.998	0.965	0.673	0.282	0.070	0.010	0.001	0.000	0.000	0.000	0.000
	5	1.000	0.991	0.837	0.474	0.163	0.032	0.003	0.000	0.000	0.000	0.000
	6	1.000	0.998	0.932	0.666	0.308	0.084	0.012	0.001	0.000	0.000	0.000
	7	1.000	1.000	0.977	0.818	0.488	0.180	0.035	0.003	0.000	0.000	0.000
	8	1.000	1.000	0.993	0.916	0.667	0.324	0.088	0.011	0.000	0.000	0.000
	9	1.000	1.000	0.998	0.967	0.814	0.500	0.186	0.033	0.002	0.000	0.000
	10	1.000	1.000	1.000	0.989	0.912	0.676	0.333	0.084	0.007	0.000	0.000
	11	1.000	1.000	1.000	0.997	0.965	0.820	0.512	0.182	0.023	0.000	0.000
	12	1.000	1.000	1.000	0.999	0.988	0.916	0.692	0.334	0.068	0.002	0.000
	13	1.000	1.000	1.000	1.000	0.997	0.968	0.837	0.526	0.163	0.009	0.000
	14	1.000	1.000	1.000	1.000	0.999	0.990	0.930	0.718	0.327	0.035	0.002

Binomial Distribution Function (*Concld.*)

						p						
	x	0.05	0.10	0.20	0.30	0.40	0.50	0.60	0.70	0.80	0.90	0.95
	15	1.000	1.000	1.000	1.000	1.000	0.998	0.977	0.867	0.545	0.115	0.013
	16	1.000	1.000	1.000	1.000	1.000	1.000	0.995	0.954	0.763	0.295	0.067
	17	1.000	1.000	1.000	1.000	1.000	1.000	0.999	0.990	0.917	0.580	0.245
	18	1.000	1.000	1.000	1.000	1.000	1.000	1.000	0.999	0.986	0.865	0.623
	19	1.000	1.000	1.000	1.000	1.000	1.000	1.000	1.000	1.000	1.000	1.000
n = 20	0	0.358	0.122	0.012	0.001	0.000	0.000	0.000	0.000	0.000	0.000	0.000
	1	0.736	0.392	0.069	0.008	0.001	0.000	0.000	0.000	0.000	0.000	0.000
	2	0.925	0.677	0.206	0.035	0.004	0.000	0.000	0.000	0.000	0.000	0.000
	3	0.984	0.867	0.411	0.107	0.016	0.001	0.000	0.000	0.000	0.000	0.000
	4	0.997	0.957	0.630	0.238	0.051	0.006	0.000	0.000	0.000	0.000	0.000
	5	1.000	0.989	0.804	0.416	0.126	0.021	0.002	0.000	0.000	0.000	0.000
	6	1.000	0.998	0.913	0.608	0.250	0.058	0.006	0.000	0.000	0.000	0.000
	7	1.000	1.000	0.968	0.772	0.416	0.132	0.021	0.001	0.000	0.000	0.000
	8	1.000	1.000	0.990	0.887	0.596	0.252	0.057	0.005	0.000	0.000	0.000
	9	1.000	1.000	0.997	0.952	0.755	0.412	0.128	0.017	0.001	0.000	0.000
	10	1.000	1.000	0.999	0.983	0.872	0.588	0.245	0.048	0.003	0.000	0.000
	11	1.000	1.000	1.000	0.995	0.943	0.748	0.404	0.113	0.010	0.000	0.000
	12	1.000	1.000	1.000	0.999	0.979	0.868	0.584	0.228	0.032	0.000	0.000
	13	1.000	1.000	1.000	1.000	0.994	0.942	0.750	0.392	0.087	0.002	0.000
	14	1.000	1.000	1.000	1.000	0.998	0.979	0.874	0.584	0.196	0.011	0.000
	15	1.000	1.000	1.000	1.000	1.000	0.994	0.949	0.762	0.370	0.043	0.003
	16	1.000	1.000	1.000	1.000	1.000	0.999	0.984	0.893	0.589	0.133	0.016
	17	1.000	1.000	1.000	1.000	1.000	1.000	0.996	0.965	0.794	0.323	0.075
	18	1.000	1.000	1.000	1.000	1.000	1.000	0.999	0.992	0.931	0.608	0.264
	19	1.000	1.000	1.000	1.000	1.000	1.000	1.000	0.999	0.988	0.878	0.642
	20	1.000	1.000	1.000	1.000	1.000	1.000	1.000	1.000	1.000	1.000	1.000
n = 25	0	0.277	0.072	0.004	0.000	0.000	0.000	0.000	0.000	0.000	0.000	0.000
	1	0.642	0.271	0.027	0.002	0.000	0.000	0.000	0.000	0.000	0.000	0.000
	2	0.873	0.537	0.098	0.009	0.000	0.000	0.000	0.000	0.000	0.000	0.000
	3	0.966	0.764	0.234	0.033	0.002	0.000	0.000	0.000	0.000	0.000	0.000
	4	0.993	0.902	0.421	0.090	0.009	0.000	0.000	0.000	0.000	0.000	0.000
	5	0.999	0.967	0.617	0.193	0.029	0.002	0.000	0.000	0.000	0.000	0.000
	6	1.000	0.991	0.780	0.341	0.074	0.007	0.000	0.000	0.000	0.000	0.000
	7	1.000	0.998	0.891	0.512	0.154	0.022	0.001	0.000	0.000	0.000	0.000
	8	1.000	1.000	0.953	0.677	0.274	0.054	0.004	0.000	0.000	0.000	0.000
	9	1.000	1.000	0.983	0.811	0.425	0.115	0.013	0.000	0.000	0.000	0.000
	10	1.000	1.000	0.994	0.902	0.586	0.212	0.034	0.002	0.000	0.000	0.000
	11	1.000	1.000	0.998	0.956	0.732	0.345	0.078	0.006	0.000	0.000	0.000
	12	1.000	1.000	1.000	0.983	0.846	0.500	0.154	0.017	0.000	0.000	0.000
	13	1.000	1.000	1.000	0.994	0.922	0.655	0.268	0.044	0.002	0.000	0.000
	14	1.000	1.000	1.000	0.998	0.966	0.788	0.414	0.098	0.006	0.000	0.000
	15	1.000	1.000	1.000	1.000	0.987	0.885	0.575	0.189	0.017	0.000	0.000
	16	1.000	1.000	1.000	1.000	0.996	0.946	0.726	0.323	0.047	0.000	0.000
	17	1.000	1.000	1.000	1.000	0.999	0.978	0.846	0.488	0.109	0.002	0.000
	18	1.000	1.000	1.000	1.000	1.000	0.993	0.926	0.659	0.220	0.009	0.000
	19	1.000	1.000	1.000	1.000	1.000	0.998	0.971	0.807	0.383	0.033	0.001
	20	1.000	1.000	1.000	1.000	1.000	1.000	0.991	0.910	0.579	0.098	0.007
	21	1.000	1.000	1.000	1.000	1.000	1.000	0.998	0.967	0.766	0.236	0.034
	22	1.000	1.000	1.000	1.000	1.000	1.000	1.000	0.991	0.902	0.463	0.127
	23	1.000	1.000	1.000	1.000	1.000	1.000	1.000	0.998	0.973	0.729	0.358
	24	1.000	1.000	1.000	1.000	1.000	1.000	1.000	1.000	0.996	0.928	0.723
	25	1.000	1.000	1.000	1.000	1.000	1.000	1.000	1.000	1.000	1.000	1.000

BINOMIAL COEFFICIENTS

n	$\binom{n}{0}$	$\binom{n}{1}$	$\binom{n}{2}$	$\binom{n}{3}$	$\binom{n}{4}$	$\binom{n}{5}$	$\binom{n}{6}$	$\binom{n}{7}$	$\binom{n}{8}$	$\binom{n}{9}$	$\binom{n}{10}$	$\binom{n}{11}$
0	1											
1	1	1										
2	1	2	1									
3	1	3	3	1								
4	1	4	6	4	1							
5	1	5	10	10	5	1						
6	1	6	15	20	15	6	1					
7	1	7	21	35	35	21	7	1				
8	1	8	28	56	70	56	28	8	1			
9	1	9	36	84	126	126	84	36	9	1		
10	1	10	45	120	210	252	210	120	45	10	1	
11	1	11	55	165	330	462	462	330	165	55	11	1
12	1	12	66	220	495	792	924	792	495	220	66	12
13	1	13	78	286	715	1287	1716	1716	1287	715	286	78
14	1	14	91	364	1001	2002	3003	3432	3003	2002	1001	364
15	1	15	105	455	1365	3003	5005	6435	6435	5005	3003	1365
16	1	16	120	560	1820	4368	8008	11440	12870	11440	8008	4368
17	1	17	136	680	2380	6188	12376	19448	24310	24310	19448	12376
18	1	18	153	816	3060	8568	18564	31824	43758	48620	43758	31824
19	1	19	171	969	3876	11628	27132	50388	75582	92378	92378	75582
20	1	20	190	1140	4845	15504	38760	77520	125970	167960	184756	167960
21	1	21	210	1330	5985	20349	54264	116280	203490	293930	352716	352716
22	1	22	231	1540	7315	26334	74613	170544	319770	497420	646646	705432

CUMULATIVE POISSON PROBABILITIES

Cumulative poisson probabilities are given by

$$\sum_{k=0}^{x} e^{-\lambda} \frac{\lambda^k}{k!}$$

The following table gives the probability that k *or more* random events are contained in an interval when the average number of such events per interval is λ.

m =	0.1	0.2	0.3	0.4	0.5	0.6	0.7	0.8	0.9	1.0
r = 0	1.0000	1.0000	1.0000	1.0000	1.0000	1.0000	1.0000	1.0000	1.0000	1.0000
1	.0952	.1813	.2592	.3297	.3935	.4512	.5034	.5507	.5934	.6321
2	.0047	.0175	.0369	.0616	.0902	.1219	.1558	.1912	.2275	.2642
3	.0002	.0011	.0036	.0079	.0144	.0231	.0341	.0474	.0629	.0803
4		.0001	.0003	.0008	.0018	.0034	.0058	.0091	.0135	.0190
5				.0001	.0002	.0004	.0008	.0014	.0023	.0037
6							.0001	.0002	.0003	.0006
7										.0001

m =	1.1	1.2	1.3	1.4	1.5	1.6	1.7	1.8	1.9	2.0
r = 0	1.0000	1.0000	1.0000	1.0000	1.0000	1.0000	1.0000	1.0000	1.0000	1.0000
1	.6671	.6988	.7275	.7534	.7769	.7981	.8173	.8347	.8504	.8647
2	.3010	.3374	.3732	.4082	.4422	.4751	.5068	.5372	.5663	.5940
3	.0996	.1205	.1429	.1665	.1912	.2166	.2428	.2694	.2963	.3233
4	.0257	.0338	.0431	.0537	.0656	.0788	.0932	.1087	.1253	.1429
5	.0054	.0077	.0107	.0143	.0186	.0237	.0296	.0364	.0441	.0527
6	.0010	.0015	.0022	.0032	.0045	.0060	.0080	.0104	0132	.0166
7	.0001	.0003	.0004	.0006	.0009	.0013	.0019	.0026	.0034	.0045
8			.0001	.0001	.0002	.0003	.0004	.0006	.0008	.0011
9							.0001	.0001	.0002	.0002

m =	2.1	2.2	2.3	2.4	2.5	2.6	2.7	2.8	2.9	3.0
r = 0	1.0000	1.0000	1.0000	1.0000	1.0000	1.0000	1.0000	1.0000	1.0000	1.0000
1	.8775	.8892	.8997	.9093	.9179	.9257	.9328	.9392	.9450	.9502
2	.6204	.6454	.6691	.6916	.7127	.7326	.7513	.7689	.7854	.8009
3	.3504	.3773	.4040	.4303	.4562	.4816	.5064	.5305	.5540	.5768
4	.1614	.1806	.2007	.2213	.2424	.2640	.2859	.3081	.3304	.3528
5	.0621	.0725	.0838	.0959	.1088	.1226	.1371	.1523	.1682	.1847
6	.0204	.0249	.0300	.0357	.0420	.0490	.0567	.0651	.0742	.0839
7	.0059	.0075	.0094	.0116	.0142	.0172	.0206	.0244	.0287	.0335
8	.0015	.0020	.0026	.0033	.0042	.0053	.0066	.0081	.0099	.0119
9	.0003	.0005	.0006	.0009	.0011	.0015	.0019	.0024	.0031	.0038
10	.0001	.0001	.0001	.0002	.0003	.0004	.0005	.0007	.0009	.0011
11					.0001	.0001	.0001	.0002	.0002	.0003
12									.0001	.0001

m =	3.1	3.2	3.3	3.4	3.5	3.6	3.7	3.8	3.9	4.0
r = 0	1.0000	1.0000	1.0000	1.0000	1.0000	1.0000	1.0000	1.0000	1.0000	1.0000
1	.9550	.9592	.9631	.9666	.9698	.9727	.9753	.9776	.9798	.9817
2	.8153	.8288	.8414	.8532	.8641	.8743	.8838	.8926	.9008	.9084
3	.5988	.6201	.6406	.6603	.6792	.6973	.7146	.7311	.7469	.7619
4	.3752	.3975	.4197	.4416	.4634	.4848	.5058	.5265	.5468	.5665

Cumulative Poisson Probabilities (*Contd.*)

m =	3.1	3.2	3.3	3.4	3.5	3.6	3.7	3.8	3.9	4.0
5	.2018	.2194	.2374	.2558	.2746	.2936	.3128	.3322	.3516	.3712
6	.0943	.1054	.1171	.1295	.1424	.1559	.1699	.1844	.1994	.2149
7	.0388	.0446	.0510	.0579	.0653	.0733	.0818	.0909	.1005	.1107
8	.0142	.0168	.0198	.0231	.0267	.0308	.0352	.0401	.0454	.0511
9	.0047	.0057	.0069	.0083	.0099	.0117	.0137	.0160	.0185	.0214
10	.0014	.0018	.0022	.0027	.0033	.0040	.0048	.0058	.0069	.0081
11	.0004	.0005	.0006	.0008	.0010	.0013	.0016	.0019	.0023	.0028
12	.0001	.0001	.0002	.0002	.0003	.0004	.0005	.0006	.0007	.0009
13				.0001	.0001	.0001	.0001	.0002	.0002	.0003
14									.0001	.0001

m =	4.1	4.2	4.3	4.4	4.5	4.6	4.7	4.8	4.9	5.0
r = 0	1.0000	1.0000	1.0000	1.0000	1.0000	1.0000	1.0000	1.0000	1.0000	1.0000
1	.9834	.9850	.9864	.9877	.9889	.9899	.9909	.9918	.9926	.9933
2	.9155	.9220	.9281	.9337	.9389	.9437	.9482	.9523	.9561	.9596
3	.7762	.7898	.8026	.8149	.8264	.8374	.8477	.8575	.8667	.8753
4	.5858	.6046	.6228	.6406	.6577	.6743	.6903	.7058	.7207	.7350
5	.3907	.4102	.4296	.4488	.4679	.4868	.5054	.5237	.5418	.5595
6	.2307	.2469	.2633	.2801	.2971	.3142	.3316	.3490	.3665	.3840
7	.1214	.1325	.1442	.1564	.1689	.1820	.1954	.2092	.2233	.2378
8	.0573	.0639	.0710	.0786	.0866	.0951	.1040	.1133	.1231	.1334
9	.0245	.0279	.0317	.0358	.0403	.0451	.0503	.0558	.0618	.0681
10	.0095	.0111	.0129	.0149	.0171	.0195	.0222	.0251	.0283	.0318
11	.0034	.0041	.0048	.0057	.0067	.0078	.0090	.0104	.0120	.0137
12	.0011	.0014	.0017	.0020	.0024	.0029	.0034	.0040	.0047	.0055
13	.0003	.0004	.0005	.0007	.0008	.0010	.0012	.0014	.0017	.0020
14	.0001	.0001	.0002	.0002	.0003	.0003	.0004	.0005	.0006	.0007
15				.0001	.0001	.0001	.0001	.0001	.0002	.0002
16									.0001	.0001

m =	5.2	5.4	5.6	5.8	6.0	6.2	6.4	6.6	6.8	7.0
r = 0	1.0000	1.0000	1.0000	1.0000	1.0000	1.0000	1.0000	1.0000	1.0000	1.0000
1	.9945	.9955	.9963	.9970	.9975	.9980	.9983	.9986	.9989	.9991
2	.9658	.9711	.9756	.9794	.9826	.9854	.9877	.9897	.9913	.9927
3	.8912	.9052	.9176	.9285	.9380	.9464	.9537	.9600	.9656	.9704
4	.7619	.7867	.8094	.8300	.8488	.8658	.8811	.8948	.9072	.9182
5	.5939	.6267	.6579	.6873	.7149	.7408	.7649	.7873	.8080	.8270
6	.4191	.4539	.4881	.5217	.5543	.5859	.6163	.6453	.6730	.6993
7	.2676	.2983	.3297	.3616	.3937	.4258	.4577	.4892	.5201	.5503
8	.1551	.1783	.2030	.2290	.2560	.2840	.3127	.3419	.3715	.4013
9	0819	.0974	.1143	.1328	.1528	.1741	.1967	.2204	.2452	.2709
10	.0397	.0488	.0591	.0708	.0839	.0984	.1142	.1314	.1498	.1695
11	.0177	.0225	.0282	.0349	.0426	.0514	.0614	.0726	.0849	.0985
12	.0073	.0096	.0125	.0160	.0201	.0250	.0307	.0373	.0448	.0534
13	.0028	.0038	.0051	.0068	.0088	.0113	.0143	.0179	.0221	.0270
14	.0010	.0014	.0020	.0027	.0036	.0048	.0063	.0080	.0102	.0128
15	.0003	.0005	.0007	.0010	.0014	.0019	.0026	.0034	.0044	.0057
16	.0001	.0002	.0002	.0004	.0005	.0007	.0010	.0014	.0018	.0024
17		.0001	.0001	.0001	.0002	.0003	.0004	.0005	.0007	.0010
18					.0001	.0001	.0001	.0002	.0003	.0004
19								.0001	.0001	.0001

Cumulative Poisson Probabilities (*Contd.*)

m =	7.2	7.4	7.6	7.8	8.0	8.2	8.4	8.6	8.8	9.0
r = 0	1.0000	1.0000	1.0000	1.0000	1.0000	1.0000	1.0000	1.0000	1.0000	1.0000
1	.9993	.9994	.9995	.9996	.9997	.9997	.9998	.9998	.9998	.9999
2	.9939	.9949	.9957	.9964	.9970	.9975	.9979	9982	.9985	.9988
3	.9745	.9781	.9812	.9839	.9862	.9882	.9900	.9914	.9927	.9938
4	.9281	.9368	.9446	.9515	.9576	.9630	.9677	.9719	.9756	.9788
5	.8445	.8605	.8751	.8883	.9004	.9113	.9211	.9299	.9379	.9450
6	.7241	.7474	.7693	.7897	.8088	.8264	.8427	.8578	.8716	.8843
7	.5796	.6080	.6354	.6616	.6866	.7104	.7330	.7543	.7744	.7932
8	.4311	.4607	.4900	.5188	.5470	.5746	.6013	.6272	.6522	.6761
9	.2973	.3243	.3518	.3796	.4075	.4353	.4631	.4906	.5177	.5443
10	.1904	.2123	.2351	.2589	.2834	.3085	.3341	.3600	.3863	.4126
11	.1133	.1293	.1465	.1648	.1841	.2045	.2257	.2478	.2706	.2940
12	.0629	.0735	.0852	.0980	.1119	.1269	.1429	.1600	.1780	.1970
13	.0327	.0391	.0464	.0546	.0638	.0739	.0850	.0971	.1102	.1242
14	.0159	.0195	.0238	.0286	.0342	.0405	.0476	.0555	.0642	.0739
15	.0073	.0092	.0114	.0141	.0173	.0209	.0251	.0299	.0353	.0415
16	.0031	.0041	.0052	.0066	.0082	.0102	.0125	.0152	.0184	.0220
17	.0013	.0017	.0022	.0029	.0037	.0047	.0059	.0074	.0091	.0111
18	.0005	.0007	.0009	.0012	.0016	.0021	.0027	.0034	.0043	.0053
19	.0002	.0003	.0004	.0005	.0006	.0009	.0011	.0015	.0019	.0024
20	.0001	.0001	.0001	.0002	.0003	.0003	.0005	.0006	.0008	.0011
21				.0001	.0001	.0001	.0002	.0002	.0003	.0004
22							.0001	.0001	.0001	.0002
23										.0001

m =	9.2	9.4	9.6	9.8	10.0	11.0	12.0	13.0	14.0	15.0
r = 0	1.0000	1.0000	1.0000	1.0000	1.0000	1.0000	1.0000	1.0000	1.0000	1.0000
1	.9999	.9999	.9999	.9999	1.0000	1.0000	1.0000	1.0000	1.0000	1.0000
2	.9990	.9991	.9993	.9994	.9995	.9998	.9999	1.0000	1.0000	1.0000
3	.9947	.9955	.9962	.9967	.9972	.9988	.9995	.9998	.9999	1.0000
4	.9816	.9840	.9862	.9880	.9897	.9951	.9977	.9990	.9995	.9998
5	.9514	.9571	.9622	.9667	.9707	.9849	.9924	.9963	.9982	.9991
6	.8959	.9065	.9162	.9250	.9329	.9625	.9797	.9893	.9945	.9972
7	.8108	.8273	.8426	.8567	.8699	.9214	.9542	.9741	.9858	.9924
8	.6990	.7208	.7416	.7612	.7798	.8568	.9105	.9460	.9684	.9820
9	.5704	.5958	.6204	.6442	.6672	.7680	.8450	.9002	.9379	.9626
10	.4389	.4651	.4911	.5168	.5421	.6595	.7576	.8342	.8906	.9301
11	.3180	.3424	.3671	.3920	.4170	.5401	.6528	.7483	.8243	.8815
12	.2168	.2374	.2588	.2807	.3032	.4207	.5384	.6468	.7400	.8152
13	.1393	.1552	.1721	.1899	.2084	.3113	.4240	.5369	.6415	.7324
14	.0844	.0958	.1081	.1214	.1355	.2187	.3185	.4270	.5356	.6368
15	.0483	.0559	.0643	.0735	.0835	.1460	.2280	.3249	.4296	.5343
16	.0262	.0309	.0362	.0421	.0487	.0926	.1556	.2364	.3306	.4319
17	.0135	.0162	.0194	.0230	.0270	.0559	.1013	.1645	.2441	.3359
18	.0066	.0081	.0098	.0119	.0143	.0322	.0630	.1095	.1728	.2511
19	.0031	.0038	.0048	.0059	.0072	.01*77	.0374	.0698	.1174	.1805
20	.0014	.0017	.0022	.0028	.0035	.0093	.0213	.0427	.0765	.1248
21	.0006	.0008	.0010	.0012	.0016	.0047	.0116	.0250	.0479	.0830
22	.0002	.0003	.0004	.0005	.0007	.0023	.0061	.0141	.0288	.0531
23	.0001	.0001	.0002	.0002	.0003	.0010	.0030	.0076	0167	.0327
24			.0001	.0001	.0001	.0005	.0015	.0040	.0093	.0195

Cumulative Poisson Probabilities (*Contd.*)

m =	9.2	9.4	9.6	9.8	10.0	11.0	12.0	13.0	14.0	15.0
25						.0002	.0007	.0020	.0050	.0122
26						.0001	.0003	.0010	.0026	.0062
27							.0001	.0005	.0013	.0033
28							.0001	.0002	.0006	.0017
29								.0001	.0003	.0009
30									.0001	.0004
31									.0001	.0002
32										.0001

m =	16.0	17.0	18.0	19.0	20.0	21.0	22.0	23.0	24.0	25.0
r = 0	1.0000	1.0000	1.0000	1.0000	1.0000	1.0000	1.0000	1.0000	1.0000	1.0000
1	1.0000	1.0000	1.0000	1.0000	1.0000	1.0000	1.0000	1.0000	1.0000	1.0000
2	1.0000	1.0000	1.0000	1.0000	1.0000	1.0000	1.0000	1.0000	1.0000	1.0000
3	1.0000	1.0000	1.0000	1.0000	1.0000	1.0000	1.0000	1.0000	1.0000	1.0000
4	.9999	1.0000	1.0000	1.0000	1.0000	1.0000	1.0000	1.0000	1.0000	1.0000
5	.9996	.9998	.9999	1.0000	1.0000	1.0000	1.0000	1.0000	1.0000	1.0000
6	.9986	.9993	.9997	.9998	.9999	1.0000	1.0000	1.0000	1.0000	1.0000
7	.9960	.9979	.9990	.9995	.9997	.9999	.9999	1.0000	1.0000	1.0000
8	.9900	.9946	.9971	.9985	.9992	.9996	.9998	.9999	1.0000	1.0000
9	.9780	.9874	.9929	.9961	.9979	.9989	.9994	.9997	.9998	.9999
10	.9567	.9739	.9846	.9911	.9950	.9972	.9985	.9992	.9996	.9998
11	.9226	.9509	.9696	.9817	.9892	.9937	.9965	.9980	.9989	.9994
12	.8730	.9153	.9451	.9653	.9786	.9871	.9924	.9956	.9975	.9986
13	.8069	.8650	.9083	.9394	.9610	.9755	.9849	.9909	.9946	.9969
14	.7255	.7991	.8574	.9016	.9339	.9566	.9722	.9826	.9893	.9935
15	.6325	.7192	.7919	.8503	.8951	.9284	.9523	.9689	.9802	.9876
16	.5333	.6285	.7133	.7852	.8435	.8889	.9231	.9480	.9656	.9777
17	.4340	.5323	.6249	.7080	.7789	.8371	.8830	.9179	.9437	.9623
18	.3407	.4360	.5314	.6216	.7030	.7730	.8310	.8772	.9129	.9395
19	.2577	.3450	.4378	.5305	.6186	.6983	.7675	.8252	.8717	.9080
20	.1878	.2637	.3491	.4394	.5297	.6157	.6940	.7623	.8197	.8664
21	.1318	.1945	.2693	.3528	.4409	.5290	.6131	.6899	.7574	.8145
22	.0892	.1385	.2009	.2745	.3563	.4423	.5284	.6106	.6861	.7527
23	.0582	.0953	.1449	.2069	.2794	.3595	.4436	.5277	.6083	.6825
24	.0367	.0633	.1011	.1510	.2125	.2840	.3626	.4449	.5272	.6061
25	.0223	.0406	.0683	.1067	.1568	.2178	.2883	.3654	.4460	.5266
26	.0131	.0252	.0446	.0731	.1122	.1623	.2229	.2923	.3681	.4471
27	.0075	.0152	.0282	.0486	.0779	.1174	.1676	.2277	.2962	.3706
28	.0041	.0088	.0173	.0313	.0525	.0825	.1225	.1726	.2323	.2998
29	.0022	.0050	.0103	.0195	.0343	.0564	.0871	.1274	.1775	.2366
30	.0011	.0027	.0059	.0118	.0218	.0374	.0602	.0915	.1321	.1821
31	.0006	.0014	.0033	.0070	.0135	.0242	.0405	.0640	.0958	.1367
32	.0003	.0007	.0018	.0040	.0081	.0152	.0265	.0436	.0678	.1001
33	.0001	.0004	.0010	.0022	.0047	.0093	.0169	.0289	.0467	.0715
34	.0001	.0002	.0005	.0012	.0027	.0055	.0105	.0187	.0314	.0498
35		.0001	.0002	.0006	.0015	.0032	.0064	.0118	.0206	.0338
36			.0001	.0003	.0008	.0018	.0038	.0073	.0132	.0225
37			.0001	.0002	.0004	.0010	.0022	.0044	.0082	.0146
38				.0001	.0002	.0005	.0012	.0026	.0050	.0092
39					.0001	.0003	.0007	.0015	.0030	.0057

Cumulative Poisson Probabilities (*Contd.*)

m =	16.0	17.0	18.0	19.0	20.0	21.0	22.0	23.0	24.0	25.0
40					.0001	.0001	.0004	.0008	.0017	.0034
41						.0001	.0002	.0004	.0010	.0020
42							.0001	.0002	.0005	.0012
43								.0001	.0003	.0007
44								.0001	.0002	.0004
45									.0001	.0002
46										.0001

m =	26.0	27.0	28.0	29.0	30.0	32.0	34.0	36.0	38.0	40.0
r = 9	1.0000	1.0000	1.0000	1.0000	1.0000	1.0000	1.0000	1.0000	1.0000	1.0000
10	.9999	.9999	1.0000	1.0000	1.0000	1.0000	1.0000	1.0000	1.0000	1.0000
11	.9997	.9998	.9999	1.0000	1.0000	1.0000	1.0000	1.0000	1.0000	1.0000
12	.9992	.9996	.9998	.9999	.9999	1.0000	1.0000	1.0000	1.0000	1.0000
13	.9982	.9990	.9994	.9997	.9998	1.0000	1.0000	1.0000	1.0000	1.0000
14	.9962	.9978	.9987	.9993	.9996	.9999	1.0000	1.0000	1.0000	1.0000
15	.9924	.9954	.9973	.9984	.9991	.9997	.9999	1.0000	1.0000	1.0000
16	.9858	.9912	.9946	.9967	.9981	.9993	.9998	.9999	1.0000	1.0000
17	.9752	.9840	.9899	.9937	.9961	.9986	.9995	.9998	1.0000	1.0000
18	.9580	.9726	.9821	.9885	.9927	.9972	.9990	.9997	.9999	1.0000
19	.9354	.9555	.9700	.9801	.9871	.9948	.9980	.9993	.9998	.9999
20	.9032	.9313	.9522	.9674	.9781	.9907	.9963	.9986	.9995	.9998
21	.8613	.8985	.9273	.9489	.9647	.9841	.9932	.9973	.9990	.9996
22	.8095	.8564	.8940	.9233	.9456	.9740	.9884	.9951	.9981	.9993
23	.7483	.8048	.8517	.8896	.9194	.9594	.9809	.9915	.9965	.9986
24	.6791	.7441	.8002	.8471	.8854	.9390	.9698	.9859	.9938	.9974
25	.6041	.6758	.7401	.7958	.8428	.9119	.9540	.9776	.9897	.9955
26	.5261	.6021	.6728	.7363	.7916	.8772	.9326	.9655	.9834	.9924
27	.4481	.5256	.6003	.6699	.7327	.8344	.9047	.9487	.9741	.9877
28	.3730	.4491	.5251	.5986	.6671	.7838	.8694	.9264	.9611	.9807
29	.3033	.3753	.4500	.5247	.5969	.7259	.8267	.8977	.9435	.9706
30	.2407	.3065	.3774	.4508	.5243	.6620	.7765	.8621	.9204	.9568
31	.1866	.2447	.3097	.3794	.4516	.5939	.7196	.8194	.8911	.9383
32	.1411	.1908	.2485	.3126	.3814	.5235	.6573	.7697	.8552	.9145
33	.1042	.1454	.1949	.2521	.3155	.4532	.5911	.7139	.8125	.8847
34	.0751	.1082	.1495	.1989	.2556	.3850	.5228	.6530	.7635	.8486
35	.0528	.0787	.1121	.1535	.2027	.3208	.4546	.5885	.7086	.8061
36	.0363	.0559	.0822	.1159	.1574	.2621	.3883	.5222	.6490	.7576
37	.0244	.0388	.0589	.0856	.1196	.2099	.3256	.4558	.5862	.7037
38	.0160	.0263	.0413	.0619	.0890	.1648	.2681	.3913	.5216	.6453
39	.0103	.0175	.0283	.0438	.0648	.1268	.2166	.3301	.4570	.5840
40	.0064	.0113	.0190	.0303	.0463	.0956	.1717	.2737	.3941	.5210
41	.0039	.0072	.0125	.0205	.0323	.0707	.1336	.2229	.3343	.4581
42	.0024	.0045	.0080	.0136	.0221	.0512	.1019	.1783	.2789	.3967
43	.0014	.0027	.0050	.0089	.0148	.0364	.0763	.1401	.2288	.3382
44	.0008	.0016	.0031	.0056	.0097	.0253	.0561	.1081	.1845	.2838
45	.0004	.0009	.0019	.0035	.0063	.0173	.0404	.0819	.1462	.2343
46	.0002	.0005	.0011	.0022	.0040	.0116	.0286	.0609	.1139	.1903
47	.0001	.0003	.0006	.0013	.0025	.0076	.0199	.0445	.0872	.1521
48	.0001	.0002	.0004	.0008	.0015	.0049	.0136	.0320	.0657	.1196
49		.0001	.0002	.0004	.0009	.0031	.0091	.0225	.0486	.0925

Cumulative Poisson Probabilities (*Contd.*)

m =	26.0	27.0	28.0	29.0	30.0	32.0	34.0	36.0	38.0	40.0
50			.0001	.0002	.0005	.0019	.0060	.0156	.0353	.0703
51			.0001	.0001	.0003	.0012	.0039	.0106	.0253	.0526
52				.0001	.0002	.0007	.0024	.0071	.0178	.0387
53					.0001	.0004	.0015	.0047	.0123	.0281
54					.0001	.0002	.0009	.0030	.0084	.0200
55						.0001	.0006	.0019	.0056	.0140
56						.0001	.0003	.0012	.0037	.0097
57							.0002	.0007	.0024	.0066
58							.0001	.0005	.0015	.0044
59							.0001	.0003	.0010	.0029
60								.0002	.0006	.0019
61								.0001	.0004	.0012
62								.0001	.0002	.0008
63									.0001	.0005
64									.0001	.0003
65										.0002
66										.0001
67										.0001

Note: For values of λ greater than 30, use the table of areas under the Normal curve to obtain approximate Poisson probabilities, putting $\mu = \lambda$ and $\sigma = \sqrt{\lambda}$.

AREAS IN UPPER TAIL OF THE NORMAL DISTRIBUTION

If $\Phi(z)$ be the cumulative distribution function of a standardised Normal variable, z then

$1-\Phi(z)=\frac{1}{\sqrt{2\pi}}\int_z^\infty e^{-z^2/2}$ is the probability that a standardised Normal variate selected at random will be greater than a value of $z\left(=\frac{x-\mu}{\sigma}\right)$.

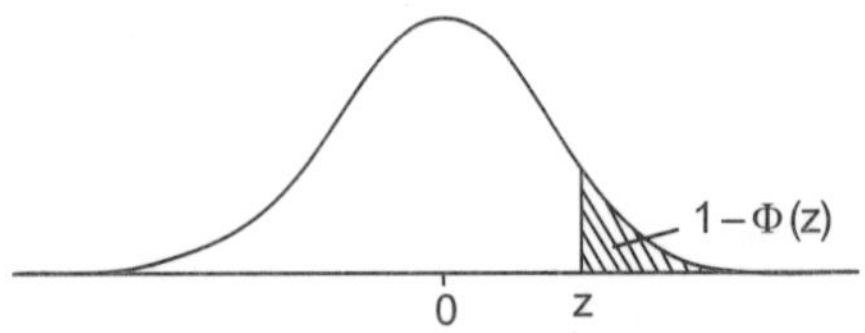

$\frac{x-u}{\sigma}$	.00	.01	.02	.03	.04	.05	.06	.07	.08	.09
0.0	.5000	4960	.4920	.4880	.4840	.4801	.4761	.4721	.4681	.4641
0.1	.4602	.4562	.4522	.4483	.4443	.4404	.4364	.4325	.4286	.4247
0.2	.4207	4168	.4129	.4090	.4052	.4013	.3974	.3936	.3897	.3859
0.3	.3821	3783	.3745	.3707	.3669	.3632	.3594	.3557	.3520	.3483
0.4	.3446	3409	.3372	.3336	.3300	.3264	.3228	.3192	.3156	.3121
0.5	.3085	3050	.3015	.2981	.2946	.2912	.2877	.2843	.2810	.2776
0.6	.2743	2709	.2676	.2643	.2611	.2578	.2546	.2514	.2483	.2451
0.7	.2420	.2389	.2358	.2327	.2296	.2266	.2236	.2206	.2177	.2148
0.8	.2119	.2090	.2061	.2033	.2005	.1977	.1949	.1922	.1894	.1867
0.9	.1841	.1814	.1788	.1762	.1736	.1711	.1685	.1660	.1635	.1611
1.0	.1587	.1562	.1539	.1515	.1492	.1469	.1446	.1423	.1401	.1379
1.1	.1357	.1335	.1314	.1292	.1271	.1251	.1230	.1210	.1190	.1170
1.2	.1151	.1131	.1112	.1093	.1075	.1056	.1038	.1020	.1003	.0985
1.3	.0968	.0951	.0934	.0918	.0901	.0885	.0869	.0853	.0838	.0823
1.4	.0808	.0793	.0778	.0764	.0749	.0735	.0721	.0708	.0694	.0681
1.5	.0668	.0655	.0643	.0630	.0618	.0606	.0594	.0582	.0571	.0559
1.6	.0548	.0537	.0526	.0516	.0505	.0495	.0485	.0475	.0465	.0455
1.7	.0446	.0436	.0427	.0418	.0409	.0401	.0392	.0384	.0375	.0367
1.8	.0359	.0351	.0344	.0336	.0329	.0322	.0314	.0307	.0301	.0294
1.9	.0287	.0281	.0274	.0268	.0262	.0256	.0250	.0244	.0239	.0233
2.0	.02275	.02222	.02169	.02118	.02068	.02018	.01970	.01923	.01876	.01831
2.1	.01786	.01743	.01700	.01659	.01618	.01578	.01539	.01500	.01463	.01426
2.2	.01390	.01355	.01321	.01287	.01255	.01222	.01191	.01160	.01130	.01101
2.3	.01072	.01044	.01017	.00990	.00964	.00939	.00914	.00889	.00866	.00842
2.4	.00820	.00798	.00776	.00755	.00734	.00714	.00695	.00676	.00657	.00639
2.5	.00621	.00604	.00587	.00570	.00554	.00539	.00523	.00508	.00494	.00480
2.6	.00466	.00453	.00440	.00427	.00415	.00402	.00391	.00379	.00368	.00357
2.7	.00347	.00336	.00326	.00317	.00307	.00298	.00289	.00280	.00272	.00264
2.8	.00256	.00248	.00240	.00233	.00226	.00219	.00212	.00205	.00199	.00193
2.9	.00187	.00181	.00175	.00169	.00164	.00159	.00154	.00149	.00144	.00139
3.0	.00135	.00131	.00126	.00122	.00118	.00114	.00111	.00107	.00104	.00100
3.1	.00097	.00094	.00090	.00087	.00084	.00082	.00079	.00076	.00074	.00071
3.2	.00069	.00066	.00064	.00062	.00060	.00058	.00056	.00054	.00052	.00050
3.3	.00048	.00047	.00045	.00043	.00042	.00040	.00039	.00038	.00036	.00035
3.4	.00034	.00032	.00031	.00030	.00029	.00028	.00027	.00026	.00025	.00024
3.5	.00023	.00022	.00022	.00021	.00020	.00019	.00019	.00018	.00017	.00017
3.6	.00016	.00015	.00015	.00014	.00014	.00013	.00013	.00012	.00012	.00011
3.7	.000108	.000104	.000100	.000096	.000092	.000088	.000085	.000082	.000078	.000075
3.8	.000072	.000069	.000067	.000064	.000062	.000059	.000057	.000054	.000052	.000050
3.9	.000048	.000046	.000044	.000042	.000041	.000039	.000037	.000036	.000034	.000033
4.0	.000032									

5.0 → 0.000 000 286 7 5.5 → 0.000 000 019 0 6.0 → 0.000 000 001 0

STANDARD NORMAL DISTRIBUTION FUNCTION

$$F(z) = \frac{1}{\sqrt{2\pi}} \int_{-\infty}^{z} e^{-t^2/2}\, dt$$

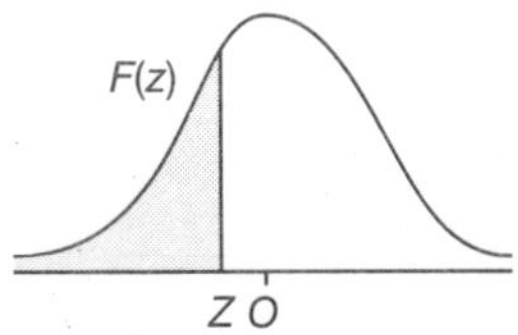

x	.00	.01	.02	.03	.04	.05	.06	.07	.08	.09
-3.9	.00005	.00005	.00004	.00004	.00004	.00004	.00004	.00004	.00003	.00003
-3.8	.00007	.00007	.00007	.00006	.00006	.00006	.00006	.00005	.00005	.00005
-3.7	.00011	.00010	.00010	.00010	.00009	.00009	.00008	.00008	.00008	.00008
-3.6	.00016	.00015	.00015	.00014	.00014	.00013	.00013	.00012	.00012	.00011
-3.5	.00023	.00022	.00022	.00021	.00020	.00019	.00019	.00018	.00017	.00017
-3.4	.00034	.00032	.00031	.00030	.00029	.00028	.00027	.00026	.00025	.00024
-3.3	.00048	.00047	.00045	.00043	.00042	.00040	.00039	.00038	.00036	.00035
-3.2	.00069	.00066	.00064	.00062	.00060	.00058	.00056	.00054	.00052	.00050
-3.1	.00097	.00094	.00090	.00087	.00084	.00082	.00079	.00076	.00074	.00071
-3.0	.00135	.00131	.00126	.00122	.00118	.00114	.00111	.00107	.00104	.00100
-2.9	.00187	.00181	.00175	.00169	.00164	.00159	.00154	.00149	.00144	.00139
-2.8	.00256	.00248	.00240	.00233	.00226	.00219	.00212	.00205	.00199	.00193
-2.7	.00347	.00336	.00326	.00317	.00307	.00298	.00289	.00280	.00272	.00264
-2.6	.00466	.00453	.00440	.00427	.00415	.00402	.00391	.00379	.00368	.00357
-2.5	.00621	.00604	.00587	.00570	.00554	.00539	.00523	.00508	.00494	.00480
-2.4	.00820	.00798	.00776	.00755	.00734	.00714	.00695	.00676	.00657	.00639
-2.3	.01072	.01044	.01017	.00990	.00964	.00939	.00914	.00889	.00866	.00842
-2.2	.01390	.01355	.01321	.01287	.01255	.01222	.01191	.01160	.01130	.01101
-2.1	.01786	.01743	.01700	.01659	.01618	.01578	.01539	.01500	.01463	.01426
-2.0	.02275	.02222	.02169	.02118	.02068	.02018	.01970	.01923	.01876	.01831
-1.9	.02872	.02807	.02743	.02680	.02619	.02559	.02500	.02442	.02385	.02330
-1.8	.03593	.03515	.03438	.03362	.03288	.03216	.03144	.03074	.03005	.02938
-1.7	.04457	.04363	.04272	.04182	.04093	.04006	.03920	.03836	.03754	.03673
-1.6	.05480	.05370	.05262	.05155	.05050	.04947	.04846	.04746	.04648	.04551
-1.5	.06681	.06552	.06426	.06301	.06178	.06057	.05938	.05821	.05705	.05592
-1.4	.08076	.07927	.07780	.07636	.07493	.07353	.07215	.07078	.06944	.06811
-1.3	.09680	.09510	.09342	.09176	.09012	.08851	.08691	.08534	.08379	.08226
-1.2	.11507	.11314	.11123	.10935	.10749	.10565	.10383	.10204	.10027	.09853
-1.1	.13567	.13350	.13136	.12924	.12714	.12507	.12302	.12100	.11900	.11702
-1.0	.15866	.15625	.15386	.15151	.14917	.14686	.14457	.14231	.14007	.13786
-0.9	.18406	.18141	.17879	.17619	.17361	.17106	.16853	.16602	.16354	.16109
-0.8	.21186	.20897	.20611	.20327	.20045	.19766	.19489	.19215	.18943	.18673
-0.7	.24196	.23885	.23576	.23270	.22965	.22663	.22363	.22065	.21770	.21476
-0.6	.27425	.27093	.26763	.26435	.26109	.25785	.25463	.25143	.24825	.24510
-0.5	.30854	.30503	.30153	.29806	.29460	.29116	.28774	.28434	.28096	.27760
-0.4	.34458	.34090	.33724	.33360	.32997	.32636	.32276	.31918	.31561	.31207
-0.3	.38209	.37828	.37448	.37070	.36693	.36317	.35942	.35569	.35197	.34827
-0.2	.42074	.41683	.41294	.40905	.40517	.40129	.39743	.39358	.38974	.38591
-0.1	.46017	.45620	.45224	.44828	.44433	.44038	.43644	.43251	.42858	.42465
-0.0	.50000	.49601	.49202	.48803	.48405	.48006	.47608	.47210	.46812	.46414

Standard Normal Distribution Function (*Contd.*)

$$F(z) = \frac{1}{\sqrt{2\pi}} \int_{-\infty}^{z} e^{-t^2/2}\, dt$$

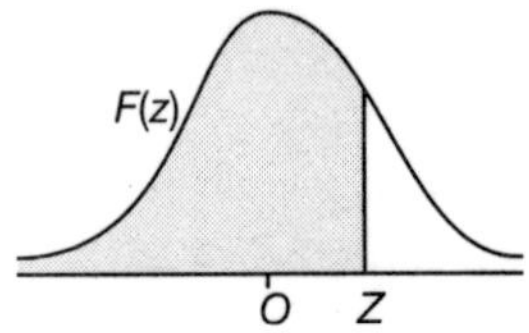

x	.00	.01	.02	.03	.04	.05	.06	.07	.08	.09
0.0	.50000	.50399	.50798	.51197	.51595	.51994	.52392	.52790	.53188	.53586
0.1	.53983	.54380	.54776	.55172	.55567	.55962	.56356	.56749	.57142	.57535
0.2	.57926	.58317	.58706	.59095	.59483	.59871	.60257	.60642	.61026	.61409
0.3	.61791	.62172	.62552	.62930	.63307	.63683	.64058	.64431	.64803	.65173
0.4	.65542	.65910	.66276	.66640	.67003	.67364	.67724	.68082	.68439	.68793
0.5	.69146	.69497	.69847	.70194	.70540	.70884	.71226	.71566	.71904	.72240
0.6	.72575	.72907	.73237	.73565	.73891	.74215	.74537	.74857	.75175	.75490
0.7	.75804	.76115	.76424	.76730	.77035	.77337	.77637	.77935	.78230	.78524
0.8	.78814	.79103	.79389	.79673	.79955	.80234	.80511	.80785	.81057	.81327
0.9	.81594	.81859	.82121	.82381	.82639	.82894	.83147	.83398	.83646	.83891
1.0	.84134	.84375	.84614	.84849	.85083	.85314	.85543	.85769	.85993	.86214
1.1	.86433	.86650	.86864	.87076	.87286	.87493	.87698	.87900	.88100	.88298
1.2	.88493	.88686	.88877	.89065	.89251	.89435	.89617	.89796	.89973	.90147
1.3	.90320	.90490	.90658	.90824	.90988	.91149	.91309	.91466	.91621	.91774
1.4	.91924	.92073	.92220	.92364	.92507	.92647	.92785	.92922	.93056	.93189
1.5	.93319	.93448	.93574	.93699	.93822	.93943	.94062	.94179	.94295	.94408
1.6	.94520	.94630	.94738	.94845	.94950	.95053	.95154	.95254	.95352	.95449
1.7	.95543	.95637	.95728	.95818	.95907	.95994	.96080	.96164	.96246	.96327
1.8	.96407	.96485	.96562	.96638	.96712	.96784	.96856	.96926	.96995	.97062
1.9	.97128	.97193	.97257	.97320	.97381	.97441	.97500	.97558	.97615	.97670
2.0	.97725	.97778	.97831	.97882	.97932	.97982	.98030	.98077	.98124	.98169
2.1	.98214	.98257	.98300	.98341	.98382	.98422	.98461	.98500	.98537	.98574
2.2	.98610	.98645	.98679	.98713	.98745	.98778	.98809	.98840	.98870	.98899
2.3	.98928	.98956	.98983	.99010	.99036	.99061	.99086	.99111	.99134	.99158
2.4	.99180	.99202	.99224	.99245	.99266	.99286	.99305	.99324	.99343	.99361
2.5	.99379	.99396	.99413	.99430	.99446	.99461	.99477	.99492	.99506	.99520
2.6	.99534	.99547	.99560	.99573	.99585	.99598	.99609	.99621	.99632	.99643
2.7	.99653	.99664	.99674	.99683	.99693	.99702	.99711	.99720	.99728	.99736
2.8	.99744	.99752	.99760	.99767	.99774	.99781	.99788	.99795	.99801	.99807
2.9	.99813	.99819	.99825	.99831	.99836	.99841	.99846	.99851	.99856	.99861
3.0	.99865	.99869	.99874	.99878	.99882	.99886	.99889	.99893	.99896	.99900
3.1	.99903	.99906	.99910	.99913	.99916	.99918	.99921	.99924	.99926	.99929
3.2	.99931	.99934	.99936	.99938	.99940	.99942	.99944	.99946	.99948	.99950
3.3	.99952	.99953	.99955	.99957	.99958	.99960	.99961	.99962	.99964	.99965
3.4	.99966	.99968	.99969	.99970	.99971	.99972	.99973	.99974	.99975	.99976
3.5	.99977	.99978	.99978	.99979	.99980	.99981	.99981	.99982	.99983	.99983
3.6	.99984	.99985	.99985	.99986	.99986	.99987	.99987	.99988	.99988	.99989
3.7	.99989	.99990	.99990	.99990	.99991	.99991	.99992	.99992	.99992	.99992
3.8	.99993	.99993	.99993	.99994	.99994	.99994	.99994	.99995	.99995	.99995
3.9	.99995	.99995	.99996	.99996	.99996	.99996	.99996	.99996	.99997	.99997

ORDINATES (Y) OF THE STANDARD NORMAL CURVE AT z

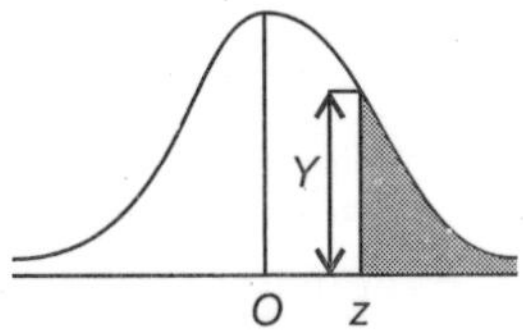

z	0.00	0.01	0.02	0.03	0.04	0.05	0.06	0.07	0.08	0.09
0.0	.3989	.3989	.3989	.3988	.3986	.3984	.3982	.3980	.3977	.3973
0.1	.3970	.3965	.3961	.3956	.3951	.3945	.3939	.3932	.3925	.3918
0.2	.3910	.3902	.3894	.3885	.3876	.3867	.3857	.3847	.3836	.3825
0.3	.3814	.3802	.3790	.3778	.3765	.3752	.3739	.3725	.3712	.3697
0.4	.3683	.3668	.3653	.3637	.3621	.3605	.3589	.3572	.3555	.3538
0.5	.3521	.3503	.3485	.3467	.3448	.3429	.3410	.3391	.3372	.3352
0.6	.3332	.3312	.3292	.3271	.3251	.3230	.3209	.3187	.3166	.3144
0.7	.3123	.3101	.3079	.3056	.3034	.3011	.2989	.2966	.2943	.2920
0.8	.2897	.2874	.2850	.2827	.2803	.2780	.2756	.2732	.2709	.2685
0.9	.2661	.2637	.2613	.2589	.2565	.2541	.2516	.2492	.2468	.2444
1.0	.2420	.2396	.2371	.2347	.2323	.2299	.2275	.2251	.2227	.2203
1.1	.2179	.2155	.2131	.2107	.2083	.2059	.2036	.2012	.1989	.1965
1.2	.1942	.1919	.1895	.1872	.1849	.1826	.1804	.1781	.1758	.1736
1.3	.1714	.1691	.1669	.1647	.1626	.1604	.1582	.1561	.1539	.1518
1.4	.1497	.1476	.1456	.1435	.1415	.1394	.1374	.1354	.1334	.1315
1.5	.1295	.1276	.1257	.1238	.1219	.1200	.1182	.1163	.1145	.1127
1.6	.1109	.1092	.1074	.1057	.1040	.1023	.1006	.0989	.0973	.0957
1.7	.0940	.0925	.0909	.0893	.0878	.0863	.0848	.0833	.0818	.0804
1.8	.0790	.0775	.0761	.0748	.0734	.0721	.0707	.0694	.0681	.0669
1.9	.0656	.0644	.0632	.0620	.0608	.0596	.0584	.0573	.0562	.0551
2.0	.0540	.0529	.0519	.0508	.0498	.0488	.0478	.0468	.0459	.0449
2.1	.0440	.0431	.0422	.0413	.0404	.0396	.0387	.0379	.0371	.0363
2.2	.0355	.0347	.0339	.0332	.0325	.0317	.0310	.0303	.0297	.0290
2.3	.0283	.0277	.0270	.0264	.0258	.0252	.0246	.0241	.0235	.0229
2.4	.0224	.0219	.0213	.0208	.0203	.0198	.0194	.0189	.0184	.0180
2.5	.0175	.0171	.0167	.0163	.0158	.0154	.0151	.0147	.0143	.0139
2.6	.0136	.0132	.0129	.0126	.0122	.0119	.0116	.0113	.0110	.0107
2.7	.0104	.0101	.0099	.0096	.0093	.0091	.0088	.0086	.0084	.0081
2.8	.0079	.0077	.0075	.0073	.0071	.0069	.0067	.0065	.0063	.0061
2.9	.0060	.0058	.0056	.0055	.0053	.0051	.0050	.0048	.0047	.0046
3.0	.0044	.0043	.0042	.0040	.0039	.0038	.0037	.0036	.0035	.0034
3.1	.0033	.0032	.0031	.0030	.0029	.0028	.0027	.0026	.0025	.0025
3.2	.0024	.0023	.0022	.0022	.0021	.0020	.0020	.0019	.0018	.0018
3.3	.0017	.0017	.0016	.0016	.0015	.0015	.0014	.0014	.0013	.0013
3.4	.0012	.0012	.0012	.0011	.0011	.0010	.0010	.0010	.0009	.0009
3.5	.0009	.0008	.0008	.0008	.0008	.0007	.0007	.0007	.0007	.0006
3.6	.0006	.0006	.0006	.0005	.0005	.0005	.0005	.0005	.0005	.0004
3.7	.0004	.0004	.0004	.0004	.0004	.0004	.0003	.0003	.0003	.0003
3.8	.0003	.0003	.0003	.0003	.0003	.0002	.0002	.0002	.0002	.0002
3.9	.0002	.0002	.0002	.0002	.0002	.0002	.0002	.0002	.0001	.0001

PERCENTAGE POINTS OF THE NORMAL DISTRIBUTION

The following table gives the 100α percentage points, z_α of a standardised normal distribution where

$$\alpha = \frac{1}{\sqrt{2\pi}} \int_{z_\alpha}^{\infty} e^{-z^2/2} dz.$$

Thus z_α is the value of a standardised normal variate which has probability α of being exceeded.

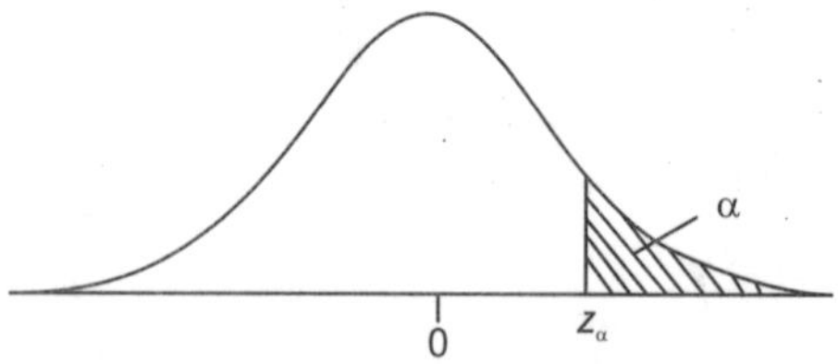

α	z_α	α	z_α	α	z_α	α	z_α	α	z_α	α	z_α
.50	0.0000	.050	1.6449	.030	1.8808	.020	2.0537	.010	23263	.050	1.6449
.45	0.1257	.048	1.6646	.029	1.8957	.019	2.0749	.009	2.3656	.010	2.3263
.40	0.2533	.046	1.6849	.028	1.9910	.018	2.0969	.008	2.4089	.001	3.0902
.35	0.3853	.044	1.7060	.027	1.9268	.017	2.1201	.007	2.4573	.000 1	3.7190
.30	0.5244	.042	1.7279	.026	1.9431	.016	2.1444	.006	2.5121	.000 01	4.2649
.25	0.6745	.040	1.7507	.025	1.9600	.015	2.1701	.005	2.5758	.025	1.9600
.20	0.8416	.038	1.7744	.024	1.9774	.014	2.1973	.004	2.6521	.005	2.5758
.15	1.0364	.036	1.7991	.023	1.9954	.013	2.2262	.003	2.7478	.000 5	3.2905
.10	1.2816	.034	1.8250	.022	2.0141	.012	2.2571	.002	2.8782	.000 05	3.8906
.05	1.6449	.032	1.8522	.021	2.0335	.011	2.2904	.001	3.0902	.000 005	4.4172

ORDINATES OF THE NORMAL DISTRIBUTION

The following table gives $\phi(z)$ for values of the standardised normal variate, z, in the interval 0.0 to 4.0, where $\phi(z) = \frac{1}{\sqrt{2\pi}} e^{-z^2/2}$.

z	.0	.1	.2	.3	.4	.5	.6	.7	.8	.9
0.0	.3989	.3970	.3910	.3814	.3683	.3521	.3332	.3123	.2897	.2661
1.0	.2420	.2179	.1942	.1714	.1497	.1295	.1109	.0940	.0790	.0656
2.0	.0540	.0440	.0355	.0283	.0224	.0175	.0136	.0104	.0079	.0060
3.0	.0044	.0033	.0024	.0017	.0012	.0009	.0006	.0004	.0003	.0002
4.0	.0001									

EXPONENTIAL FUNCTION e^x

For any negative exponential distribution, the tabulated function may be used to find the proportion of the distribution in excess of x times the mean. For an example, in random sampling of an exponential variate with a mean of 8, the probability that a single value will exceed 6 is 0.4724 since 6 is 0.75 times the distribution mean. Further, the 1% point of the distribution is 4.61 times the mean.

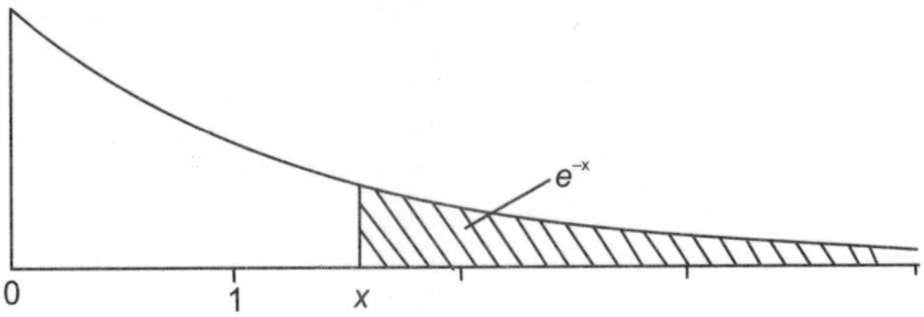

x	.0	.1	.2	.3	.4	.5	.6	.7	.8	.9
1.0	.3679	.3329	.3012	.2725	.2466	.2231	.2019	.1827	.1653	.1496
2.0	.1353	.1225	.1108	.1003	.0907	.0821	.0743	.0672	.0608	.0550
3.0	.0498	.0450	.0408	.0369	.0334	.0302	.0273	.0247	.0224	.0202
4.0	.0183	.0166	.0150	.0136	.0123	.0111	.0101	$.0^{2}910$	$.0^{2}823$	$.0^{2}745$
5.0	$.0^{2}674$	$.0^{2}610$	$.0^{2}552$	$.0^{2}499$	$.0^{2}452$	$.0^{2}409$	$.0^{2}370$	$.0^{2}335$	$.0^{2}303$	$.0^{2}274$
6.0	$.0^{2}248$	$.0^{2}224$	$.0^{2}203$	$.0^{2}184$	$.0^{2}166$	$.0^{2}150$	$.0^{2}136$	$.0^{2}123$	$0^{2}111$	$.0^{2}101$
7.0	$.0^{3}912$	$.0^{3}825$	$.0^{3}747$	$.0^{3}676$	$.0^{3}611$	$.0^{3}553$	$.0^{3}500$	$.0^{3}453$	$.0^{3}410$	$.0^{3}371$
8.0	$.0^{3}335$	$.0^{3}304$	$.0^{3}275$	$.0^{3}249$	$.0^{3}225$	$.0^{3}203$	$.0^{3}184$	$.0^{3}167$	$.0^{3}151$	$.0^{3}136$
9.0	$.0^{3}123$	$.0^{3}112$	$0^{3}101$	$.0^{4}914$	$.0^{4}827$	$.0^{4}749$	$.0^{4}677$	$.0^{4}613$	$,0^{4}555$	$.0^{4}502$
10.0	$.0^{4}454$	$.0^{4}411$	$.0^{4}372$	$.0^{4}336$	$.0^{4}304$	$.0^{4}275$	$.0^{4}249$	$.0^{4}225$	$.0^{4}204$	$.0^{4}185$
11.0	$.0^{4}167$	$0^{4}151$	$.0^{4}137$	$.0^{4}124$	$.0^{4}112$	$.0^{4}101$	$.0^{5}917$	$.0^{5}829$	$.0^{5}750$	$.0^{5}679$
12.0	$.0^{5}614$	$.0^{5}556$	$.0^{5}502$	$.0^{5}455$	$.0^{5}412$	$.0^{5}373$	$.0^{5}337$	$.0^{5}305$	$.0^{5}276$	$.0^{5}250$
13.0	$.0^{5}226$	$.0^{5}205$	$.0^{5}185$	$.0^{5}167$	$.0^{5}152$	$.0^{5}137$	$.0^{5}124$	$.0^{5}112$	$.0^{5}102$	$.0^{6}919$
14.0	$.0^{6}832$	$.0^{6}752$	$.0^{6}681$	$.0^{6}616$	$.0^{6}557$	$.0^{6}504$	$.0^{6}456$	$.0^{6}413$	$.0^{6}374$	$.0^{6}338$
15.0	$.0^{6}306$	$.0^{6}277$	$.0^{6}250$	$.0^{6}227$	$.0^{6}205$	$.0^{6}186$	$.0^{6}168$	$.0^{6}152$	$.0^{6}137$	$.0^{6}124$
16.0	$.0^{6}113$	$.0^{6}102$	$.0^{7}921$	$.0^{7}834$	$.0^{7}754$	$.0^{7}683$	$.0^{7}618$	$.0^{7}559$	$.0^{7}506$	$.0^{7}458$
17.0	$.0^{7}414$	$.0^{7}375$	$.0^{7}339$	$.0^{7}307$	$.0^{7}278$	$.0^{7}251$	$.0^{7}227$	$.0^{7}206$	$0^{7}186$	$0^{7}168$
18.0	$.0^{7}152$	$.0^{7}138$	$.0^{7}125$	$.0^{7}113$	$.0^{7}102$	$.0^{8}924$	$.0^{8}836$	$.0^{8}756$	$.0^{8}684$	$.0^{8}619$
19.0	$.0^{8}560$	$.0^{8}507$	$.0^{8}459$	$,0^{8}415$	$,0^{8}376$	$.0^{8}340$	$.0^{8}307$	$.0^{8}278$	$.0^{8}252$	$.0^{8}228$
20.0	$.0^{8}206$									

x	.00	.01	.02	.03	.04	.05	.06	.07	.08	.09
0	1.0000	.9900	.9802	.9704	.9608	.9512	.9418	.9324	.9231	.9139
.1	.9048	.8958	.8869	.8781	.8694	.8607	.8521	.8437	.8353	.8270
.2	.8187	.8106	.8025	.7945	.7866	.7788	.7711	.7634	.7558	.7483
.3	.7408	.7334	.7261	.7189	.7118	.7047	.6977	.6907	.6839	.6771
.4	.6703	.6636	.6570	.6505	.6440	.6376	.6313	.6250	.6188	.6126
.5	.6065	.6005	.5945	.5886	.5827	.5770	.5712	.5655	.5599	.5543
.6	.5488	.5434	.5379	.5326	.5273	.5220	.5169	.5117	.5066	.5016
.7	.4966	.4916	.4868	.4819	.4771	.4724	.4677	.4630	.4584	.4538
.8	.4493	.4449	.4404	.4360	.4317	.4274	.4232	.4190	.4148	.4107
.9	.4066	.4025	.3985	.3946	.3906	.3867	.3829	.3791	.3753	.3716
1.0	.3679	.3642	.3606	.3570	.3535	.3499	.3465	.3430	.3396	.3362
1.1	.3329	.3296	.3263	.3230	.3198	.3166	.3135	.3104	.3073	.3042
1.2	.3012	.2892	.2952	.2923	.2894	.2865	.2837	.2808	.2780	.2753

Exponential Function e^{-x} (*Contd.*)

x	.00	.01	.02	.03	.04	.05	.06	.07	.08	.09
1.3	.2725	.2698	.2671	.2645	.2618	.2592	.2567	.2541	.2516	.2491
1.4	.2466	.2441	.2417	.2393	.2369	.2346	.2322	.2299	.2276	.2254
1.5	.2231	.2209	.2187	.2165	.2144	.2122	.2101	.2080	.2060	.2039
1.6	.2019	.1999	.1979	.1959	.1940	.1920	.1901	.1882	.1864	.1845
1.7	.1827	.1809	.1791	.1773	.1755	.1738	.1720	.1703	.1686	.1670
1.8	.1653	.1637	.1620	.1604	.1588	.1572	.1557	.1541	.1526	.1511
1.9	.1496	.1481	.1466	.1451	.1437	.1423	.1409	.1395	.1381	.1367
2.0	.1353	.1340	.1327	.1313	.1300	.1287	.1275	.1262	.1249	.1237
2.1	.1225	.1212	.1200	.1188	.1177	.1165	.1153	.1142	.1130	.1119
2.2	.1108	.1097	.1086	.1075	.1065	.1054	.1044	.1035	.1023	.1013
2.3	.1003	.0993	.0983	.0973	.0963	.0954	.0944	.0935	.0926	.0916
2.4	.0907	.0898	.0889	.0880	.0872	.0863	.0854	.0846	.0837	.0829
2.5	.0821	.0813	.0805	.0797	.0789	.0781	.0773	.0765	.0758	.0750
2.6	.0743	.0735	.0728	.0721	.0714	.0707	.0699	.0693	.0686	.0679
2.7	.0672	.0665	.0659	.0652	.0646	.0639	.0633	.0627	.0620	.0614
2.8	.0608	.0602	.0596	.0590	.0584	.0578	.0573	.0567	.0561	.0556
2.9	.0550	.0545	.0539	.0534	.0529	.0523	.0518	.0513	.0508	.0503
3.0	.0498	.0493	.0488	.0483	.0478	.0474	.0469	.0464	.0460	.0455
3.1	.0450	.0446	.0442	.0437	.0433	.0429	.0424	.0420	.0416	.0412
3.2	.0408	.0404	.0400	.0396	.0392	.0388	.0384	.0380	.0376	.0373
3.3	.0369	.0365	.0362	.0358	.0354	.0351	.0347	.0344	.0340	.0337
3.4	.0334	.0330	.0327	.0324	.0321	.0317	.0314	.0311	.0308	.0305
3.5	.0302	.0299	.0296	.0293	.0290	.0287	.0284	.0282	.0279	.0276
3.6	.0273	.0271	.0268	.0265	.0263	.0260	.0257	.0255	.0252	.0250
3.7	.0247	.0245	.0242	.0240	.0238	.0235	.0233	.0231	.0228	.0226
3.8	.0224	.0221	.0219	.0217	.0215	.0213	.0211	.0209	.0207	.0204
3.9	.0202	.0200	.0198	.0196	.0194	.0193	.0191	.0189	.0187	.0185
4.0	.0183	.0181	.0180	.0178	.0176	.0174	.0172	.0171	.0169	.0167
4.1	.0166	.0164	.0162	.0161	.0159	.0158	.0156	.0155	.0153	.0151
4.2	.0150	.0148	.0147	.0146	.0144	.0143	.0141	.0140	.0138	.0137
4.3	.0136	.0134	.0133	.0132	.0130	.0129	.0128	.0127	.1025	.0124
4.4	.0123	.0122	.0120	.0119	.0118	.0117	.0116	.0114	.0113	.0112
4.5	.0111	.0110	.0109	.0108	.0107	.0106	.0105	.0104	.0103	.0102
4.6	.0101	.0100	.0099	.0098	.0097	.0096	.0095	.0094	.0093	.0092
4.7	.0091	.0090.	.0089	.0088	.0087	.0087	.0086	.0085	.0084	.0083
4.8	.0082	.0081	.0081	.0080	.0079	.0078	.0078	.0077	.0076	.0075
4.9	.0074	.0074	.0073	.0072	.0072	.0071	.0070	.0069	.0069	.0068
5.0	.0067									

Note: To keep each entry compact, the superscript digit refers to the number of zeros which precede the first significant figure. Thus $0.0^3123 = 0.000\ 123$.

PERCENTAGE POINTS OF THE t DISTRIBUTION

The following table gives the value of $t_{\alpha\nu}$ – the 100α percentage point of the t distribution for ν degrees of freedom. The values of t are obtained by solving the equation:

$$\alpha = \Gamma[\tfrac{1}{2}(\nu+1)][\Gamma(\tfrac{1}{2}\nu)]^{-1}\,(\nu\pi)^{-1/2}\int_t^{\infty}(1+x^2/\nu)^{-(\nu+1)/2}\,dx$$

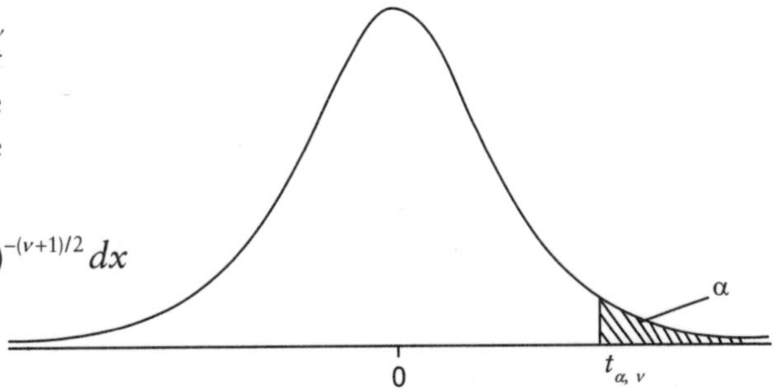

α =	0.10	0.05	0.025	0.01	0.005	0.001	0.0005
ν = 1	3.078	6.314	12.706	31.821	63.657	318.31	636.62
2	1.886	2.920	4.303	6.965	9.925	22.326	31.598
3	1.638	2.353	3.182	4.541	5.841	10.213	12.924
4	1.533	2.132	2.776	3.747	4.604	7.173	8.610
5	1.476	2.015	2.571	3.365	4.032	5.893	6.869
6	1.440	1.943	2.447	3.143	3.707	5.208	5.959
7	1.415	1.895	2.365	2.998	3.499	4.785	5.408
8	1.397	1.860	2.306	2.896	3.355	4.501	5.041
9	1.383	1.833	2.262	2.821	3.250	4.297	4.781
10	1.372	1.812	2.228	2.764	3.169	4.144	4.587
11	1.363	1.796	2.201	2.718	3.106	4.025	4.437
12	1.356	1.782	2.179	2.681	3.055	3.930	4.318
13	1.350	1.771	2.160	2.650	3.012	3.852	4.221
14	1.345	1.761	2.145	2.624	2.977	3.787	4.140
15	1.341	1.753	2.131	2.602	2.947	3.733	4.073
16	1.337	1.746	2.120	2.583	2.921	3.686	4.015
17	1.333	1.740	2.110	2.567	2.898	3.646	3.965
18	1.330	1.734	2.101	2.552	2.878	3.610	3.922
19	1.328	1.729	2.093	2.539	2.861	3.579	3.883
20	1.325	1.725	2.086	2.528	2.845	3.552	3.850
21	1.323	1.721	2.080	2.518	2.831	3.527	3.819
22	1.321	1.717	2.074	2.508	2.819	3.505	3.792
23	1.319	1.714	2.069	2.500	2.807	3.485	3.767
24	1.318	1.711	2.064	2.492	2.797	3.467	3.745
25	1.316	1.708	2.060	2.485	2.787	3.450	3.725
26	1.315	1.706	2.056	2.479	2.779	3.435	3.707
27	1.314	1.703	2.052	2.473	2.771	3.421	3.690
28	1.313	1.701	2.048	2.467	2.763	3.408	3.674
29	1.311	1.699	2.045	2.462	2.756	3.396	3.659
30	1.310	1.697	2.042	2.457	2.750	3.385	3.646
40	1.303	1.684	2.021	2.423	2.704	3.307	3.551
60	1.296	1.671	2.000	2.390	2.660	3.232	3.460
120	1.289	1.658	1.980	2.358	2.617	3.160	3.373
∞	1.282	1.645	1.960	2.326	2.576	3.090	3.291

Note: The above table is for one tail only, that is, for positive values of t. For $|t|$ the column headings for α should be doubled.

PERCENTAGE POINTS OF THE χ^2 DISTRIBUTION

The following table gives $\chi^2_{\alpha\nu}$ – the 100α percentage point of the χ^2 distribution for ν degrees of freedom.

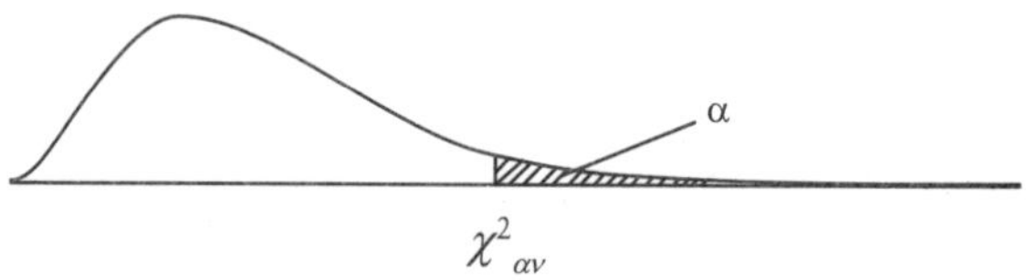

α =	.995	.99	.98	.975	.95	.90	.80	.75	.70	.50
ν = 1	$.0^4393$	$.0^3157$	$.0^3628$	$.0^3982$	.00393	.0158	.0642	.102	.148	.455
2	.0100	.0201	.0404	.0506	.103	.211	.446	.575	.713	1.386
3	.0717	.115	.185	.216	.352	.584	1.005	1.213	1.424	2.366
4	.207	.297	.429	.484	.711	1.064	1.649	1.923	2.195	3.357
5	.412	.554	.752	.831	1.145	1.610	2.343	2.675	3.000	4.351
6	.676	.872	1.134	1.237	1.635	2.204	3.070	3.455	3.828	5.348
7	.989	1.239	1.564	1.690	2.167	2.833	3.822	4.255	4.671	6.346
8	1.344	1.646	2.032	2.180	2.733	3.490	4.594	5.071	5.527	7.344
9	1.735	2.088	2.532	2.700	3.325	4.168	5.380	5.899	6.393	8.343
10	2.156	2.558	3.059	3.247	3.940	4.865	6.179	6.737	7.267	9.342
11	2.603	3.053	3.609	3.816	4.575	5.578	6.989	7.584	8.148	10.341
12	3.074	3.571	4.178	4.404	5.226	6.304	7.807	8.438	9.034	11.340
13	3.565	4.107	4.765	5.009	5.892	7.042	8.634	9.299	9.926	12.340
14	4.075	4.660	5.368	5.629	6.571	7.790	9.467	10.165	10.821	13.339
15	4.601	5.229	5.985	6.262	7.261	8.547	10.307	11.036	11.721	14.339
16	5.142	5.812	6.614	6.908	7.962	9.312	11.152	11.912	12.624	15.338
17	5.697	6.408	7.255	7.564	8.672	10.085	12.002	12.792	13.531	16.338
18	6.265	7.015	7.906	8.231	9.390	10.865	12.857	13.675	14.440	17.338
19	6.844	7.633	8.567	8.907	10.117	11.651	13.716	14.562	15.352	18.338
20	7.434	8.260	9.237	9.591	10.851	12.443	14.578	15.452	16.266	19.337
21	8.034	8.897	9.915	10.283	11.591	13.240	15.445	16.344	17.182	20.337
22	8.643	9.542	10.600	10.982	12.338	14.041	16.314	17.240	18.101	21.337
23	9.260	10.196	11.293	11.688	13.091	14.848	17.187	18.137	19.021	22.337
24	9.886	10.856	11.992	12.401	13.848	15.659	18.062	19.037	19.943	23.337
25	10.520	11.524	12.697	13.120	14.611	16.473	18.940	19.939	20.867	24.337
26	11.160	12.198	13.409	13.844	15.379	17.292	19.820	20.843	21.792	25.336
27	11.808	12.879	14.125	14.573	16.151	18.114	20.703	21.749	22.719	26.336
28	12.461	13.565	14.847	15.308	16.928	18.939	21.588	22.657	23.647	27.336
29	13.121	14.256	15.574	16.047	17.708	19.768	22.475	23.567	24.577	28.336
30	13.787	14.953	16.306	16.791	18.493	20.599	23.364	24.478	25.508	29.336
40	20.706	22.164	23.838	24.433	26.509	29.051	32.345	33.660	34.872	39.335
50	27.991	29.707	31.664	32.357	34.764	37.689	41.449	42.942	44.313	49.335
60	35.535	37.485	39.699	40.482	43.188	46.459	50.641	52.294	53.809	59.335
70	43.275	45.442	47.893	48.758	51.739	55.329	59.898	61.698	63.346	69.334
80	51.171	53.539	56.213	57.153	60.391	64.278	69.207	71.145	72.915	79.334
90	59.196	61.754	64.634	65.646	69.126	73.291	78.558	80.625	82.511	89.334
100	67.327	70.065	73.142	74.222	77.929	82.358	87.945	90.133	92.129	99.334

Percentage Points of the χ^2 Distribution (*Contd.*)

.30	.25	.20	.10	.05	.025	.02	.01	.005	.001	$= \alpha$
1.074	1.323	1.642	2.706	3.841	5.024	5.412	6.635	7.879	10.827	$\nu = 1$
2.408	2.773	3.219	4.605	5.991	7.378	7.824	9.210	10.597	13.815	2
3.665	4.108	4.642	6.251	7.815	9.348	9.837	11.345	12.838	16.268	3
4.878	5.385	5.989	7.779	9.488	11.143	11.668	13.277	14.860	18.465	4
6.064	6.626	7.289	9.236	11.070	12.832	13.388	15.086	16.750	20.517	5
7.231	7.841	8.558	10.645	12.592	14.449	15.033	16.812	18.548	22.457	6
8.383	9.037	9.803	12.017	14.067	16.013	16.622	18.475	20.278	24.322	7
9.524	10.219	11.030	13.362	15.507	17.535	18.168	20.090	21.955	26.125	8
10.656	11.389	12.242	14.684	16.919	19.023	19.679	21.666	23.589	27.877	9
11.781	12.549	13.442	15.987	18.307	20.483	21.161	23.209	25.188	29.588	10
12.899	13.701	14.631	17.275	19.675	21.920	22.618	24.725	26.757	31.264	11
14.011	14.845	15.812	18.549	21.026	23.337	24.054	26.217	28.300	32.909	12
15.119	15.984	16.985	19.812	22.362	24.736	25.472	27.688	29.819	34.528	13
16.222	17.117	18.151	21.064	23.685	26.119	26.873	29.141	31.319	36.123	14
17.322	18.245	19.311	22.307	24.996	27.488	28.259	30.578	32.801	37.697	15
18.418	19.369	20.465	23.542	26.296	28.845	29.633	32.000	34.267	39.252	16
19.511	20.489	21.615	24.769	27.587	30.191	30.995	33.409	35.718	40.790	17
20.601	21.605	22.760	25.989	28.869	31.526	32.346	34.805	37.156	42.312	18
21.689	22.718	23.900	27.204	30.144	32.852	33.687	36.191	38.582	43.820	19
22.775	23.828	25.038	28.412	31.410	34.170	35.020	37.566	39.997	45.315	20
23.858	24.935	26.171	29.615	32.671	35.479	36.343	38.932	41.401	46.797	21
24.939	26.039	27.301	30.813	33.924	36.781	37.659	40.289	42.796	48.268	22
26.018	27.141	28.429	32.007	35.172	38.076	38.968	41.638	44.181	49.728	23
27.096	28.241	29.553	33.196	36.415	39.364	40.270	42.980	45.558	51.179	24
28.172	29.339	30.675	34.382	37.652	40.646	41.566	44.314	46.928	52.620	25
29.246	30.434	31.795	35.563	38.885	41.923	42.856	45.642	48.290	54.052	26
30.319	31.528	32.912	36.741	40.113	43.194	44.140	46.963	49.645	55.476	27
31.391	32.620	34.027	37.916	41.337	44.461	45.419	48.278	50.993	56.893	28
32.461	33.711	35.139	39.087	42.557	45.722	46.693	49.588	52.336	58.302	29
33.530	34.800	36.250	40.256	43.773	46.979	47.962	50.892	53.672	59.703	30
44.165	45.616	47.269	51.805	55.759	59.342	60.436	63.691	66.766	73.402	40
54.723	56.334	58.164	63.167	67.505	71.420	72.613	76.154	79.490	86.661	50
65 227	66.981	68.972	74.397	79.082	83.298	84.580	88.379	91.952	99.607	60
75.689	77.577	79.715	85.527	90.531	95.023	96.388	100.425	104.215	112.317	70
86.120	88.130	90.405	96.578	101.880	106.629	108.069	112.329	116.321	124.839	80
96.524	98.650	101.054	107.565	113.145	118.136	119.648	124.116	128.299	137.208	90
106.906	109.141	111.667	118.498	124.342	129.561	131.142	135.807	140.170	149.449	100

Note: For values of $\nu > 30$, approximate values of χ^2 may be obtained from the expression $\nu\left[1-\frac{2}{9\nu}\pm\frac{x}{\sigma}\sqrt{\frac{2}{9\nu}}\right]^3$ where $\frac{x}{\sigma}$ is the normal deviate cutting off the corresponding tails of a normal distribution. If $\frac{x}{\sigma}$ is taken at the 0.02 level, so that 0.01 of the normal distribution is in each tail, the expression yields χ^2 at the 0.99 and 0.01 points.

For very large values of ν, it is sufficiently accurate to compute $\sqrt{2\chi^2}$, the distribution of which is approximately normal around a mean of $\sqrt{2\nu-1}$ and with a standard deviation of 1.

PERCENTAGE POINTS OF THE *F* DISTRIBUTION

The following table gives the values of $F_{a;\ v1,\ v2}$ the 100α percentage point of the F distribution having v_1 degrees of freedom in the numerator and v_2 degrees of freedom in the denominator. For each pair of values of v_1 and v_2, $F_{a;\ v1,\ v2}$ is tabulated for α = 0.05, 0.025, 0.01, 0.001, the 0.025 values being bracketed.

The lower percentage points of the distribution may be obtained from the relation:

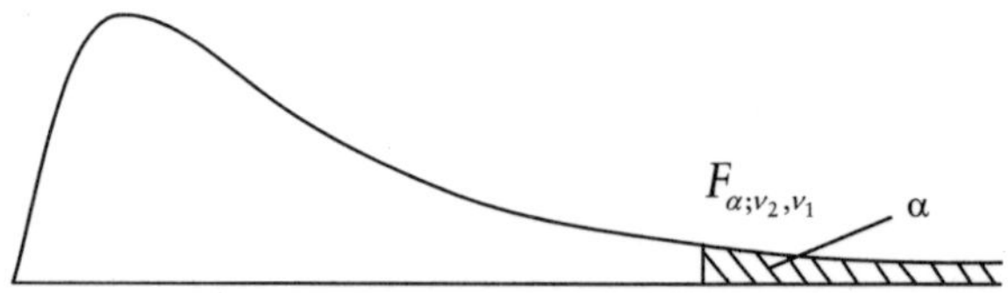

$$F_{1-\alpha;v_1,v_2} = 1 / F_{\alpha;v_2,v_1}$$

v_2 \ v_1	1	2	3	4	5	6	7	8	10	12	24	∞
1	161.4	199.5	215.7	224.6	230.2	234.0	236.8	238.9	241.9	243.9	249.0	254.3
	(648)	(800)	(864)	(900)	(922)	(937)	(948)	(957)	(969)	(977)	(997)	(1018)
	4052	5000	5403	5625	5764	5859	5928	5981	6056	6106	6235	6366
	4053*	5000*	5405*	5625*	5764*	5859*	5929*	5981*	6056*	6107*	6235*	6366*
2	18.5	19.0	19.2	19.2	19.3	19.3	19.4	19.4	19.4	19.4	19.5	19.5
	(38.5)	(39.0)	(39.2)	(39.2)	(39.3)	(39.3)	(39.4)	(39.4)	(39.4)	(39.4)	(39.5)	(39.5)
	98.5	99.0	99.2	99.2	99.3	99.3	99.4	99.4	99.4	99.4	99.5	99.5
	998.5	999.0	999.2	999.2	999.3	999.3	999.4	999.4	999.4	999.4	999.5	999.5
3	10.13	9.55	9.28	9.12	9.01	8.94	8.89	8.85	8.79	8.74	8.64	8.53
	(17.4)	(16.0)	(15.4)	(15.1)	(14.9)	(14.7)	(14.6)	(14.5)	(14.4)	(14.3)	(14.1)	(13.9)
	34.1	30.8	29.5	28.7	28.2	27.9	27.7	27.5	27.2	27.1	26.6	26.1
	167.0	148.5	141.1	137.1	134.6	132.8	131.5	130.6	129.2	128.3	125.9	123.5
4	7.71	6.94	6.59	6.39	6.26	6.16	6.09	6.04	5.96	5.91	5.77	5.63
	(12.22)	(10.65)	(9.98)	(9.60)	(9.36)	(9.20)	(9.07)	(8.98)	(8.84)	(8.75)	(8.51)	(8.26)
	21.2	18.0	16.7	16.0	15.5	15.2	15.0	14.8	14.5	14.4	13.9	13.5
	74.14	61.25	56.18	53.44	51.71	50.53	49.66	49.00	48.05	47.41	45.77	44.05
5	6.61	5.79	5.41	5.19	5.05	4.95	4.88	4.82	4.74	4.68	4.53	4.36
	(10.01)	(8.43)	(7.76)	(7.39)	(7.15)	(6.98)	(6.85)	(6.76)	(6.62)	(6.52)	(6.28)	(6.02)
	16.26	13.27	12.06	11.39	10.97	10.67	10.46	10.29	10.05	9.89	9.47	9.02
	47.18	37.12	33.20	31.09	29.75	28.83	28.16	27.65	26.92	26.42	25.14	23.79
6	5.99	5.14	4.76	4.53	4.39	4.28	4.21	4.15	4.06	4.00	3.84	3.67
	(8.81)	(7.26)	(6.60)	(6.23)	(5.99)	(5.82)	(5.70)	(5.60)	(5.46)	(5.37)	(5.12)	(4.85)
	13.74	10.92	9.78	9.15	8.75	8.47	8.26	8.10	7.87	7.72	7.31	6.88
	35.51	27.00	23.70	21.92	20.80	20.03	19.46	19.03	18.41	17.99	16.90	15.75
7	5.59	4.74	4.35	4.12	3.97	3.87	3.79	3.73	3.64	3.57	3.41	3.23
	(8.07	(6.54)	(5.89)	(5.52)	(5.29)	(5.12)	(4.99)	(4.90)	(4.76)	(4.67)	(4.42)	(4.14)
	12.25	9.55	8.45	7.85	7.46	7.19	6.99	6.84	6.62	6.47	6.07	5.65
	29.25	21.69	18.77	17.20	16.21	15.52	15.02	14.63	14.08	13.71	12.73	11.70
8	5.32	4.46	4.07	3.84	3.69	3.58	3.50	3.44	3.35	3.28	3.12	2.93
	(7.57)	(6.06)	(5.42)	(5.05)	(4.82)	(4.65)	(4.53)	(4.43)	(4.30)	(4.20)	(3.95)	(3.67)
	11.26	8.65	7.59	7.01	6.63	6.37	6.18	6.03	5.81	5.67	5.28	4.86
	25.42	18.49	15.83	14.39	13.48	12.86	12.40	12.05	11.54	11.19	10.30	9.34
9	5.12	4.26	3.86	3.63	3.48	3.37	3.29	3.23	3.14	3.07	2.90	2.71
	(7.21)	(5.71)	(5.08)	(4.72)	(4.48)	(4.32)	(4.2(5)	(4.10)	(3.96)	(3.87)	(3.61)	(3.33)
	10.56	8.02	6.99	6.42	6.06	5.80	5.61	5.47	5.26	5.11	4.73	4.31
	22.86	16.39	13.90	12.56	11.71	11.13	10.69	10.37	9.87	9.57	8.72	7.81

Percentage Points of the *F* Distribution (*Contd.*)

v_2 \ v_1	1	2	3	4	5	6	7	8	10	12	24	∞
10	4.96	4.10	3.71	3.48	3.33	3.22	3.14	3.07	2.98	2.91	2.74	2.54
	(6.94)	(5.46)	(4.83)	(4.47)	(4.24)	(4.07)	(3.95)	(3.85)	(3.72)	(3.62)	(3.37)	(3.08)
	10.04	7.56	6.55	5.99	5.64	5.39	5.20	5.06	4.85	4.71	4.33	3.91
	21.04	14.91	12.55	11.28	10.48	9.93	9.52	9.20	8.74	8.44	7.64	6.76
11	4.84	3.98	3.59	3.36	3.20	3.09	3.01	2.95	2.85	2.79	2.61	2.40
	(6.72)	(5.26)	(4.63)	(4.28)	(4.04)	(3.88)	(3.76)	(3.66)	(3.53)	(3.43)	(3.17)	(2.88)
	9.65	7.21	6.22	5.67	5.32	5.07	4.89	4.74	4.54	4.40	4.02	3.60
	19.69	13.81	11.56	10.35	9.58	9.05	8.66	8.35	7.92	7.63	6.85	6.00
12	4.75	3.89	3.49	3.26	3.11	3.00	2.91	2.85	2.75	2.69	2.51	2.30
	(6.55)	(5.10)	(4.47)	(4.12)	(3.89)	(3.73)	(3.61)	(3.51)	(3.37)	(3.28)	(3.02)	(2.72)
	9.33	6.93	5.95	5.41	5.06	4.82	4.64	4.50	4.30	4.16	3.78	3.36
	18.64	12.97	10.80	9.63	8.89	8.38	8.00	7.71	7.29	7.00	6.25	5.42
13	4.67	3.81	3.41	3.18	3.03	2.92	2.83	2.77	2.67	2.60	2.42	2.21
	(6.41)	(4.97)	(4.35)	(4.00)	(3.77)	(3.60)	(3.48)	(3.39)	(3.25)	(3.15)	(2.89)	(2.60)
	9.07	6.70	5.74	5.21	4.86	4.62	4.44	4.30	4.10	3.96	3.59	3.17
	17.82	12.31	10.21	9.07	8.35	7.86	7.49	7.21	6.80	6.52	5.78	4.97
14	4.60	3.74	3.34	3.11	2.96	2.85	2.76	2.70	2.60	2.53	2.35	2.13
	(6.30)	(4.86)	(4.24)	(3.89)	(3.66)	(3.50)	(3.38)	(3.29)	(3.15)	(3.05)	(2.79)	(2.49)
	8.86	6.51	5.56	5.04	4.70	4.46	4.28	4.14	3.94	3.80	3.43	3.00
	17.14	11.78	9.73	8.62	7.92	7.44	7.08	6.80	6.40	6.13	5.41	4.60
16	4.49	3.63	3.24	3.01	2.85	2.74	2.66	2.59	2.49	2.42	2.24	2.01
	(6.12)	(4.69)	(4.08)	(3.73)	(3.50)	(3.34)	(3.22)	(3.12)	(2.99)	(2.89)	(2.63)	(2.32)
	8.53	6.23	5.29	4.77	4.44	4.20	4.03	3.89	3.69	3.55	3.18	2.75
	16.12	10.97	9.01	7.94	7.27	6.80	6.46	6.19	5.81	5.55	4.85	4.06
18	4.41	3.55	3.16	2.93	2.77	2.66	2.58	2.51	2.41	2.34	2.15	1.92
	(5.98)	(4.56)	(3.95)	(3.61)	(3.38)	(3.22)	(3.10)	(3.01)	(2.87)	(2.77)	(2.50)	(2.19)
	8.29	6.01	5.09	4.58	4.25	4.01	3.84	3.71	3.51	3.37	3.00	2.57
	15.38	10.39	8.49	7.46	6.81	6.35	6.02	5.76	5.39	5.13	4.45	3.67
20	4.35	3.49	3.10	2.87	2.71	2.60	2.51	2.45	2.35	2.28	2.08	1.84
	(5.87)	(4.46)	(3.86)	(3.51)	(3.29)	(3.13)	(3.01)	(2.91)	(2.77)	(2.68)	(2.41)	(2.09)
	8.10	5.85	4.94	4.43	4.10	3.87	3.70	3.56	3.37	3.23	2.86	2.42
	14.82	9.95	8.10	7.10	6.46	6.02	5.69	5.44	5.08	4.82	4.15	3.38
22	4.30	3.44	3.05	2.82	2.66	2.55	2.46	2.40	2.30	2.23	2.03	1.78
	(5.79)	(4.38)	(3.78)	(3.44)	(3.22)	(3.05)	(2.93)	(2.84)	(2.70)	(2.60)	(2.33)	(2.00)
	7.95	5.72	4.82	4.31	3.99	3.76	3.59	3.45	3.26	3.12	2.75	2.31
	14.38	9.61	7.80	6.81	6.19	5.76	5.44	5.19	4.83	4.58	3.92	3.15
24	4.26	3.40	3.01	2.78	2.62	2.51	2.42	2.36	2.25	2.18	1.98	1.73
	(5.72)	(4.32)	(3.72	(3.38)	(3.15)	(2.99)	(2.87)	(2.78)	(2.64)	(2.54)	(2.27)	(1.94)
	7.82	5.61	4.72	4.22	3.90	3.67	3.50	3.36	3.17	3.03	2.66	2.21
	14.03	9.34	7.55	6.59	5.98	5.55	5.23	4.99	4.64	4.39	3.74	2.97
26	4.23	3.37	2.98	2.74	2.59	2.47	2.39	2.32	2.22	2.15	1.95	1.69
	(5.66)	(4.27)	(3.67)	(3.33)	(3.10)	(2.94)	(2.82)	(2.73)	(2.59)	(2.49)	(2.22)	(1.88)
	7.72	5.53	4.64	4.14	3.82	3.59	3.42	3.29	3.09	2.96	2.58	2.13
	13.74	9.12	7.36	6.41	5.80	5.38	5.07	4.83	4.48	4.24	3.59	2.82
28	4.20	3.34	2.95	2.71	2.56	2.45	2.36	2.29	2.19	2.12	1.91	1.65
	(5.61)	(4.22)	(3.63)	(3.29)	(3.06)	(2.90)	(2.78)	(2.69)	(2.55)	(2.45)	(2.17)	(1.83)
	7.64	5.45	4.57	4.07	3.75	3.53	3.36	3.23	3.03	2.90	2.52	2.06
	13.50	8.93	7.19	6.25	5.66	5.24	4.93	4.69	4.35	4.11	3.46	2.69

Percentage Points of the *F* Distribution (*Contd.*)

ν_1 / ν_2	1	2	3	4	5	6	7	8	10	12	24	∞
30	4.17	3.32	2.92	2.69	2.53	2.42	2.33	2.27	2.16	2.09	1.89	1.62
	(5.57)	(4.18)	(3.59)	(3.25)	(3.03)	(2.87)	(2.75)	(2.65)	(2.51)	(2.41)	(2.14)	(1.79)
	7.56	5.39	4.51	4.02	3.70	3.47	3.30	3.17	2.98	2.84	2.47	2.01
	13.29	8.77	7.05	6.12	5.53	5.12	4.82	4.58	4.24	4.00	3.36	2.59
40	4.08	3.23	2.84	2.61	2.45	2.34	2.25	2.18	2.08	2.00	1.79	1.51
	(5.42)	(4.05)	(3.46)	(3.13)	(2.90)	(2.74)	(2.62)	(2.53)	(2.39)	(2.29)	(2.01)	(1.64)
	7.31	5.18	4.31	3.83	3.51	3.29	3.12	2.99	2.80	2.66	2.29	1.80
	12.61	8.25	6.59	5.70	5.13	4.73	4.44	4.21	3.87	3.64	3.01	2.23
60	4.00	3.15	2.76	2.53	2.37	2.25	2.17	2.10	1.99	1.92	1.70	1.39
	(5.29)	(3.93)	(3.34)	(3.01)	(2.79)	(2.63)	(2.51)	(2.41)	(2.27)	(2.17)	(1.88)	(1.48)
	7.08	4.98	4.13	3.65	3.34	3.12	2.95	2.82	2.63	2.50	2.12	1.60
	11.97	7.77	6.17	5.31	4.76	4.37	4.09	3.86	3.54	3.32	2.69	1.89
120	3.92	3.07	2.68	2.45	2.29	2.18	2.09	2.02	1.91	1.83	1.61	1.25
	(5.15)	(3.80)	(3.23)	(2.89)	(2.67)	(2.52)	(2.39)	(2.30)	(2.16)	(2.05)	(1.76)	(1.31)
	6.85	4.79	3.95	3.48	3.17	2.96	2.79	2.66	2.47	2.34	1.95	1.38
	11.38	7.32	5.78	4.95	4.42	4.04	3.77	3.55	3.24	3.02	2.40	1.54
∞	3.84	3.00	2.60	2.37	2.21	2.10	2.01	1.94	1.83	1.75	1.52	1.00
	(5.02)	(3.69)	(3.12)	(2.79)	(2.57)	(2.41)	(2.29)	(2.19)	(2.05)	(1.94)	(1.64)	(1.00)
	6.63	4.61	3.78	3.32	3.02	2.80	2.64	2.51	2.32	2.18	1.79	1.00
	10.83	6.91	5.42	4.62	4.10	3.74	3.47	3.27	2.96	2.74	2.13	1.00

Note: Entries marked by * must be multiplied by 100.

PERCENTAGE POINTS OF THE CORRELATION COEFFICIENT

For a total correlation, v is 2 less than the number of pairs in the sample; for a partial correlation, the number of eliminated variates should also be subtracted. The probabilities at the head of the columns refer to the two-tail test of significance and give the chance that $|r|$ will be greater than the tabulated values given that the true product moment correlation, ρ, is zero. For a single-tail test the probabilities should be halved.

The following table gives the values of the correlation coefficient for different levels of significance.

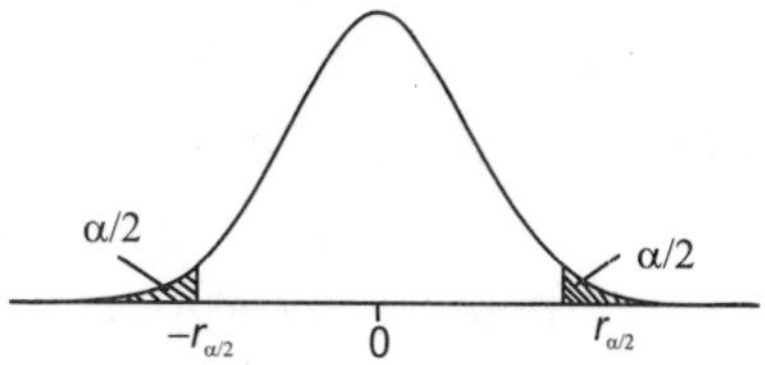

	0.1	0.05	0.02	0.01	0.001
$\nu = 1$	.98769	.99692	.999507	.999877	.9999988
2	.90000	.95000	.98000	.990000	.99900
3	.8054	.8783	.93433	.95873	.99116
4	.7293	.8114	.8822	.91720	.97406
5	.6694	.7545	.8329	.8745	.95074
6	.6215	.7067	.7887	.8343	.92493
7	.5822	.6664	.7498	.7977	.8982
8	.5494	.6319	.7155	.7646	.8721
9	.5214	.6021	.6851	.7348	.8471
10	.4973	.5760	.6581	.7079	.8233
11	.4762	.5529	.6339	.6835	.8010
12	.4575	.5324	.6120	.6614	.7800
13	.4409	.5139	.5923	.6411	.7603
14	.4259	.4973	.5742	.6226	.7420
15	.4124	.4821	.5577	.6055	.7246
16	.4000	.4683	.5425	.5897	.7084
17	.3887	.4555	.5285	.5751	.6932
18	.3783	.4438	.5155	.5614	.6787
19	.3687	.4329	.5034	.5487	.6652
20	.3598	.4227	.4921	.5368	.6524
25	.3233	.3809	.4451	.4869	.5974
30	.2960	.3494	.4093	.4487	.5541
35	.2746	.3246	.3810	.4182	.5189
40	.2573	.3044	.3578	.3932	.4896
45	.2428	.2875	.3384	.3721	.4648
50	.2306	.2732	.3218	.3541	.4433
60	.2108	.2500	.2948	.3248	.4078
70	.1954	.2319	.2737	.3017	.3799
80	.1829	.2172	.2565	.2830	.3568
90	.1726	.2050	.2422	.2673	.3375
100	.1638	.1946	.2301	.2540	.3211

TUKEY'S WHOLLY SIGNIFICANT DIFFERENCE AT 5% LEVEL

This table gives the values of the WSD/s $(\bar{x})$ where:

WSD = 'Wholly Significant Difference' between means at 5% level

s = Estimate of standard deviation

$s\ (\bar{x})$ = Estimate of standard error of the means = $s / \sqrt{n}$

k = Number of means under test

ν = Degrees of freedom associated with s^2, the estimate of error variance

n = Number of readings in each mean

k / ν	2	3	4	5	6	7	8	9	10	12	15	20	30	60
1	21.96	28.80	34.56	39.60	44.28	48.78	53.10	55.62	57.96	62.46	69.12	77.58	91.98	116.64
2	6.83	8.54	9.88	10.98	11.95	12.87	13.73	14.27	14.76	15.68	17.02	18.67	21.41	26.11
3	4.89	6.00	6.84	7.50	8.07	8.60	9.08	9.41	9.69	10.22	10.98	11.89	13.40	15.96
4	4.19	5.10	5.76	6.27	6.70	7.10	7.45	7.70	7.92	8.31	8.88	9.55	10.62	12.47
5	3.86	4.66	5.24	5.68	6.04	6.37	6.66	6.88	7.06	7.39	7.86	8.41	9.28	10.77
6	3.64	4.38	4.91	5.31	5.63	5.92	6.17	6.37	6.53	6.82	7.23	7.70	8.45	9.73
7	3.50	4.20	4.70	5.06	5.35	5.62	5.85	6.03	6.17	6.44	6.81	7.24	7.91	9.05
8	3.41	4.08	4.55	4.89	5.17	5.41	5.62	5.79	6.00	6.18	6.53	6.92	7.53	8.57
9	3.34	3.98	4.44	4.76	5.03	5.26	5.46	5.62	5.75	5.98	6.31	6.68	7.25	7.99
10	3.28	3.91	4.35	4.66	4.91	5.13	5.32	5.48	5.61	5.83	6.14	6.49	7.02	7.94
12	3.19	3.80	4.22	4.52	4.75	4.96	5.13	5.28	5.40	5.61	5.90	6.22	6.70	7.54
15	3.11	3.69	4.09	4.37	4.60	4.79	4.95	5.09	5.20	5.39	5.66	5.95	6.39	7.14
20	3.04	3.60	3.98	4.25	4.45	4.63	4.78	4.91	5.02	5.19	5.44	5.71	6.11	6.79
30	2.97	3.51	3.87	4.12	4.32	4.48	4.61	4.74	4.84	5.00	5.23	5.47	5.83	6.44
60	2.90	3.41	3.76	4.00	4.18	4.33	4.45	4.57	4.66	4.81	5.02	5.24	5.56	6.09
∞	2.82	3.32	3.66	3.88	4.04	4.18	4.29	4.40	4.49	4.63	4.82	5.01	5.29	5.76

PERCENTAGE POINTS OF SPEARMAN'S RANK CORRELATION COEFFICIENT

The Spearman's rank correlation coefficient is given by $r_s = 1 - \frac{6\Sigma d^2}{n(n^2-1)}$ where n is the number of observations in each of two series and d is the difference between the ranks of the corresponding observations in each series.

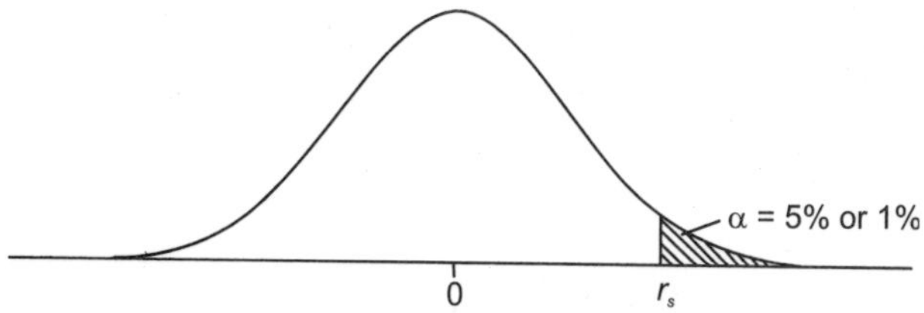

An observed value of r_s greater than or equal to that tabulated indicates a significant positive correlation between the ranks of the two series of observations at the level indicated (5% or 1%). For a two-sided test the probabilities should be doubled. For a significant negative correlation between ranks, an observed r_s should be negative but numerically greater than or equal to the tabulated value.

	(One-tailed test)	
α =	5%	1%
n = 4	1.000	—
5	0.900	1.000
6	0.829	0.943
7	0.714	0.893
8	0.643	0.833
9	0.600	0.783
10	0.564	0.746
12	0.506	0.712
14	0.456	0.645
16	0.425	0.601
18	0.399	0.564
20	0.377	0.534
22	0.359	0.508
24	0.343	0.485
26	0.329	0.465
28	0.317	0.448
30	0.306	0.432

When n is large (greater than about 10 say), the significance of an observed value of r_s under the null hypothesis may be tested using:

$$t \approx r_s\sqrt{\frac{n-2}{1-r_s^2}}$$

where t is Student's t with $n - 2$ degrees of freedom.

PERCENTAGE POINTS OF KENDALL'S RANK CORRELATION COEFFICIENT

The Kendall's rank correlation coefficient is given by $r_K = \dfrac{S}{\frac{n}{2}(n-1)}$ where n is the number of observations in each of two series, and S is the net score for the $\frac{n}{2}(n-1)$ pairs among the n ranks of the second set.

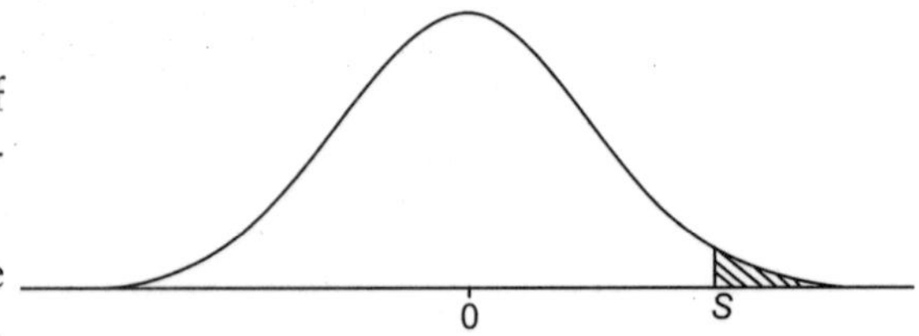

Note that the tabulation here is in terms of the numerator S which, for a given n, has a one-to-one correspondence with r_K.

The paired observations of ranks are rearranged so that one set is in ascending order. The corresponding ranks of the second set are examined two at a time, scoring +1 if the pair is in ascending order and –1 if it is in descending order. S is the net score for the $\frac{n}{2}(n-1)$ pairs amongst the n ranks of the second set.

The table gives the probability that the net total score, S, will be greater than or equal to any possible positive values when the ranks of the two sets are not correlated. Since the distribution of S, and also of r_K, is symmetrical, a negative correlation of ranks can be tested against the null hypothesis by interpreting the tabulated S values as negative. For a two-tailed test, the tabulated probability corresponding to the observed value of S should be doubled (except for $S = 0$).

S	4	5	8	9
0	.625	.592	.548	.540
2	.375	.408	.452	.460
4	.167	.242	.360	.381
6	.042	.117	.274	.306
8		.042	.199	.238
10		.0083	.138	.179
12			.089	.130
14			.054	.090
16			.031	.060
18			.016	.038
20			.0071	.022
22			.0028	.012
24			.0009	.0063
26			.0002	.0029
28				.0012
30				.0004

S	6	7	10
1	.500	.500	.500
3	.360	.386	.431
5	.235	.281	.364
7	.136	.191	.300
9	.068	.119	.242
11	.028	.068	.190
13	.0083	.035	.146
15	.0014	.015	.108
17		.0054	.078
19		.0014	.054
21		.0002	.036
23			.023
25			.014
27			.0083
29			.0046
31			.0023
33			.0011
35			.0005

For values of n greater than 10, it is sufficiently accurate to use the fact that r_K is approximately normal with mean of zero and variance of:

$$\frac{2(2n+5)}{9n(n-1)}$$

Thus r_K can be expressed as a standardised normal variate by writing:

$$z = \frac{\dfrac{S}{\frac{n}{2}(n-1)} - 0}{\sqrt{\left(\dfrac{2(2n+5)}{9n(n-1)}\right)}} = S\sqrt{\frac{18}{n(n-1)(2n+5)}}$$

PERCENTAGE POINTS OF NAIR'S 'STUDENTISED' EXTREME DEVIATE FROM THE MEAN

For a random sample of size n from a normal population, the lowest observation is $x_{(1)}$, the highest is $x_{(n)}$ and the arithmetic mean is $\bar{x}$. s is an independent estimate, external to the sample, of the population standard deviation and having ν degrees of freedom.

The maximum value of $\dfrac{|x_{(1)} - \bar{x}|}{s}$ and $\dfrac{|x_{(n)} - \bar{x}|}{s}$ gives a test criterion for identifying a single possible extreme value.

The following table gives the upper 5% and 1% points of the distribution of the statistic for n and ν. An observed value of the statistic greater than or equal to that tabulated suggests that $x_{(1)}$ or $x_{(n)}$ as appropriate is not from the same population as the remainder of the sample.

5% Significance Level

n =	3	4	5	6	7	8	9
ν = 10	2.02	2.29	2.49	2.63	2.75	2.85	2.93
11	1.99	2.26	2.44	2.58	2.70	2.79	2.87
12	1.97	2.22	2.40	2.54	2.65	2.75	2.83
13	1.95	2.20	2.38	2.51	2.62	2.71	2.79
14	1.93	2.18	2.35	2.48	2.59	2.68	2.76
15	1.92	2.16	2.33	246	2.56	2.65	2.73
16	1.90	2.14	2.31	2.44	2.54	2.63	2.70
17	1.89	2.13	2.30	2.42	2.52	2.61	2.68
18	1.88	2.12	2.28	2.41	2.51	2.59	2.66
19	1.87	2.11	2.27	2.39	2.49	2.58	2.65
20	1.87	2.10	2.26	2.38	2.48	2.56	2.63
24	1.84	2.07	2.23	2.35	2.44	2.52	2.59
30	1.82	2.04	2.20	2.31	2.40	2.48	2.55
40	1.80	2.02	2.17	2.28	2.37	2.44	2.51
60	1.78	1.99	2.14	2.25	2.33	2.41	2.47
120	1.76	1.97	2.11	2.21	2.30	2.37	2.43
∞	1.74	1.94	2.08	2.18	2.27	2.33	2.39

1% Significance Level

n =	3	4	5	6	7	8	9
ν = 10	2.76	3.05	3.25	3.39	3.50	3.59	3.67
11	2.71	3.00	3.19	3.33	3.44	3.53	3.61
12	2.67	2.95	3.14	3.28	3.39	3.48	3.55
13	2.63	2.91	3.10	3.24	3.34	3.43	3.51
14	2.60	2.87	3.06	3.20	3.30	3.39	3.47
15	2.57	2.84	3.02	3.16	3.27	3.35	3.43
16	2.55	2.81	3.00	3.13	3.24	3.32	3.39
17	2.53	2.79	2.97	3.10	3.21	3.29	3.36
18	2.50	2.77	2.95	3.08	3.'18	3.27	3.34
19	2.49	2.75	2.92	3.06	3.16	3.24	3.31
20	2.47	2.73	2.91	3.04	3.14	3.22	3.29
24	2.43	2.68	2.85	2.97	3.07	3.15	3.22
30	2.38	2.62	2.79	2.91	3.01	3.08	3.15
40	2.34	2.57	2.73	2.85	2.94	3.02	3.08
60	2.30	2.52	2.68	2.79	2.88	2.95	3.01
120	2.25	2.48	2.62	2.73	2.82	2.89	2.95
∞	2.22	2.43	2.57	2.68	2.76	2.83	2.88

UPPER PERCENTAGE POINTS OF DIXON'S RANK DIFFERENCE RATIO

The following table gives the upper 100 α percentage points of three rank difference ratios, each ratio being appropriate to a particular range of sample size, α = 0.10, 0.05 and 0.01. No estimate of the population standard deviation is necessary.

If a random sample of size n from a normal population is ranked in ascending order of magnitude resulting in $x_{(1)}, x_{(2)}, \ldots x_{(n)}$, then an observed value of the rank difference ratio greater than or equal to that tabulated indicates that the particular extreme value of x, that is $x_{(1)}$ or $x_{(n)}$, is not consistent with the remainder of the sample.

Rank difference ratio is given by

Rank Difference Ratio		
$\dfrac{x_{(2)} - x_{(1)}}{x_{(n)} - x_{(1)}}$	or	$\dfrac{x_{(n)} - x_{(n-1)}}{x_{(n)} - x_{(1)}}$
$\dfrac{x_{(3)} - x_{(1)}}{x_{(n-1)} - x_{(1)}}$	or	$\dfrac{x_{(n)} - x_{(n-2)}}{x_{(n)} - x_{(2)}}$
$\dfrac{x_{(3)} - x_{(1)}}{x_{(n-2)} - x_{(1)}}$	or	$\dfrac{x_{(n)} - x_{(n-2)}}{x_{(n)} - x_{(3)}}$

	Significance Level		
α =	.010	0.05	0.01
n = 3	0.886	0.941	0.988
4	0.679	0.765	0.889
5	0.557	0.642	0.780
6	0.482	0.560	0.698
7	0.434	0.507	0.637
8	0.650	0.710	0.829
9	0.594	0.657	0.776
10	0.551	0.612	0.726
11	0.517	0.576	0.679
12	0.490	0.546	0.642
13	0.467	0.521	0.615
14	0.448	0.501	0.593
15	0.472	0.525	0.616
16	0.454	0.507	0.595
17	0.438	0.490	0.577
18	0.424	0.475	0.561
19	0.412	0.462	0.547
20	0.401	0.450	0.535

PERCENTAGE POINTS OF *D* IN THE ONE-SAMPLE KOLMOGOROV-SMIRNOV DISTRIBUTION

D is the maximum value of $| F_{OBS}(x) - F_{EXP}(x) |$ where $F_{OBS}(x)$ and $F_{EXP}(x)$ are observed and expected (theoretical) cumulative probability distribution functions evaluated at each of a set of values of the observed variable *x*.

The following table gives the values of D_α, the 100α percentage point of the *D* distribution for $\alpha = 0.20, 0.15, 0.10, 0.05$ and 0.01.

n \ α	.20	.15	.10	.05	.01
1	.900	.925	.950	.975	.995
2	.684	.726	.776	.842	.929
3	.565	.597	.642	.708	.828
4	.494	.525	.564	.624	.733
5	.446	.474	.510	.565	.669
6	.410	.436	.470	.521	.618
7	.381	.405	.438	.486	.577
8	.358	.381	.411	.457	.543
9	.339	.360	.388	.432	.514
10	.322	.342	.368	.410	.490
11	.307	.326	.352	.391	.468
12	.295	.313	.338	.375	.450
13	.284	.302	.325	.361	.433
14	.274	.292	.314	.349	.418
15	.266	.283	.304	.338	.404
16	.258	.274	.295	.328	.392
17	.250	.266	.286	.318	.381
18	.244	.259	.278	.309	.371
19	.237	.252	.272	.301	.363
20	.231	.246	.264	.294	.356
25	.21	.22	.24	.27	.32
30	.19	.20	.22	.24	.29
35	.18	.19	.21	.23	.27
Over 35	$\frac{1.07}{\sqrt{n}}$	$\frac{1.14}{\sqrt{n}}$	$\frac{1.22}{\sqrt{n}}$	$\frac{1.36}{\sqrt{n}}$	$\frac{1.63}{\sqrt{n}}$

Values of *D* greater than or equal to those tabulated for the given sample size *n* and significance level α suggest that the sample has not been drawn from a population with the properties of the assumed theoretical distribution.

PERCENTAGE POINTS OF D IN THE TWO-SAMPLE KOLMOGOROV-SMIRNOV DISTRIBUTION

In the two-sample case, given two random samples of size n_1 and n_2 respectively from the same distribution of a continuous variate, x, D is defined as the maximum absolute value of $F_1(x) - F_2(x)$.

$F_1(x)$ and $F_2(x)$ are the two empirical cumulative relative frequency distribution functions evaluated at each of a set of values of the observed variable, x. It is recommended that the number of values of x, and hence the number of values of $|F_1(x) - F_2(x)|$, should be as large as feasible given the observed data, otherwise the test may be less powerful.

The following table gives the values of D_α, the 100α upper percentage points of the Kolmogorov–Smirnov two–sample test statistic for $\alpha = 0.05$, 0.025, 0.01 and 0.001.

		n_2											
n_1	α	2	3	4	5	6	7	8	9	10	11	12	13
2	0.05	–	–	–	–	–	–	1.000	1.000	1.000	1.000	1.000	1.000
2	0.025	–	–	–	–	–	–	–	–	–	–	1.000	1.000
2	0.01	–	–	–	–	–	–	–	–	–	–	–	–
2	0.001	–	–	–	–	–	–	–	–	–	–	–	–
3	0.05	–	–	–	1.000	1.000	1.000	.875	.889	.900	.909	.833	.846
3	0.025	–	–	–	–	1.000	1.000	1.000	1.000	1.000	.909	.917	.923
3	0.01	–	–	–	–	–	–	–	1.000	1.000	1.000	1.000	1.000
3	0.001	–	–	–	–	–	–	–	–	–	–	–	–
4	0.05	–	–	1.000	1.000	.833	.857	.875	.778	.750	.750	.750	.750
4	0.025	–	–	–	1.000	1.000	1.000	.875	.889	.900	.818	.833	.846
4	0.01	–	–	–	–	1.000	1.000	1.000	1.000	.900	.909	.917	.923
4	0.001	–	–	–	–	–	–	–	–	–	–	–	1.000
5	0.05	–	1.000	1.000	1.000	.800	.800	.750	.778	.800	.709	.717	.692
5	0.025	–	–	1.000	1.000	1.000	.857	.800	.800	.800	.800	.750	.723
5	0.01	–	–	–	1.000	1.000	1.000	.875	.889	.900	.818	.833	.800
5	0.001	–	–	–	–	–	–	–	1.000	1.000	1.000	1.000	1.000
6	0.05	–	1.000	.833	.800	.833	.714	.708	.722	.667	.652	.667	.667
6	0.025	–	1.000	1.000	1.000	1.000	.833	.750	.778	.733	.727	.750	.692
6	0.01	–	–	1.000	1.000	1.000	.857	.833	.833	.800	.818	.833	.769
6	0.001	–	–	–	–	–	–	1.000	1.000	1.000	1.000	.917	.923
7	0.05	–	1.000	.857	.800	.714	.857	.714	.667	.657	.623	.631	.615
7	0.025	–	1.000	1.000	.857	.833	.857	.732	.714	.700	.675	.667	.637
7	0.01	–	–	1.000	1.000	.857	.857	.857	.778	.757	.766	.714	.714
7	0.001	–	–	–	–	–	1.000	1.000	1.000	.900	.909	.857	.857
8	0.05	1.000	.875	.875	.750	.708	.714	.750	.639	.600	.602	.625	.596
8	0.025	–	1.000	.875	.800	.750	.732	.750	.667	.675	.659	.667	.625
8	0.01	–	–	1.000	.875	.833	.857	.875	.764	.750	.727	.708	.692
8	0.001	–	–	–	–	1.000	1.000	1.000	.889	.875	.875	.833	.846

Percentage Points of *D* in the Two-sample Kolmogorov-Smirnov Distribution (*Contd.*)

						n_2							
14	15	16	17	18	19	20	21	22	23	24	25	α	n_1
.929	.933	.938	.941	.944	.947	.950	.905	.909	.913	.917	.920	0.05	2
1.000	1.000	1.000	1.000	1.000	1.000	1.000	.952	.955	.957	.958	.960	0.025	2
–	–	–	–	–	1.000	1.000	1.000	1.000	1.000	1.000	1.000	0.01	2
–	–	–	–	–	–	–	–	–	–	–	–	0.001	2
.857	.800	.813	.824	.833	.789	.800	.810	.773	.783	.792	.800	0.05	3
.929	.867	.875	.882	.889	.895	.850	.857	.864	.870	.833	.840	0.025	3
1.000	.933	.938	.941	.944	.947	.950	.905	.909	.913	.917	.920	0.01	3
–	–	–	–	–	–	–	1.000	1.000	1.000	1.000	1.000	0.001	3
.750	.733	.750	.706	.694	.697	.750	.702	.705	.696	.708	.680	0.05	4
.786	.750	.813	.765	.750	.750	.800	.750	.750	.750	.750	.750	0.025	4
.857	.867	.875	.882	.833	.842	.850	.857	.818	.826	.833	.840	0.01	4
1.000	1.000	1.000	1.000	1.000	1.000	.950	.952	.955	.957	.958	.960	0.001	4
.657	.733	.675	.647	.667	.642	.650	.657	.636	.626	.633	.640	0.05	5
.729	.733	.738	.706	.722	.695	.750	.705	.709	.696	.675	.720	0.025	5
.800	.800	.800	.800	.778	.747	.800	.762	.755	.757	.750	.760	0.01	5
1.000	.933	.938	.941	.944	.895	.900	.905	.909	.913	.875	.880	0.001	5
.643	.633	.625	.608	.667	.614	.600	.595	.591	.580	.625	.587	0.05	6
.690	.700	.667	.657	.722	.667	.650	.643	.652	.623	.667	.640	0.025	6
.762	.767	.750	.716	.778	.728	.733	.714	.697	.703	.708	.713	0.01	6
.929	.933	.875	.833	.889	.842	.833	.833	.833	.826	.833	.833	0.001	6
.643	.590	.571	.571	.571	.571	.564	.619	.545	.553	.548	.554	0.05	7
.714	.648	.652	.647	.635	.632	.614	.667	.623	.609	.607	.600	0.025	7
.786	.714	.688	.706	.690	.684	.664	.714	.669	.671	.667	.657	0.01	7
.857	.857	.857	.824	.802	.805	.800	.810	.812	.783	.792	.777	0.001	7
.571	.558	.625	.566	.556	.539	.550	.530	.534	.533	.542	.520	0.05	8
.625	.617	.625	.588	.597	.592	.600	.577	.580	.576	.583	.560	0.025	8
.679	.675	.688	.647	.653	.645	.650	.637	.636	.625	.667	.625	0.01	8
.804	.808	.813	.816	.778	.770	.775	.750	.750	.745	.792	.750	0.001	8

Percentage Points of D in the Two-sample Kolmogorov-Smirnov Distribution (*Contd.*)

n_1	α	n_2 = 2	3	4	5	6	7	8	9	10	11	12	13
9	0.05	1.000	.889	.778	.778	.722	.667	.639	.667	.589	.596	.583	.556
9	0.025	–	1.000	.889	.800	.778	.714	.667	.778	.667	.636	.639	.615
9	0.01	–	1.000	1.000	.889	.833	.778	.764	.778	.700	.707	.694	.667
9	0.001	–	–	–	1.000	1.000	1.000	.889	.889	.889	.818	.806	.778
10	0.05	1.000	.900	.750	.800	.667	.657	.600	.589	.700	.545	.550	.538
10	0.025	–	1.000	.900	.800	.733	.700	.675	.667	.700	.618	.600	.592
10	0.01	–	1.000	.900	.900	.800	.757	.750	.700	.800	.700	.667	.646
10	0.001	–	–	–	1.000	1.000	.900	.875	.889	.900	.809	.800	.769
11	0.05	1.000	.909	.750	.709	.652	.623	.602	.596	.545	.636	.545	.524
11	0.025	–	.909	.818	.800	.727	.675	.659	.636	.618	.636	.576	.587
11	0.01	–	1.000	.909	.818	.818	.766	.727	.707	.700	.727	.652	.636
11	0.001	–	–	–	1.000	1.000	.909	.875	.818	.809	.818	.750	.755
12	0.05	1.000	.833	.750	.717	.667	.631	.625	.583	.550	.545	.583	.519
12	0.025	1.000	.917	.833	.750	.750	.667	.667	.639	.600	.576	.667	.538
12	0.01	–	1.000	.917	.833	.833	.714	.708	.694	.667	.652	.667	.609
12	0.001	–	–	–	1.000	.917	.857	.833	.806	.800	.750	.833	.750
13	0.05	1.000	.846	.750	.692	.667	.615	.596	.556	.538	.524	.519	.538
13	0.025	1.000	.923	.846	.723	.692	.637	.625	.615	.592	.587	.538	.615
13	0.01	–	1.000	.923	.800	.769	.714	.692	.667	.646	.636	.609	.692
13	0.001	–	–	1.000	1.000	.923	.857	.846	.778	.769	.755	.750	.769
14	0.05	.929	.857	.750	.657	.643	.643	.571	.556	.529	.532	.512	.489
14	0.025	1.000	.929	.786	.729	.690	.714	.625	.603	.586	.565	.560	.549
14	0.01	–	1.000	.857	.800	.762	.786	.679	.667	.643	.623	.619	.571
14	0.001	–	–	1.000	1.000	.929	.857	.804	.778	.757	.747	.714	.709
15	0.05	.933	.800	.733	.733	.633	.590	.558	.556	.533	.509	.517	.492
15	0.025	1.000	.867	,750	.733	.700	.648	.617	.600	.600	.570	.550	.533
15	0.01	–	.933	.867	.800	.767	.714	.675	.667	.667	.618	.600	.590
15	0.001	–	–	1.000	.933	.933	.857	.808	.778	.767	.727	.717	.703
16	0.05	.938	.813	.750	.675	.625	.571	.625	.542	.525	.506	.500	.486
16	0.025	1.000	.875	.813	.738	.667	.652	.625	.590	.563	.545	.542	.534
16	0.01	–	.938	.875	.800	.750	.688	.688	.653	.625	.602	.604	.582
16	0.001	–	–	1.000	.938	.875	.857	.813	.764	.738	.722	.708	.688
17	0.05	.941	.824	.706	.647	.608	.571	.566	.536	.524	.497	.490	.475
17	0.025	1.000	.882	.765	.706	.657	.647	.588	.588	.565	.545	.529	.516
17	0.01	–	.941	.882	.800	.716	.706	.647	.647	.624	.588	.583	.575
17	0.001	–	–	1.000	.941	.833	.824	.816	.765	.741	.706	.691	.688
18	0.05	.944	.833	.694	.667	.667	.571	.556	.556	.511	.490	.500	.470
18	0.025	1.000	.889	.750	.722	.722	.635	.597	.611	.556	.540	.556	.513
18	0.01	–	.944	.833	.778	.778	.690	.653	.667	.600	.596	.583	.560
18	0.001	–	–	1.000	.944	.889	.802	.778	.778	.733	.707	.694	.667

Percentage Points of D in the Two-sample Kolmogorov-Smirnov Distribution (*Contd.*)

						n_2							
14	15	16	17	18	19	20	21	22	23	24	25	α	n_1
.556	.556	.542	.536	.556	.520	.517	.524	.510	.512	.514	.507	0.05	9
.603	.600	.590	.588	.611	.573	.556	.571	.556	.556	.556	.547	0.025	9
.667	.667	.653	.647	.667	.626	.617	.619	.616	.609	.611	.600	0.01	9
.778	.778	.764	.765	.778	.737	.739	.730	.727	.734	.722	.720	0.001	9
.529	.533	.525	.524	.511	.495	.550	.500	.491	.496	.492	.500	0.05	10
.586	.600	.563	.565	.556	.542	.600	.552	.536	.539	.533	.540	0.025	10
.643	.667	.625	.624	.600	.595	.650	.600	.591	.596	.583	.600	0.01	10
.757	.767	.738	.741	.733	.700	.750	.710	.700	.696	.692	.700	0.001	10
.532	.509	.506	.497	.490	.488	.486	.485	.500	.470	.470	.469	0.05	11
.565	,570	.545	.545	.540	.531	.527	.532	.545	.518	.519	.509	0.025	11
.623	.618	.602	.588	.596	.584	.577	.580	.591	.561	.568	.560	0.01	11
.747	.727	.722	.706	.707	.699	.700	.680	.727	.684	.667	.669	0.001	11
.512	.517	.500	.490	.500	.474	.483	.476	.470	.453	.500	.460	0.05	12
.560	.550	.542	.529	.556	.526	.517	.512	.508	.496	.542	.500	0.025	12
.619	.600	.604	.583	.583	.570	.583	.560	.561	.540	.583	.550	0.01	12
.714	.717	.708	.691	.694	.684	.683	.667	.659	.659	.667	.640	0.001	12
.489	.492	.486	.475	.470	.462	.462	.462	.455	.452	.449	.446	0.05	13
.549	.533	.534	.516	.513	.510	.500	.502	.493	.488	.484	.486	0.025	13
.571	.590	.582	.575	.560	.559	.550	.549	.545	.538	.532	.529	0.01	13
.709	.703	.688	.688	.667	.664	.650	.656	.647	.639	.638	.615	0.001	13
.571	.467	.473	.466	.460	.455	.450	.476	.448	.441	.435	.429	0.05	14
.571	.524	.518	.513	.500	.500	.493	.500	.481	.478	.476	.474	0.025	14
.643	.586	.563	.563	.556	.556	.543	.548	.532	.528	.524	.520	0.01	14
.786	.667	.679	.668	.659	.662	.643	.643	.636	.627	.625	.626	0.001	14
.467	.533	.475	.455	.456	.446	.450	.438	.436	.432	.433	.427	0.05	15
.524	.600	.496	.506	.500	.495	.500	.486	.467	.472	.467	.467	0.025	15
.586	.600	.554	.557	.544	.533	.533	.533	.524	.519	.517	.520	0.01	15
.667	.733	.675	.647	.644	.632	.650	.629	.621	.609	.617	.613	0.001	15
.473	.475	.500	.456	.444	.438	.438	.432	.426	.427	.438	.418	0.05	16
.518	.496	.563	.500	.486	.477	.488	.467	.466	.459	.479	.453	0.025	16
.563	.554	.625	.526	.535	.526	.525	.515	.511	.508	.521	.498	0.01	16
.679	.675	.688	.640	.646	.632	.625	.619	.614	.601	.604	.595	0.001	16
.466	.455	.456	.471	.435	.437	.429	.423	.420	.417	.412	.407	0.05	17
.513	.506	.500	.529	.484	.467	.471	.465	.455	.458	.449	.447	0.025	17
.563	.557	.526	.588	.536	.514	.515	.504	.500	.501	.498	.487	0.01	17
.668	.647	.640	.706	.611	.619	.615	.608	.602	.593	.588	.586	0.001	17
.460	.456	.444	.435	.500	.415	.422	.421	.414	.411	.417	.400	0.05	18
.500	.500	.486	.484	.500	.465	.461	.460	.449	.444	.458	.436	0.025	18
.556	.544	.535	.536	.556	.515	.506	.500	.495	.493	.500	.480	0.01	18
.659	.644	.646	.611	.667	.620	.594	.595	.591	.585	.583	.571	0.001	18

Percentage Points of *D* in the Two-sample Kolmogorov-Smirnov Distribution (*Contd.*)

n_1	α	n_2 = 2	3	4	5	6	7	8	9	10	11	12	13
19	0.05	.947	.789	.697	.642	.614	.571	.539	.520	.495	.488	.474	.462
19	0.025	1.000	.895	.750	.695	.667	.632	.592	.573	.542	.531	.526	.510
19	0.01	1.000	.947	.842	.747	.728	.684	.645	.626	.595	.584	.570	.559
19	0.001	–	–	1.000	.895	.842	.80f	i .770	.737	.700	.699	.684	.664
20	0.05	.950	.800	.750	.650	.600	.564	.550	.517	.550	.486	.483	.462
20	0.025	1.000	.850	.800	.750	.650	.614	.600	.556	.600	.527	.517	.500
20	0.01	1.000	.950	.850	.800	.733	.664	.650	.617	.650	.577	.583	.550
20	0.001	–	–	.950	.900	.833	.800	.775	.739	.750	.700	.683	.650
21	0.05	.905	.810	.702	.657	.595	.619	.530	.524	.500	.485	.476	.462
21	0.025	.952	.857	.750	.705	.643	.667	.577	.571	.552	.532	.512	.502
21	0.01	1.000	.905	.857	.762	.714	.714	.637	.619	.600	.580	.560	.549
21	0.001	–	1.000	.952	.905	.833	.810	.750	.730	.710	.680	.667	.656
22	0.05	.909	.773	.705	.636	.591	545	.534	.510	.491	.500	.470	.455
22	0.025	.955	.864	.750	.709	.652	.623	.580	.556	.536	.545	.508	.493
22	0.01	1.000	.909	.818	.755	.697	.669	.636	.616	.591	.591	.561	.545
22	0.001	–	1.000	.955	.909	.833	.812	.750	.727	.700	.727	.659	.647
23	0.05	.913	.783	.696	.626	.580	.552	.533	.512	.496	.470	.453	.452
23	0.025	.957	.870	.750	.696	.623	.609	.576	.556	.539	.518	.496	.488
23	0.01	1.000	.913	.826	.757	.703	.671	.625	.609	.596	.561	.540	.538
23	0.001	–	1.000	.957	.913	.826	.782	.745	.734	.696	.684	.659	.639
24	0.05	.917	.792	.708	.633	.625	.548	.542	.514	.492	.470	.500	.449
24	0.025	.958	.833	.750	.675	.667	.607	.583	.556	.533	.519	.542	.484
24	0.01	1.000	.917	.833	.750	.708	.661	.667	.611	.583	.568	.583	.532
24	0.001	–	1.000	.958	.875	.833	.792	.792	.722	.692	.667	.667	.638
25	0.05	.920	.800	.680	.640	.587	.554	.520	.507	.500	.469	.460	.446
25	0.025	.960	.840	.750	.720	.640	.600	.560	.547	.540	.509	.500	.486
25	0.01	1.000	.920	.840	.760	.713	.657	.625	.600	.600	.560	.550	.529
25	0.001	–	1.000	.960	.880	.833	.777	.750	.720	.700	.669	.640	.615

Percentage Points of D in the Two-sample Kolmogorov-Smirnov Distribution (*Contd.*)

n_2													
14	15	16	17	18	19	20	21	22	23	24	25	α	n_1
.455	.446	.438	.437	.415	.474	.421	.409	.404	.405	.401	.394	0.05	19
.500	.495	.477	.467	.465	.526	.445	.451	.443	.435	.436	.432	0.025	19
.556	.533	.526	.514	.515	.526	.492	.499	.488	.478	.478	.472	0.01	19
.662	.632	.632	.619	.620	.632	.592	.594	.579	.579	.572	.564	0.001	19
.450	.450	.438	.429	.422	.421	.450	.412	.400	.400	.400	.400	0.05	20
.493	.500	.488	.471	.461	.445	.500	.429	.436	.433	.433	.430	0.025	20
.543	.533	.525	.515	.506	.492	.550	.474	.482	.476	.475	.470	0.01	20
.643	.650	.625	.615	.594	.592	.650	.569	.577	.570	.567	.560	0.001	20
.476	.438	.432	.423	.421	.409	.412	.429	.396	.391	.393	.385	0.05	21
.500	.486	.467	.465	.460	.451	.429	.476	.439	.427	.423	.419	0.025	21
.548	.533	.515	.504	.500	.499	.474	.524	.483	.470	.470	.465	0.01	21
.643	.629	.619	.608	.595	.594	.569	.619	.578	.557	.560	.552	0.001	21
.448	.436	.426	.420	.414	.404	.400	.396	.409	.383	.386	.380	0.05	22
.481	.467	.466	.455	.449	.443	.436	.439	.455	.423	.420	.415	0.025	22
.532	.524	.511	.500	.495	.488	.482	.483	.500	.468	.458	.455	0.01	22
.636	.621	.614	.602	.591	.579	.577	.578	.591	.557	.553	.544	0.001	22
.441	.432	.427	.417	.411	.405	.400	.391	.383	.435	.371	.376	0.05	23
.478	.472	.459	.458	.444	.435	.433	.427	.423	.435	.409	.412	0.025	23
.528	.519	.508	.501	.493	.478	.476	.470	.468	.478	.451	.456	0.01	23
.627	.609	.601	.593	.585	.579	.570	.557	.557	.565	.536	.543	0.001	23
.435	.433	.438	.412	.417	.401	.400	.393	.386	.371	.417	.375	0.05	24
.476	.467	.479	.449	.458	.436	.433	.423	.420	.409	.458	.397	0.025	24
.524	.517	.521	.498	.500	.478	.475	.470	.458	.451	.500	.437	0.01	24
.625	.617	.604	.588	.583	.572	.567	.560	.553	.536	.583	.520	0.001	24
.429	.427	.418	.407	.400	.394	.400	.385	.380	.376	.375	.400	0.05	25
.474	.467	.453	.447	.436	.432	.430	.419	.415	.412	.397	.440	0.025	25
.520	.520	.498	.487	.480	.472	.470	.465	.455	.456	.437	.480	0.01	25
.626	.613	.595	.586	.571	.564	.560	.552	.544	.543	.520	.560	0.001	25

α	0.05	0.025	0.01	0.001
D_α	$1.36\sqrt{\frac{n_1+n_2}{n_1 n_2}}$	$1.48\sqrt{\frac{n_1+n_2}{n_1 n_2}}$	$1.63\sqrt{\frac{n_1+n_2}{n_1 n_2}}$	$1.95\sqrt{\frac{n_1+n_2}{n_1 n_2}}$

Values of D equal to or greater than that tabulated for a given pair of sample sizes n_1 and n_2 and for a specified significance level α suggest that the two samples have not come from the same population. The tabulated values of α relate to a two-sided alternative hypothesis regarding the difference between the sampled populations.

For values of n_1 and n_2 greater than 25, it is sufficiently informative in relation to *statistical* significance to calculate the critical values from the formulae given at the foot of the table.

LOWER PERCENTAGE POINTS OF THE WILCOXON SIGNED-RANK DISTRIBUTION

For a random sample from a symmetric continuous distribution with mean of μ_0, subtract μ_0 from each sample member, discard all zero values and rank the *absolute* values of the remaining differences in ascending order. For tied ranks, use the average of the ranks if there had been no ties. Let w_+ and w_- be the sum of the ranks which correspond to positive and negative differences respectively.

The following table gives the lower percentage points of W_+, the sum of positive ranks.

α	0.05	Nearer exact probability	0.025	Nearer exact probability	0.01	Nearer exact probability	0.005	Nearer exact probability	0.001
n = 5	1	.0625	–		–		–		–
6	2	.0469	1	.0312	–		–		–
7	4	.0547	2	.0234	0	.0078	–		–
8	6	.0547	4	.0273	2	.0117	0	.0039	–
9	8	.0488	6	.0273	3	.0098	2	.0059	–
10	11	.0527	8	.0244	5	.0098	3	.0049	0
11	14	.0508	11	.0269	7	.0093	5	.0049	1
12	17	.0461	14	.0261	10	.0105	7	.0046	2
13	21	.0471	17	.0239	13	.0107	10	.0052	4
14	26	.0520	21	.0247	16	.0101	13	.0054	6
15	30	.0473	25	.0240	20	.0108	16	.0051	8
16	36	.0523	30	.0253	24	.0107	20	.0055	11
17	41	.0492	35	.0253	28	.0101	23	.0047	14
18	47	.0494	40	.0241	33	.0104	28	.0052	18
19	54	.0521	46	.0247	38	.0102	32	.0047	21
20	60	.0487	52	.0242	43	.0096	38	.0053	26
21	68	.0516	59	.0251	49	.0097	43	.0051	30
22	75	.0492	66	.0250	56	.0104	49	.0052	35
23	83	.0490	73	.0242	62	.0098	55	.0051	40
24	92	.0505	81	.0245	69	.0097	61	.0048	45
25	101	.0507	90	.0258	77	.0101	68	.0048	51
26	110	.0497	98	.0247	85	.0102	76	.0051	58
27	120	.0502	107	.0246	93	.0100	84	.0052	64
28	130	.0496	117	.0252	102	.0102	92	.0051	71
29	141	.0504	127	.0253	111	.0101	100	.0049	79
30	152	.0502	137	.0249	120	.0098	109	.0050	86

A test of the hypothesis that a sample has come from a continuous symmetric population with specified mean value μ_0 (or that the mean difference of matched pairs is μ_0) against the alternative hypothesis that $\mu < \mu_0$, would lead to rejection if the calculated value of w_+ is less than or equal to the tabulated value of W_+. To test the hypothesis against the alternative $\mu > \mu_0$, reject if w_- is less than or equal to the tabulated value. For the two-sided alternative $\mu \neq \mu_0$, use the smaller value of w_+ and w_-; in this case the tabulated value of α should be doubled.

Because the distribution of ranks is discrete, the possible values of W are not usually associated with round number values of probability (as usually quoted for α). The above table gives the exact probability of observing a given value of W_+ or smaller. This exact probability is the one *nearer* to the given α, not the largest one which is less than or equal to α. Where this probability is unacceptably larger than required, reduction of the tabulated value of W_+ by 1 will bring the significance level below the designated value of α. For $\alpha = 0.001$, the probabilities of the given W_+ are all less than or equal to 0.001.

Note: As n increases, the distribution of W_+ (and W_-) tends to normality with mean of $\frac{n(n+1)}{4}$ and standard deviation of $\sqrt{\frac{n(n+1)(2n+1)}{24}}$. For $n > 30$, the use of the approximation will give results accurate enough for most practical purposes.

LOWER PERCENTAGE POINTS OF THE MANN-WHITNEY *U*-DISTRIBUTION

The following table gives the critical values of the Mann-Whitney statistic U for one-tailed tests for $\alpha = 0.05$, 0.025, 0.01 and 0.001. For two-tailed tests the probabilities should be doubled.

n_1	α	n_2: 3	4	5	6	7	8	9	10	11	12	13	14	15	16	17	18	19	20
1	0.05	–	–	–	–	–	–	–	–	–	–	–	–	–	–	–	–	0	0
1	0.025	–	–	–	–	–	–	–	–	–	–	–	–	–	–	–	–	–	–
1	0.01	–	–	–	–	–	–	–	–	–	–	–	–	–	–	–	–	–	–
1	0.001	–	–	–	–	–	–	–	–	–	–	–	–	–	–	–	–	–	–
2	0.05	–	–	0	0	0	1	1	1	1	2	2	3	3	3	3	4	4	4
2	0.025	–	–	–	–	–	0	0	0	0	1	1	1	1	1	2	2	2	2
2	0.01	–	–	–	–	–	–	–	–	–	–	0	0	0	0	0	0	1	1
2	0.001	–	–	–	–	–	–	–	–	–	–	–	–	–	–	–	–	–	–
3	0.05	0	0	1	2	2	3	4	4	5	5	6	7	7	8	9	9	10	11
3	0.025	–	–	0	1	1	2	2	3	3	4	4	5	5	6	6	7	7	8
3	0.01	–	–	–	–	0	0	1	1	1	2	2	2	3	3	4	4	4	5
3	0.001	–	–	–	–	–	–	–	–	–	–	–	–	–	–	0	0	0	0
4	0.05	–	1	2	3	4	5	6	7	8	9	10	11	12	14	15	16	17	18
4	0.025	–	0	1	2	3	4	4	5	6	7	8	9	10	11	11	12	13	13
4	0.01	–	–	0	1	1	2	3	3	4	5	5	6	7	7	8	9	9	10
4	0.001	–	–	–	–	–	–	–	0	0	0	1	1	1	2	2	3	3	3
5	0.05	–	–	4	5	6	8	9	11	12	13	15	16	18	19	20	22	23	25
5	0.025	–	–	2	3	5	6	7	8	9	11	12	13	14	15	17	18	19	20
5	0.01	–	–	1	2	3	4	5	6	7	8	9	10	11	12	13	14	15	16
5	0.001	–	–	–	–	0	0	1	1	2	2	3	3	4	5	5	6	7	7
6	0.05	–	–	–	7	8	10	12	14	16	17	19	21	23	25	26	28	30	32
6	0.025	–	–	–	5	6	8	10	11	13	14	16	17	19	21	22	24	25	27
6	0.01	–	–	–	3	4	6	7	8	9	11	12	13	15	16	18	19	20	22
6	0.001	–	–	–	0	1	2	2	3	4	4	5	6	7	8	9	10	11	12
7	0.05	–	–	–	–	11	13	15	17	19	21	24	26	28	30	33	35	37	39
7	0.025	–	–	–	–	8	10	12	14	16	18	20	22	24	26	28	30	32	34
7	0.01	–	–	–	–	6	8	9	11	12	14	16	17	19	21	23	24	26	28
7	0.001	–	–	–	–	2	3	3	5	6	7	8	9	10	11	13	14	15	16
8	0.05	–	–	–	–	–	15	18	20	23	26	28	31	33	36	39	41	44	47
8	0.025	–	–	–	–	–	13	15	17	19	22	24	26	29	31	34	36	38	41
8	0.01	–	–	–	–	–	10	11	13	15	17	20	22	24	26	28	30	32	34
8	0.001	–	–	–	–	–	5	5	6	8	9	11	12	14	15	17	18	20	21
9	0.05	–	–	–	–	–	–	21	24	27	30	33	36	39	42	45	48	51	54
9	0.025	–	–	–	–	–	–	17	20	23	26	28	31	34	37	39	42	45	48
9	0.01	–	–	–	–	–	–	14	16	18	21	23	26	28	31	33	36	38	40
9	0.001	–	–	–	–	–	–	7	8	10	12	14	15	17	19	21	23	25	26
10	0.05	–	–	–	–	–	–	–	27	31	34	37	41	44	48	51	55	58	62
10	0.025	–	–	–	–	–	–	–	23	26	29	33	36	39	42	45	48	52	55
10	0.01	–	–	–	–	–	–	–	19	22	24	27	30	33	36	38	41	44	47
10	0.001	–	–	–	–	–	–	–	10	12	14	17	19	21	23	25	27	29	32
11	0.05	–	–	–	–	–	–	–	–	34	38	42	46	50	54	57	61	65	69
11	0.025	–	–	–	–	–	–	–	–	30	33	37	40	44	47	51	55	58	62
11	0.01	–	–	–	–	–	–	–	–	25	28	31	34	37	41	44	47	50	53
11	0.001	–	–	–	–	–	–	–	–	15	17	20	22	24	27	29	32	34	37
12	0.05	–	–	–	–	–	–	–	–	–	42	47	51	55	60	64	68	72	77
12	0.025	–	–	–	–	–	–	–	–	–	37	41	45	49	53	57	61	65	69
12	0.01	–	–	–	–	–	–	–	–	–	31	35	38	42	46	49	53	56	60
12	0.001	–	–	–	–	–	–	–	–	–	20	23	25	28	31	34	37	40	42
13	0.05	–	–	–	–	–	–	–	–	–	–	51	56	61	65	70	75	80	84
13	0.025	–	–	–	–	–	–	–	–	–	–	45	50	54	59	63	67	72	76

Lower Percentage Points of the Mann-Whitney *U*-Distribution (*Contd.*)

n_1	α	n_2 3	4	5	6	7	8	9	10	11	12	13	14	15	16	17	18	19	20
13	0.01	–	–	–	–	–	–	–	–	–	–	39	43	47	51	55	59	63	67
13	0.001	–	–	–	–	–	–	–	–	–	–	26	29	32	35	38	42	45	48
14	0.05	–	–	–	–	–	–	–	–	–	–	–	61	66	71	77	82	87	92
14	0.025	–	–	–	–	–	–	–	–	–	–	–	55	59	64	67	74	78	83
14	0.01	–	–	–	–	–	–	–	–	–	–	–	47	51	56	60	65	69	73
14	0.001	–	–	–	–	–	–	–	–	–	–	–	32	36	39	43	46	50	54
15	0.05	–	–	–	–	–	–	–	–	–	–	–	–	72	77	83	88	94	100
15	0.025	–	–	–	–	–	–	–	–	–	–	–	–	64	70	75	80	85	90
15	0.01	–	–	–	–	–	–	–	–	–	–	–	–	56	61	66	70	75	80
15	0.001	–	–	–	–	–	–	–	–	–	–	–	–	40	43	47	51	55	59
16	0.05	–	–	–	–	–	–	–	–	–	–	–	–	–	83	89	95	101	107
16	0.025	–	–	–	–	–	–	–	–	–	–	–	–	–	75	81	86	92	98
16	0.01	–	–	–	–	–	–	–	–	–	–	–	–	–	66	71	76	82	87
16	0.001	–	–	–	–	–	–	–	–	–	–	–	–	–	48	52	56	60	65
17	0.05	–	–	–	–	–	–	–	–	–	–	–	–	–	–	96	102	109	115
17	0.025	–	–	–	–	–	–	–	–	–	–	–	–	–	–	87	93	99	105
17	0.01	–	–	–	–	–	–	–	–	–	–	–	–	–	–	77	82	88	93
17	0.001	–	–	–	–	–	–	–	–	–	–	–	–	–	–	57	61	66	70
18	0.05	–	–	–	–	–	–	–	–	–	–	–	–	–	–	–	109	116	123
18	0.025	–	–	–	–	–	–	–	–	–	–	–	–	–	–	–	99	106	112
18	0.01	–	–	–	–	–	–	–	–	–	–	–	–	–	–	–	88	94	100
18	0.001	–	–	–	–	–	–	–	–	–	–	–	–	–	–	–	66	71	76
19	0.05	–	–	–	–	–	–	–	–	–	–	–	–	–	–	–	–	123	130
19	0.025	–	–	–	–	–	–	–	–	–	–	–	–	–	–	–	–	113	119
19	0.01	–	–	–	–	–	–	–	–	–	–	–	–	–	–	–	–	101	107
19	0.001	–	–	–	–	–	–	–	–	–	–	–	–	–	–	–	–	77	82
20	0.05	–	–	–	–	–	–	–	–	–	–	–	–	–	–	–	–	–	138
20	0.025	–	–	–	–	–	–	–	–	–	–	–	–	–	–	–	–	–	127
20	0.01	–	–	–	–	–	–	–	–	–	–	–	–	–	–	–	–	–	114
20	0.001	–	–	–	–	–	–	–	–	–	–	–	–	–	–	–	–	–	88

Given two random samples of size n_1 and n_2 ($n_1 \le n_2$), rank the $n_1 + n_2$ observations in increasing order of size. If there are ties, give each tied rank the average of the ranks they would each have had if there were no ties.

Denote the sum of the ranks of the sample of size n_1 by R_1 and that obtained from n_2 by R_2.

Calculate $u_1 = n_1 n_2 + ½n_1 (n_1 + 1) - R_1$

and $u_2 = n_1 n_2 + ½n_2 (n_2 + 1) - R_2$ although since $u_1 + u_2 = n_1 n_2$, this can more easily be obtained as $n_1n_2 - u_1$.

For an alternative hypothesis $\mu_1 < \mu_2$, the observed value, μ_1, should be less than the tabulated value of U in order to reject the hypothesis that $\mu_1 = \mu_2$ at the one-sided α level of significance.

To reject equality of population means in favour of the one-sided alternative $\mu_1 > \mu_2$, μ_2 should be less than U.

For the two-sided alternative $\mu_1 \ne \mu_2$, reject the equality of means hypothesis if the smaller of u_1 and u_2 is less than the tabulated value of U though the significance level of this test will be double the quoted value of α.

Note that the quoted significance levels, α, are not generally exact, especially for small n_1 and n_2. The exact Type I error, in all cases, will not exceed the quoted value of α, being the highest feasible value less than α. For example, for $n_1 = 4$ and $n_2 = 6$, the exact probability of observing a U-value of 3 or less is 0.033, not 0.05. The tabulation is thus conservative regarding Type I errors.

Note: For values of $n_2 > 20$, the sampling distribution of U rapidly tends to normality with mean $= \frac{1}{2}n_1n_2$ and standard deviation $= \sqrt{\frac{n_1n_2(n_1+n_2+1)}{12}}$. Thus we have that the statistic $\dfrac{U - \frac{1}{2}n_1n_2}{\sqrt{\frac{n_1n_2(n_1+n_2+1)}{12}}}$ will be distributed like the unit normal variate, Z.

PERCENTAGE POINTS OF FRIEDMAN'S DISTRIBUTION

The following table gives the probabilities associated with values as large as, or larger than, the possible observed values of Friedman's statistic χ_r^2 given by $\chi_r^2 = \frac{12}{nk(k+1)} \sum_{j=1}^{k} R_j^2 - 3n(k+1)$

$k = 3$

$n = 2$		$n = 3$		$n = 4$		$n = 5$	
χ_r^2	p	χ_r^2	p	χ_r^2	p	χ_r^2	p
0	1.000	.000	1.000	.0	1.000	.0	1.000
1	.833	.667	.944	.5	.931	.4	.954
3	.5	2.000	.528	1.5	.653	1.2	.691
4	.167	2.667	.361	2.0	.431	1.6	.522
–	–	4.667	.194	3.5	.273	2.8	.367
–	–	6.000	.028	4.5	.125	3.6	.182
–	–	–	–	6.0	.069	4.8	.124
–	–	–	–	6.5	.042	5.2	.093
–	–	–	–	8.0	.0046	6.4	.039
–	–	–	–	–	–	7.6	.024
–	–	–	–	–	–	8.4	.0085
–	–	–	–	–	–	10.0	.00077

$k = 3$

$n = 6$		$n = 7$		$n = 8$		$n = 9$	
χ_r^2	p	χ_r^2	p	χ_r^2	p	χ_r^2	p
.00	1.000	.000	1.000	.00	1.000	.000	1.000
.33	.956	.286	.964	.25	.967	.222	.971
1.00	.740	.857	.768	.75	.794	.667	.814
1.33	.570	1.143	.620	1.00	.654	.889	.865
2.33	.430	2.000	.486	1.75	.531	1.556	.569
3.00	.252	2.571	.305	2.25	.355	2.000	.398
4.00	.184	3.429	.237	3.00	.285	2.667	.328
4.33	.142	3.714	.192	3.25	.236	2.889	.278
5.33	.072	4.571	.112	4.00	.149	3.556	.187
6.33	.052	5.429	.085	4.75	.120	4.222	.154
7.00	.029	6.000	.052	5.25	.079	4.667	.107
8.33	.012	7.143	.027	6.25	.047	5.556	.069
9.00	.0081	7.714	.021	6.75	.038	6.000	.057
9.33	.0055	8.000	.016	7.00	.030	6.222	.048
10.33	.0017	8.857	.0084	7.75	.018	6.889	.031
12.00	.00013	10.286	.0036	9.00	.0099	8.000	.019
		10.571	.0027	9.25	.0080	8.222	.016
		11.143	.0012	9.75	.0048	8.667	.010
		12.286	.00032	10.75	.0024	9.556	.0060
		14.000	.000021	12.00	.0011	10.667	.0035
				12.25	.00086	10.889	.0029
				13.00	.00026	11.556	.0013
				14.25	.000061	12.667	.00066

Percentage Points of Friedman's Distribution (*Contd.*)

$n = 6$		$n = 7$		$n = 8$		$n = 9$	
χ_r^2	p	χ_r^2	p	χ_r^2	p	χ_r^2	p
				16.00	.0000036	13.556	.00035
						14.000	.00020
						14.222	.000097
						14.889	.000054
						16.222	.000011
						18.000	.0000006

$k = 4$

$n = 2$		$n = 3$		$n = 4$			
χ_r^2	p	χ_r^2	p	χ_r^2	p	χ_r^2	p
.0	1.000	.2	1.000	.0	1.000	5.7	.141
.6	.958	.6	.958	.3	.992	6.0	.105
1.2	.834	1.0	.910	.6	.928	6.3	.094
1.8	.792	1.8	.727	.9	.900	6.6	.077
2.4	.625	2.2	.608	1.2	.800	6.9	.068
3.0	.542	2.6	.524	1.5	.754	7.2	.054
3.6	.458	3.4	.446	1.8	.677	7.5	.052
4.2	.375	3.8	.342	2.1	.649	7.8	.036
4.8	.208	4.2	.300	2.4	.524	8.1	.033
5.4	.167	5.0	.207	2.7	.508	8.4	.019
6.0	.042	5.4	.175	3.0	.432	8.7	.014
		5.8	.148	3.3	.389	9.3	.012
		6.6	.075	3.6	.355	9.6	.0069
		7.0	.054	3.9	.324	9.9	.0062
		7.4	.033	4.5	.242	10.2	.0027
		8.2	.017	4.8	.200	10.8	.0016
		9.0	.0017	5.1	.190	11.1	.00094
				5.4	.158	12.0	.000072

Given k matched samples, each therefore of size n, and with the observations lying on at least an ordinal scale, Friedman's distribution may be used to test the hypothesis that the k groups have a common mean value. This is equivalent to a standard balanced analysis of variance but without the requirement that each population is normally distributed.

The data are laid out in n rows (equivalent to blocks) and k columns and the observations in each row are converted to their equivalent ranks between 1 and k. If there are ties, the appropriate average rank is used. The total, R_j, of the ranks in each column is used to form the sum of these squared rank totals.

Note: For larger values of n and k than tabulated, the associated probability may be found from a table of the χ^2 distribution since Friedman's statistic χ_r^2 tends towards a χ^2 distribution with $(k - 1)$ degrees of freedom as n and k increase.

UPPER TAILS OF THE KRUSKAL-WALLIS DISTRIBUTION

Given a set of random samples, not necessarily of equal size, obtained from each of k populations, the hypothesis of equality of the population means can be tested. The measured variable should produce at least ordinal values so that the responses can be ranked without many ties. The structure of the test is that of a one-way analysis of variance but without the need to assume a normal distribution.

To apply the test, convert the whole data set to one sequence of ascending order of ranks. The smallest observation would thus have rank 1, the next larger one than this would be assigned the rank 2 and so on. It is convenient to imagine the ranks laid out in k columns corresponding to the original data.

The total number of readings $n = \sum_{j=1}^{k} n_j$ where n_j is the number of observations (ranks) in column j.

The statistic H is calculated from $H = \frac{12}{n(n+1)} \sum_{j=1}^{k} \frac{R_j^2}{n_j} - 3(n+1)$

The following table shows for k = 3, the probability that, for group sample sizes of n_1, n_2 and n_3, the statistic H takes certain preferred values or higher ones in the upper tail of the distribution. Thus for sample sizes 4, 4 and 2, a value of H = 5.2364 or higher would occur with probability of 0.052 just by chance when there is no actual difference between the k population means. The next possible higher value of H is 5,4545 with probability 0.046.

Sample sizes				
n_1	n_2	n_3	H	P
2	1	1	2.7000	.500
2	2	1	3.6000	.200
2	2	2	3.7143	.200
			4.5714	.067
3	1	1	3.2000	.300
3	2	1	3.8571	.133
			4.2857	.100
3	2	2	4.4643	.105
			4.5000	.067
			4.7143	.048
			5.3572	.029
3	3	1	4.0000	.129
			4.5714	.100
			5.1429	.043
3	3	2	4.2500	.121
			4.5556	.100
4	3	2	4.4444	.102
			4.5111	.098
			5.4000	.051
			5.4444	.046
			6.3000	.011
			6.4444	.008
4	3	3	4.7000	.101
			4.7091	.092
			5.7273	.050
			5.7909	.046
			6.7091	.013
			6.7455	.010
4	4	1	4.0667	.102
			4.1667	.082
			4.8667	.054
			4.9667	.048
			6.1667	.022
			6.6667	.010
4	4	2	4.4455	.103
			4.5545	.098

Upper Tails of the Kruskal-Wallis Distribution (*Contd.*)

Sample sizes				
n_1	n_2	n_3	H	P
			5.1389	.061
			5.3611	.032
			6.2500	.011
3	3	3	4.6222	.100
			5.0667	.086
			5.6000	.050
			5.6889	.029
			6.4889	.011
			7.2000	.004
4	1	1	3.5714	.200
4	2	1	4.0179	.114
			4.5000	.076
			4.8214	.057
4	2	2	4.1667	.105
			4.4583	.100
			5.1250	.052
			5.3333	.033
			6.0000	.014
4	3	1	3.8889	.129
			4.0556	.093
			5.0000	.057
			5.2083	.050
			5.8333	.021
5	2	2	4.2933	.122
			4.3733	.090
			5.0400	.056
			5.1600	.034
			6.1333	.013
			6.5333	.008
5	3	1	3.8400	.123
			4.0178	.095
			4.8711	.052
			4.9600	.048
			6.4000	.012
5	3	2	4.4945	.101
			4.6509	.091
			5.1055	.052
			5.2509	.049
			6.8218	.010
			6.9091	.009

Sample sizes				
n_1	n_2	n_3	H	P
			5.2364	.052
			5.4545	.046
			6.8727	.011
			7.0364	.006
4	4	3	4.4773	.102
			4.5455	.099
			5.5758	.051
			5.5985	.049
			7.1364	.011
			7.1439	.010
4	4	4	4.5001	.104
			4.6539	.097
			5.6538	.054
			5.6923	.049
			7.5385	.011
			7.6538	.008
5	1	1	3.8571	.143
5	2	1	4.0500	.119
			4.2000	.095
			4.4500	.071
			5.0000	.048
			5.2500	.036
5	4	4	4.5527	.102
			4.6187	.100
			5.6176	.050
			5.6571	.049
			7.7440	.011
			7.76.4	.009
5	5	1	4.0364	.105
			4.1091	.086
			4.9091	.053
			5.1273	.046
			6.8364	.011
			7.3091	.009
5	5	2	4.5077	.100
			4.6231	.097
			5.2462	.051
			5.3385	.047
			7.2692	.010
			7.3385	.010

Upper Tails of the Kruskal-Wallis Distribution (*Contd.*)

Sample sizes				
n_1	n_2	n_3	H	P
5	3	3	4.4121	.109
			4.5333	.097
			5.5152	.051
			5.6485	.049
			6.9818	.011
			7.0788	.009
5	4	1	3.9600	.102
			3.9873	.098
			4.8600	.056
			4.9855	.044
			6.8400	.011
			6.9545	.008
5	4	2	4.5182	.101
			4.5409	.098
			5.2682	.050
			5.2727	.049
			7.1182	.010
			7.2045	.009
5	4	3	4.5231	.103
			4.5487	.099
			5.6308	.050
			5.6564	.049
			7.3949	.011
			7.4449	.010
5	5	3	4.5363	.102
			4.5451	.100
			5.6264	.051
			5.7055	.046
			7.5429	.010
			7.5780	.010
5	5	4	4.5200	.101
			4.5229	.099
			5.6429	.050
			5.6657	.049
			7.7914	.010
			7.8229	.010
5	5	5	4.5000	.102
			4.5600	.100
			5.6600	.051
			5.7800	.049
			7.9800	.010
			8.0000	.009

Note: As n tends to infinity, the distribution of H tends to that of χ^2 with $(k - 1)$ df. This can be used as a reasonable approximation for H, given values of n_i and k outside the tabulated range.

CONTROL CHART FACTORS FOR SAMPLE MEAN ($\bar{X}$)

To obtain the limits,

either multiply σ by the appropriate value of $A_{0.025}$ and $A_{0.001}$

or multiply $\bar{R}$ by the appropriate value of $A'_{0.025}$ and $A'_{0.001}$

then add the result to and subtract it from the centre line of the chart ($\bar{\bar{X}}$ or μ).

No. in sample	*For inner limits* $A_{0.025}$	*For outer limits* $A_{0.001}$	*For inner limits* $A'_{0.025}$	*For outer limits* $A'_{0.001}$
2	1.386	2.185	1.229	1.937
3	1.132	1.784	0.668	1.054
4	0.980	1.545	0.476	0.750
5	0.876	1.382	0.377	0.594
6	0.800	1.262	0.316	0.498
7	0.741	1.168	0.274	0.432
8	0.693	1.092	0.244	0.384
9	0.653	1.030	0.220	0.347
10	0.620	0.977	0.202	0.317
11	0.591	0.932	0.186	0.294
12	0.566	0.892	0.174	0.274
13	0.544	0.857		
14	0.524	0.826		
15	0.506	0.798		
16	0.490	0.773		
17	0.475	0.750		
18	0.462	0.728		
19	0.450	0.709		
20	0.438	0.691		
21	0.428	0.674		
22	0.418	0.659		
23	0.409	0.644		
24	0.400	0.631		
25	0.392	0.618		
26	0.384	0.606		
27	0.377	0.595		
28	0.370	0.584		
29	0.364	0.574		
30	0.358	0.564		

Note: Samples containing more than 12 individuals should not be used when utilising the range as a substitute for the standard deviation in calculating the X control limits.

CONTROL CHART FACTORS FOR SAMPLE RANGE USING $\bar{R}$

To obtain the limits, multiply $\bar{R}$ by the appropriate value of D' (the subscript of each D' value is the probability of a sample range of n readings being *larger* than the calculated limit).

No. in sample n	*For lower limits* *Action* $D'_{0.999}$	*Warning* $D'_{0.975}$	*For upper limits* *Warning* $D'_{0.025}$	*Action* $D'_{0.001}$
2	0.00	0.04	2.81	4.12
3	0.04	0.18	2.17	2.98
4	0.10	0.29	1.93	2.57
5	0.16	0.37	1.81	2.34
6	0.21	0.42	1.72	2.21
7	0.26	0.46	1.66	2.11
8	0.29	0.50	1.62	2.04
9	0.32	0.52	1.58	1.99
10	0.35	0.54	1.56	1.93
11	0.38	0.56	1.53	1.91
12	0.40	0.58	1.51	1.87

CONTROL CHART FACTORS FOR SAMPLE RANGE USING σ

To obtain the limits, multiply σ by the appropriate value of D (the subscript of each D value is the probability of a sample range of n readings being *larger* than the calculated limit).

To obtain the expected value of sample range ($\bar{R}$) multiply σ by the appropriate value of d_2.

No. in sample n	*For lower limits* *Action* $D'_{0.999}$	*Warning* $D'_{0.975}$	*For upper limits* *Warning* $D'_{0.025}$	*Action* $D'_{0.001}$	d_2
2	0.00	0.04	3.17	4.65	1.128
3	0.06	0.30	3.68	5.05	1.693
4	0.20	0.59	3.98	5.30	2.059
5	0.37	0.85	4.20	5.45	2.326
6	0.54	1.06	4.36	5.60	2.534
7	0.69	1.25	4.49	5.70	2.704
8	0.83	1.41	4.61	5.80	2.847
9	0.96	1.55	4.70	5.90	2.970
10	1.08	1.67	4.79	5.95	3.078
11	1.20	1.78	4.86	6.05	3.173
12	1.30	1.88	4.92	6.10	3.258

CONTROL CHART CONSTANTS

No. of Obser-vations in samples	*Chart for averages*			*Chart for Standard deviations*					*Chart for ranges*				
	Factors for control limits			*Factors for central line*	*Factors for control limits*				*Factors for central line*	*Factors for control limits*			
n	A	A_1	A_2	C_2	B_1	B_2	B_3	B_4	d_2	D_1	D_2	D_3	D_4
2	2.121	3.760	1.880	0.5642	0	1.843	0	3.276	1.128	0	3.686	0	3.262
3	1.732	2.394	1.023	0.7236	0	1.858	0	2.568	1.663	0	4.358	0	2.575
4	1.500	1.880	0.729	0.7979	0	1.808	0	.266	2.059	0	4.698	0	2.282
5	.342	1.596	0.577	0.8407	0	1.756	0	.089	2.326	0	4.918	0	2.115
6	1.225	1.410	0.438	0.8686	0.026	1.711	0.030	1.970	2.534	0	5.078	0	2.004
7	1.134	1.277	0.419	0.8882	0.105	1.672	0.118	1.882	2.704	0.205	5.203	0.076	1.924
8	1.061	1.175	0.373	0.9027	0.167	1.638	0.185	1.815	2.847	0.387	5.307	0.136	1.864
9	1.000	1.094	0.337	0.9139	0.219	1.609	0.239	1.760	2.970	0.546	5.394	0.184	1.816
10	0.949	1.028	0.308	0.9227	0.262	1.584	0.284	1.716	3.078	0.687	5.469	0.223	1.777
11	0.905	0.973	0.285	0.9300	0.299	1.561	0.321	1.679	3.173	0.812	5.534	0.256	1.744
12	0.866	0.925	0.266	0.9359	0.331	1.541	0.354	1.646	3.258	0.924	5.592	0.284	1.716
13	0.832	0.884	0.249	0.9410	0.359	1.523	0.382	1.618	3.336	1.026	5.646	0.308	1.692
14	0.802	0.848	0.235	0.9453	0.384	1.507	0.406	1.594	3.407	1.121	5.693	0.329	1.671
15	0.775	0.816	0.223	0.9490	0.406	1.492	0.428	1.572	3.472	1.207	5.737	0.348	1.652

Statistics	*Standard Giver*		*Analysis of Past Data*	
	Central line	*Limits*	*Central line*	*Limits*
$\bar{X}$	$\bar{X}$	$\bar{X} \pm A\sigma'$	$\bar{X}$	$\bar{X} \pm A_1\sigma$ or $\bar{X} \pm A_2R$
σ	$C_2\sigma'$	$B_1\sigma_1'$, $B_2\sigma'$	$\bar{\sigma}$	$B_2\bar{\sigma}, B_4\bar{\sigma}$
R	$d_2\sigma$	$D_1\sigma'$, $D_2\sigma'$	$\bar{R}$	$D_3\bar{R}, D_1\bar{R}$

VALUES OF e^{-m} (FOR COMPUTING POISSON PROBABILITIES) ($o < m < 1$)

m	0	1	2	3	4	5	6	7	8	9
0.0	1.0000	.9900	.9802	.9704	.9608	.9512	.9418	.9324	.9231	.9139
0.1	0.9048	.8958	.8860	.8781	.8694	.8607	.8521	.8437	.8353	.8270
0.2	0.8187	.8106	.8025	.7945	.7866	.7788	.7711	.7634	.7558	.7483
0.3	0.7408	.7334	.7261	.7189	.7118	.7047	.6977	.6907	.6839	.6771
0.4	0.6703	.6636	.6570	.6505	.6440	.6376	.6313	.6250	.6188	.6126
0.5	0.6065	.6005	.5945	.5886	.5827	.5770	.5712	.5655	.5599	.5543
0.6	0.5488	.5434	.5379	.5326	.5278	.5220	.5160	.5117	.5066	.5016
0.7	0.4966	.4916	.4868	.4810	.4771	.4724	.4670	.4630	.4584	.4538
0.8	0.4493	.4449	.4404	.4360	.4317	.4274	.4232	.4190	.4148	.4107
0.9	0.4066	.4025	.3985	.3946	.3906	.3867	.3829	.3791	.3753	.3716

(m = 1, 2, 3, ...10)

m	1	2	3	4	5	6	7	8	9	10
e^{-m}	.36788	.13534	.04979	.01832	.00698	.00279	.00092	.000395	.000123	.000045

Note: To obtain values of e^{-m} for other values of m, use the laws of exponents.

Example: $e^{-2.35} = (e^{-2.00})\,(e^{-0.35}) = (.13534)\,(.7047) = .095374$.

TOLERANCE FACTORS FOR THE NORMAL DISTRIBUTION

For a normal distribution with unknown mean and unknown standard deviation, a random sample of size n can be used to state with a confidence probability of 95% or 99% that the tolerance interval $\bar{x} \pm ks$ will contain a proportion of at least $1 - \alpha$ of the population.

$\bar{x}$ is the observed sample mean and s is the sample standard deviation $\left[= \sqrt{\frac{\Sigma(x - \bar{x})^2}{n-1}} \right]$.

The tabulated values give the multiplier, k.

95% Confidence Probability

	Proportion $(1-\alpha)$		
n	0.90	0.95	0.99
2	32.019	37.764	48.430
3	8.380	9.916	12.861
4	5.369	6.370	8.299
5	4.275	5.079	6.634
6	3.712	4.414	5.775
7	3.369	4.007	5.248
8	3.136	3.732	4.891
9	2.967	3.532	4.631
10	2.839	3.379	4.433
11	2.737	3.259	4.277
12	2.655	3.162	4.150
13	2.587	3.081	4.044
14	2.529	3.012	3.955
15	2.480	2.954	3.878
16	2.437	2.903	3.812
17	2.400	2.858	3.754
18	2.366	2.819	3.702
19	2.337	2.784	3.656
20	2.310	2.752	3.615
25	2.208	2.631	3.457
30	2.140	2.549	3.350
35	2.090	2.490	3.272
40	2.052	2.445	3.213
45	2.021	2.408	3.165
50	1.996	2.379	3.126
55	1.976	2.354	3.094
60	1.958	2.333	3.066
65	1.943	2.315	3.042
70	1.929	2.299	3.021
75	1.917	2.285	3.002
80	1.907	2.272	2.986
85	1.897	2.261	2.971
90	1.889	2.251	2.958
95	1.881	2.241	2.945
100	1.874	2.233	2.934
150	1.825	2.175	2.859
200	1.798	2.143	2.816
250	1.780	2.121	2.788
300	1.767	2.106	2.767
400	1.749	2.084	2.739
500	1.737	2.070	2.721
600	1.729	2.060	2.707
700	1.722	2.052	2.697
800	1.717	2.046	2.688
900	1.712	2.040	2.682
1000	1.709	2.036	2.676
∞	1.645	1.960	2.576

99% Confidence Probability

	Proportion $(1-\alpha)$		
n	0.90	0.95	0.99
2	160.193	188.491	242.300
3	18.930	22.401	29.055
4	9.398	11.150	14.527
5	6.612	7.855	10.260
6	5.337	6.345	8.301
7	4.613	5.488	7.187
8	4.147	4.936	6.468
9	3.822	4.550	5.966
10	3.582	4.265	5.594
11	3.397	4.045	5.308
12	3.250	3.870	5.079
13	3.130	3.727	4.893
14	3.029	3.608	4.737
15	2.945	3.507	4.605
16	2.872	3.421	4.492
17	2.808	3.345	4.393
18	2.753	3.279	4.307
19	2.703	3.221	4.230
20	2.659	3.168	4.161
25	2.494	2.972	3.904
30	2.385	2.841	3.733
35	2.306	2.748	3.611
40	2.247	2.677	3.518
45	2.200	2.621	3.444
50	2.162	2.576	3.385
55	2.130	2.538	3.335
60	2.103	2.506	3.293
65	2.080	2.478	3.257
70	2.060	2.454	3.225
75	2.042	2.433	3.197
80	2.026	2.414	3.173
85	2.012	2.397	3.150
90	1.999	2.382	3.130
95	1.987	2.368	3.112
100	1.977	2.355	3.096
150	1.905	2.270	2.983
200	1.865	2.222	2.921
250	1.839	2.191	2.880
300	1.820	2.169	2.850
400	1.794	2.138	2.809
500	1.777	2.117	2.783
600	1.764	2.102	2.763
700	1.755	2.091	2.748
800	1.747	2.082	2.736
900	1.741	2.075	2.726
1000	1.736	2.068	2.718
∞	1.645	1.960	2.576

SAMPLE SIZE FOR TWO-SIDED DISTRIBUTION-FREE TOLERANCE LIMITS

The following table gives the size of a random sample to be taken from any underlying distribution in order to ensure that there is a given confidence probability of $100(1 - \gamma)\%$ that at least a proportion $(1 - \alpha)$ of the members of the distribution will lie between the smallest and largest values subsequently observed in the sample.

Proportion $1 - \alpha$	*Confidence Probability: $100(1 - \gamma)\%$* 50%	70%	90%	95%	99%	99.5%
0.995	336	488	777	947	1325	1483
0.99	168	244	388	473	662	740
0.95	34	49	77	93	130	146
0.90	17	24	38	46	64	72
0.85	11	16	25	30	42	47
0.80	9	12	18	22	31	34
0.75	7	10	15	18	24	27
0.70	6	8	12	14	20	22
0.60	4	6	9	10	14	16
0.50	3	5	7	8	11	12

SAMPLE SIZE FOR ONE-SIDED DISTRIBUTION-FREE TOLERANCE LIMITS

The following table gives the size of a random sample to be taken from any underlying distribution in order to ensure that there is a given confidence probability of $100(1 - \gamma)\%$ that at least a proportion $(1 - \alpha)$ of the members of the distribution will be greater than the smallest (or less than the largest) value subsequently observed in the sample.

Proportion $1 - \alpha$	*Confidence Probability: $100(1 - \gamma)\%$* 50	70	95	99	99.5
0.995	139	241	598	919	1379
0.99	69	120	299	459	688
0.95	14	24	59	90	135
0.90	7	12	29	44	66
0.85	5	8	19	29	43
0.80	4	6	14	21	31
0.75	3	5	11	17	25
0.70	2	4	9	13	20
0.60	2	3	6	10	14
0.50	1	2	5	7	10

NUMBER OF RUNS ON EITHER SIDE OF THE MEAN, 5.0% POINT

Table for Testing Randomness of Grouping in a Sequence of Alternatives
(Probability of an equal or smaller number of runs than that listed is $P = 0.05$)
n = cases on one side of average: m = cases on other side of average
m is always taken as the smaller number of cases, n the larger.

m =	6	7	8	9	10	11	12	13	14	15	16	17	18	19	20
n = 6	3														
7	4	4													
8	4	4	5												
9	4	5	5	6											
10	5	5	6	6	6										
11	5	5	6	6	7	7									
12	5	6	6	7	7	8	8								
13	5	6	6	7	8	8	9	9							
14	5	6	7	7	8	8	9	9	10						
15	6	6	7	8	8	9	9	10	10	11					
16	6	6	7	8	8	9	10	10	11	11	11				
17	6	7	7	8	9	9	10	10	11	11	12	12			
18	6	7	8	8	9	10	10	11	11	12	12	13	13		
19	6	7	8	8	9	10	10	11	12	12	13	13	14	14	
20	6	7	8	9	9	10	11	11	12	12	13	13	14	14	15

NUMBER OF RUNS ON EITHER SIDE OF THE MEAN, 0.5% POINT

Table for Testing Randomness of Grouping in a Sequence of Alternatives
(Probability of an equal or smaller number of runs than that listed is $P = 0.005$)
n = cases on one side of average: m = cases on other side of average
m is always taken as the smaller number of cases, n the larger.

m =	6	7	8	9	10	11	12	13	14	15	16	17	18	19	20
n = 6	2														
7	2	3													
8	3	3	3												
9	3	3	3	4											
10	3	3	4	4	5										
11	3	4	4	5	5	5									
12	3	4	4	5	5	6	6								
13	3	4	5	5	5	6	6	7							
14	4	4	5	5	6	6	7	7	7						
15	4	4	5	6	6	7	7	7	8	8					
16	4	5	5	6	6	7	7	8	8	9	9				
17	4	5	5	6	7	7	8	8	8	9	9	10			
18	4	5	6	6	7	7	8	8	9	9	10	10	11		
19	4	5	6	6	7	8	8	9	9	10	10	10	11	11	
20	4	5	6	7	7	8	8	9	9	10	10	11	11	12	12

NUMBER OF RUNS ABOVE AND BELOW THE MEDIAN

Limiting values for the total number of runs above and below the median of a set of values for $P = 0.005$ and $P = 0.05$ where P *is* the probability of an equal or smaller total number of runs.

$m = n$	0.005	0.05
10	4	6
11	5	7
12	6	8
13	7	9
14	7	10
15	8	11
16	9	11
17	10	12
18	10	13
19	11	14
20	12	15
21	13	16
22	14	17
23	14	17
24	15	16
25	16	19
26	17	20
27	18	21
28	18	22
29	19	23
30	20	24
31	21	25
32	22	25
33	23	26
34	23	27
35	24	28
36	25	29
37	26	30
38	27	31
39	28	32
40	29	33
41	29	34
42	30	35
43	31	35
44	32	36
45	33	37
46	34	38
47	35	39
48	35	40
49	36	41
50	37	42
51	38	43
52	39	44
53	40	45
54	41	45
55	42	46
56	42	47
57	43	48
58	44	49
59	45	50
60	46	51
61	47	52
62	48	53
63	49	54
64	49	55
65	50	56
66	51	57
67	52	58
68	53	58
69	54	59
70	55	60
71	56	61
72	57	62
73	57	63
74	58	64
75	59	65
76	60	66
77	61	67
78	62	68
79	63	69
80	64	70
81	65	71
82	66	71
83	66	72
84	67	73
85	68	74
86	69	75
87	70	76
88	71	77
89	72	78
90	73	79
91	74	80
92	75	81
93	75	82
94	76	83
95	77	84
96	78	85
97	79	86
98	80	87
99	81	87
100	82	88

LENGTHS OF RUNS ON EITHER SIDE OF THE MEDIAN

Probability of getting at least one run of specified size or more.

N	0.05	0.01	0.001
10	5	—	—
20	7	8	9
30	8	9	—
40	9	10	12
50	10	11	—

Note: Larger runs than these suggest the existence of non-random influences.

CRITICAL VALUES OF LENGTHS OF RUNS UP AND DOWN

	Probability equal to or less than 0.0032		*Probability equal to or less than 0.0583*	
N	*Run*	*Probability of an equal or greater run*	*Run*	*Probability of an equal or greater run*
4	–	–	3	0.0833
5	–	–	4	0.0167
6	5	0.0028	4	0.0306
7	6	0.0004	4	0.0444
8	6	0.0007	4	0.0583
9	6	0.0011	5	0.0099
10	6	0.0014	5	0.0123
11	6	0.0018	5	0.0147
12	6	0.0021	5	0.0170
13	6	0.0025	5	0.0194
14	6	0.0028	5	0.0217
15	6	0.0032*	5	0.0239*
20	7	0.0006*	5	0.0355*
40	7	0.0015*	6	0.0118*
60	7	0.0023*	6	0.0186*
80	7	0.0032*	6	0.0254*
100	8	0.0005*	6	0.0322*
200	8	0.0010*	7	0.0085*
500	8	0.0024*	7	0.0215*
1,000	9	0.0005*	7	0.0428*
5,000	9	0.0025*	8	0.0245*

Note: *Probabilities based on approximation of the exact distribution by the Poisson exponential.

DERIVATION OF SINGLE SAMPLING PLANS

Values of np_1 and c for constructing single sampling plans whose OC Curve is required to pass through the two points $(p_1, 1 - \alpha)$ and (p_2, β).

Here p_1 is the fraction nonconforming for which the risk of wrongful rejection is to be α, and p_2 is the fraction non-conforming for which the risk of wrongful acceptance is to be β $(p_2 > p_1)$. To construct the plan, find the tabular value of p_2/p_1 in the column for the given α and β which is equal to or just greater than the given value of the ratio. The sample size is found by dividing the value of np_1 corresponding to the selected ratio by p_1. The acceptance number is the value of c corresponding to the selected value of the ratio.

	Values of p_2/p_1 for					*Values of p_2/p_1 for*			
c	$\alpha = 0.05$ $\beta = 0.10$	$\alpha = 0.05$ $\beta = 0.05$	$\alpha = 0.05$ $\beta = 0.01$	np_1	c	$\alpha = 0.01$ $\beta = 0.10$	$\alpha = 0.01$ $\beta = 0.05$	$\alpha = 0.01$ $\beta = 0.01$	np_1
0	44.890	58.404	89.781	0.052	0	229.105	298.073	458.210	0.010
1	10.946	13.349	18.681	0.355	1	26.184	31.933	44.686	0.149
2	6.509	7.699	10.280	0.818	2	12.206	14.439	19.278	0.436
3	4.890	5.675	7.352	1.366	3	8.115	9.418	12.202	0.823
4	4.057	4.646	5.890	1.970	4	6.249	7.156	9.072	1.279
5	3.549	4.023	5.017	2.613	5	5.195	5.889	7.343	1.785
6	3.206	3.604	4.435	3.286	6	4.520	5.082	6.253	2.330
7	2.957	3.303	4.019	3.981	7	4.050	4.524	5.506	2.906
8	2.768	3.074	3.707	4.695	8	3.705	4.115	4.962	3.507
9	2.618	2.895	3.462	5.426	9	3.440	3.803	4.548	4.130
10	2.497	2.750	3.265	6.169	10	3.229	3.555	4.222	4.771
11	2.397	2.630	3.104	6.924	11	3.058	3.354	3.959	5.428
12	2.312	2.528	2.968	7.690	12	2.915	3.188	3.742	6.099
13	2.240	2.442	2.852	8.464	13	2.795	3.047	3.559	6.782
14	2.177	2.367	2.752	9.246	14	2.692	2.927	3.403	7.477
15	2.122	2.302	2.665	10.035	15	2.603	2.823	3.269	8.181
16	2.073	2.244	2.588	10.831	16	2.524	2.732	3.151	8.895
17	2.029	2.192	2.520	11.633	17	2.455	2.652	3.048	9.616
18	1.990	2.145	2.458	12.442	18	2.393	2.580	2.956	10.346
19	1.954	2.103	2.403	13.254	19	2.337	2.516	2.874	11.082
20	1.922	2.065	2.352	14.072	20	2.287	2.458	2.799	11.825
21	1.892	2.030	2.307	14.894	21	2.241	2.405	2.733	12.574
22	1.865	1.999	2.265	15.719	22	2.200	2.357	2.671	13.329
23	1.840	1.969	2.223	16.548	23	2.162	2.313	2.615	14.088
24	1.817	1.942	2.191	17.382	24	2.126	2.272	2.564	14.853
25	1.795	1.917	2.158	18.218	25	2.094	2.235	2.516	15.623
26	1.775	1.893	2.127	19.058	26	2.064	2.200	2.472	16.397
27	1.757	1.871	2.098	19.900	27	2.035	2.168	2.431	17.175
28	1.739	1.850	2.071	20.746	28	2.009	2.138	2.393	17.957
29	1.723	1.831	2.046	21.594	29	1.985	2.110	2.358	18.742
30	1.707	1.813	2.023	22.444	30	1.962	2.083	2.324	19.532
31	1.692	1.796	2.001	23.298	31	1.940	2.059	2.293	20.324
32	1.679	1.780	1.980	24.152	32	1.920	2.035	2.264	21.120
33	1.665	1.764	1.960	25.010	33	1.900	2.013	2.236	21.919
34	1.653	1.750	1.941	25.870	34	1.882	1.992	2.210	22.721
35	1.641	1.736	1.923	26.731	35	1.865	1.973	2.185	23.525
36	1.630	1.723	1.906	27.594	36	1.848	1.954	2.162	24.333
37	1.619	1.710	1.890	28.460	37	1.833	1.936	2.139	25.143
38	1.609	1.698	1.875	29.327	38	1.818	1.920	2.118	25.955
39	1.599	1.687	1.860	30.196	39	1.804	1.903	2.098	26.770
40	1.590	1.676	1.846	31.066	40	1.790	1.887	2.079	27.587
41	1.581	1.666	1.833	31.938	41	1.777	1.873	2.060	28.406
42	1.572	1.656	1.820	32.812	42	1.765	1.859	2.043	29.228
43	1.564	1.646	1.807	33.686	43	1.753	1.845	2.026	30.051
44	1.556	1.637	1.796	34.563	44	1.742	1.832	2.010	30.877
45	1.548	1.628	1.784	35.441	45	1.731	1.820	1.994	31.704
46	1.541	1.619	1.773	36.320	46	1.720	1.808	1.980	32.534
47	1.534	1.611	1.763	37.200	47	1.710	1.796	1.965	33.365
48	1.527	1.603	1.752	38.082	48	1.701	1.785	1.952	34.198
49	1.521	1.596	1.743	38.965	49	1.691	1.775	1.938	35.032

CONSTRUCTION OF O.C. CURVES FOR SINGLE SAMPLING PLANS

Values of np_1 for which the probability of acceptance of c or fewer non-conforming items in a sample of n is $P(A)$.

To find the non-conforming fraction p, corresponding to a probability of acceptance $P(A)$ in a single sampling plan with sample size n and acceptance number c, divide by n the entry in the row for the given c and the column for the given $P(A)$. Plot $P(A)$ against p.

$P(A)$ =	0.995	0.990	0.975	0.950	0.900	0.750	0.500	0.250	0.100	0.050	0.025	0.010	0.005
c = 0	0.0050	0.010	0.025	0.051	0.105	0.288	0.693	1.386	2.303	2.996	3.689	4.605	5.298
1	0.103	0.149	0.242	0.355	0.532	0.961	1.678	2.693	3.890	4.744	5.572	6.638	7.430
2	0.338	0.436	0.619	0.818	1.102	1.727	2.674	3.920	5.322	6.296	7.224	8.406	9.274
3	0.672	0.823	1.090	1.366	1.745	2.535	3.672	5.109	6.681	7.754	8.768	10.045	10.978
4	1.078	1.279	1.623	1.970	2.433	3.369	4.671	6.274	7.994	9.154	10.242	11.605	12.594
5	1.537	1.785	2.202	2.613	3.152	4.219	5.670	7.423	9.275	10.513	11.668	13.108	14.150
6	2.037	2.330	2.814	3.286	3.895	5.083	6.670	8.558	10.532	11.842	13.060	14.571	15.660
7	2.571	2.906	3.454	3.981	4.656	5.956	7.669	9.684	11.771	13.148	14.422	16.000	17.134
8	3.132	3.507	4.115	4.695	5.432	6.838	8.669	10.802	12.995	14.434	15.763	17.403	18.578
9	3.717	4.130	4.795	5.426	6.221	7.726	9.669	11.914	14.206	15.705	17.085	18.783	19.998
10	4.321	4.771	5.491	6.169	7.021	8.620	10.668	13.020	15.407	16.962	18.390	20.145	21.398
11	4.943	5.428	6.201	6.924	7.829	9.519	11.668	14.121	16.598	18.208	19.682	21.490	22.779
12	5.580	6.099	6.922	7.690	8.646	10.422	12.668	15.217	17.782	19.442	20.962	22.821	24.145
13	6.231	6.782	7.654	8.464	9.470	11.329	13.668	16.310	18.958	20.668	22.230	24.139	25.496
14	6.893	7.477	8.396	9.246	10.300	12.239	14.668	17.400	20.128	21.886	23.490	25.446	26.836
15	7.566	8.181	9.144	10.035	11.135	13.152	15.668	18.486	21.292	23.098	24.741	27.743	28.166
16	8.249	8.895	9.902	10.831	11.976	14.068	16.668	19.570	22.452	24.302	25.984	28.031	29.484
17	8.942	9.616	10.666	11.633	12.822	14.986	17.668	20.652	23.606	25.500	27.220	29.310	30.792
18	9.644	10.346	11.438	12.442	13.672	15.907	18.668	21.731	24.756	26.692	28.448	30.581	32.092
19	10.353	11.082	12.216	13.254	14.525	16.830	19.668	22.808	25.902	27.879	29.671	31.845	33.383
20	11.069	11.825	12.999	14.072	15.383	17.755	20.668	23.883	27.045	29.062	30.888	33.103	34.668
21	11.791	12.574	13.787	14.894	16.244	18.682	21.668	24.956	28.184	30.241	32.102	34.355	35.947
22	12.520	13.329	14.580	15.719	17.108	19.610	22.668	26.028	29.320	31.416	33.309	35.601	37.219
23	13.255	14.088	15.377	16.548	17.975	20.540	23.668	27.098	30.453	32.586	34.512	36.841	38.485
24	13.995	14.853	16.178	17.382	18.844	21.471	24.668	28.167	31.584	33.752	35.710	38.077	39.745
25	14.740	15.623	16.984	18.218	19.717	22.404	25.667	29.234	32.711	34.916	36.905	39.308	41.000
26	15.490	16.397	17.793	19.058	20.592	23.338	26.667	30.300	33.836	36.077	38.096	40.535	42.252
27	16.245	17.175	18.606	19.900	21.469	24.273	27.667	31.365	34.959	37.234	39.284	41.757	43.497
28	17.004	17.957	19.422	20.746	22.348	25.209	28.667	32.428	36.080	38.389	40.468	42.975	44.738
29	17.767	18.742	20.241	21.594	23.229	26.147	29.667	33.491	37.198	39.541	41.649	44.190	45.976
30	18.534	19.532	21.063	22.444	24.113	27.086	30.667	34.552	38.315	40.690	42.827	45.401	47.210
31	19.305	20.324	21.888	23.298	24.998	28.025	31.667	35.613	39.430	41.838	44.002	46.609	48.440
32	20.079	21.120	22.716	24.152	25.885	28.966	32.667	36.672	40.543	42.982	45.174	47.813	49.666
33	20.856	21.919	23.546	25.010	26.774	29.907	33.667	37.731	41.654	44.125	46.344	49.015	50.888
34	21.638	22.721	24.379	25.870	27.664	30.849	34.667	38.788	42.764	45.266	47.512	50.213	52.108
35	22.422	23.525	25.214	26.731	28.556	31.792	35.667	39.845	43.872	46.404	48.676	51.409	53.324
36	23.208	24.333	26.052	27.594	29.450	32.736	36.667	40.901	44.978	47.540	49.840	52.601	54.538
37	23.998	25.143	26.891	28.460	30.345	33.681	37.667	41.957	46.083	48.676	51.000	53.791	55.748
38	24.791	25.955	27.733	29.327	31.241	34.626	38.667	43.011	47.187	49.808	52.158	54.979	56.956
39	25.586	26.770	28.576	30.196	32.139	35.572	39.667	44.065	48.289	50.940	53.314	56.164	58.160
40	26.384	27.587	29.422	31.066	33.038	36.519	40.667	45.118	49.390	52.069	54.469	57.347	59.363
41	27.184	28.406	30.270	31.938	33.938	37.466	41.667	46.171	50.490	53.197	55.622	58.528	60.563
42	27.986	29.228	31.120	32.812	34.839	38.414	42.667	47.223	51.589	54.324	56.772	59.717	61.761
43	28.791	30.051	31.970	33.686	35.742	39.363	43.667	48.274	52.686	55.449	57.921	60.884	62.956
44	29.598	30.877	32.824	34.563	36.646	40.312	44.667	49.325	53.782	56.572	59.068	62.059	64.150
45	30.408	31.704	33.678	35.441	37.550	41.262	45.667	50.375	54.878	57.695	60.214	63.231	65.340
46	31.219	32.534	34.534	36.320	38.456	42.212	46.667	51.425	55.972	58.816	61.358	64.402	66.529
47	32.032	33.365	35.392	37.200	39.363	43.163	47.667	52.474	57.064	59.936	62.500	65.571	67.716
48	32.848	34.198	36.250	38.082	40.270	44.115	48.667	53.522	58.158	61.054	63.641	66.738	68.901
49	33.664	35.032	37.111	38.965	41.179	45.067	49.667	54.571	59.249	62.171	64.780	67.903	70.084

RANDOM NUMBERS

03 47 43 73 86	36 96 47 36 61	46 98 63 71 62	33 26 16 80 45	60 11 14 10 95
97 74 24 67 62	42 81 14 57 20	42 53 32 37 32	27 07 36 07 51	24 51 79 89 73
16 76 62 27 66	56 50 26 71 07	32 90 79 78 53	13 55 38 58 59	88 97 54 14 10
12 56 85 99 26	96 96 68 27 31	05 03 72 93 15	57 12 10 14 21	88 26 49 81 76
55 59 56 35 64	38 54 82 46 22	31 62 43 09 90	06 18 44 32 53	23 83 01 30 30
16 22 77 94 39	49 54 43 54 82	17 37 93 23 78	87 35 20 96 43	84 26 34 91 64
84 42 17 53 31	57 24 55 06 88	77 04 74 47 67	21 76 33 50 25	83 92 12 06 76
63 01 63 78 59	16 95 55 67 19	98 10 50 71 75	12 86 73 58 07	44 39 52 38 79
33 21 12 34 29	78 64 56 07 82	52 42 07 44 38	15 51 00 13 42	99 66 02 79 54
57 60 86 32 44	09 47 27 96 54	49 17 46 09 62	90 52 84 77 27	08 02 73 43 28
18 18 07 92 46	44 17 16 58 09	79 83 86 19 62	06 76 50 03 10	55 23 64 05 05
26 62 38 97 75	84 16 07 44 99	83 11 46 32 24	20 14 85 88 45	10 93 72 88 71
23 42 40 64 74	82 97 77 77 81	07 45 32 14 08	32 98 94 07 72	93 85 79 10 75
52 36 28 19 95	50 92 26 11 97	00 56 76 31 38	80 22 02 53 53	86 60 42 04 53
37 85 94 35 12	83 39 50 08 30	42 34 07 96 88	54 42 06 87 98	35 85 29 48 39
70 29 17 12 13	40 33 20 38 26	13 89 51 03 74	17 76 37 13 04	07 74 21 19 30
56 62 18 37 35	96 83 50 87 75	97 12 25 93 47	70 33 24 03 54	97 77 46 44 80
99 49 57 22 77	88 42 95 45 72	16 64 36 16 00	04 43 18 66 79	94 77 24 2190
16 08 15 04 72	33 27 14 34 09	45 59 34 68 49	12 72 07 34 45	99 27 72 95 14
31 16 93 32 43	50 27 89 87 19	20 15 37 00 49	52 85 66 60 44	38 68 88 11 80
68 34 30 13 70	55 74 30 77 40	44 22 78 84 26	04 33 46 09 52	68 07 97 06 57
74 57 25 65 76	59 29 97 68 60	71 91 38 67 54	13 58 18 24 76	15 54 55 95 52
27 42 37 86 53	48 55 90 65 72	96 57 69 36 10	96 46 92 42 45	97 60 49 04 91
00 39 68 29 61	66 37 32 20 30	77 84 57 03 29	10 45 65 04 26	11 04 96 67 24
29 94 98 94 24	68 49 69 10 82	53 75 91 93 30	34 25 20 57 27	40 48 73 51 92
16 90 82 66 59	83 62 64 11 12	67 19 00 72 74	60 57 2129 68	02 02 37 03 31
11 27 94 75 06	06 09 19 74 66	02 94 37 34 02	76 70 90 30 86	38 45 94 30 38
35 24 10 16 20	33 32 51 26 38	79 78 45 04 91	16 92 53 56 16	02 75 50 95 98
38 23 16 86 38	42 38 97 01 50	87 75 66 81 41	40 01 74 91 62	48 51 84 08 32
31 96 25 91 47	96 44 33 49 13	34 86 82 53 91	00 52 43 48 85	27 55 26 89 62
66 67 40 67 14	64 05 71 95 86	11 05 65 09 68	76 83 20 37 90	57 16 00 11 66
14 90 84 45 11	75 73 88 05 90	52 27 41 14 86	22 98 12 22 08	07 52 74 95 80
68 05 51 18 00	33 96 02 75 19	07 60 62 93 55	59 33 82 43 90	49 37 38 44 59
20 46 78 73 90	97 51 40 14 02	04 02 33 3108	39 54 16 49 36	47 95 93 13 30
64 19 58 97 79	15 06 15 93 20	01 90 10 75 06	40 78 78 89 62	02 67 74 17 33
05 26 93 70 60	22 35 85 15 13	92 03 51 59 77	59 56 78 06 83	52 91 05 70 74
07 97 10 88 23	09 98 42 99 64	61 71 62 99 15	06 51 29 16 93	58 05 77 08 51
68 71 86 85 85	54 87 66 47 54	73 32 08 11 12	44 95 92 63 16	29 56 24 29 48
26 99 61 65 53	58 37 78 80 70	42 10 50 67 53	32 17 55 85 74	94 44 67 16 94
14 65 52 68 75	87 59 36 22 41	26 78 63 06 55	13 08 27 01 50	15 29 39 39 43
17 53 77 58 71	71 41 61 50 72	12 41 94 96 26	44 95 27 36 99	02 96 74 30 83
90 26 59 21 19	23 52 23 33 12	95 93 02 18 39	07 02 18 36 07	25 99 32 70 23
41 23 52 55 99	31 04 49 69 96	10 47 48 45 88	13 4143 89 20	97 17 14 49 17
60 20 50 81 69	31 99 73 68 68	35 81 33 03 76	24 30 12 48 60	18 99 10 72 34
91 25 38 05 90	94 58 28 41 36	45 37 59 03 09	90 35 57 29 12	82 62 54 65 60
34 50 57 74 37	98 80 33 00 91	09 77 93 19 82	74 94 80 04 04	45 07 31 66 49
85 22 04 30 43	73 81 53 94 79	33 62 46 86 28	08 31 54 46 31	53 94 13 38 47
09 79 13 77 48	73 82 97 22 21	05 03 27 24 83	72 89 44 05 60	35 80 39 94 88
88 75 80 18 14	22 95 75 42 49	39 32 82 22 49	02 48 07 70 37	16 04 61 67 87
90 96 23 70 00	39 00 03 06 90	55 85 78 38 36	94 37 30 69 32	90 89 00 76 33

Random Numbers (*Contd.*)

53 74 23 99 67	61 32 28 69 84	94 62 67 86 24	98 33 41 19 95	47 53 53 38 09
63 38 06 86 54	99 00 65 26 94	02 82 90 23 07	79 62 67 80 60	75 91 12 81 19
35 30 58 21 46	06 72 17 10 94	25 21 31 75 96	49 28 24 00 49	55 65 79 78 07
63 43 36 82 69	65 51 18 37 88	61 38 44 12 45	32 92 85 88 65	54 34 81 85 35
98 25 37 55 26	01 91 82 81 46	74 71 12 94 97	24 02 71 37 07	03 92 18 66 75
02 63 21 17 69	71 50 80 89 56	38 15 70 11 48	43 40 45 86 98	00 83 26 91 03
64 55 22 21 82	48 22 28 06 00	61 54 13 43 91	82 78 12 23 29	06 66 24 12 27
85 07 26 13 89	01 10 07 82 04	59 63 69 36 03	69 11 15 83 80	13 29 54 19 28
58 54 16 24 15	51 54 44 82 00	62 61 65 04 69	38 18 65 18 97	85 72 13 49 21
34 85 27 84 87	61 48 64 56 26	90 18 48 13 26	37 70 15 42 57	65 65 80 39 07
03 92 18 27 46	57 99 16 96 56	30 33 72 85 22	84 64 38 56 98	99 01 30 98 64
62 95 30 27 59	37 75 41 66 48	86 97 80 61 45	23 53 04 01 63	45 76 08 64 27
08 45 93 15 22	60 21 75 46 91	98 77 27 85 42	28 88 61 08 84	69 62 03 42 73
07 08 55 18 40	45 44 75 13 90	24 94 96 61 02	57 55 66 83 15	73 42 37 11 61
01 85 89 95 66	51 10 19 34 88	15 84 97 19 75	12 76 39 43 78	64 63 91 08 25
72 84 71 14 35	19 11 58 49 26	50 11 17 17 76	86 31 57 20 18	95 60 78 46 75
88 78 28 16 84	13 52 53 94 53	75 45 69 30 96	73 89 65 70 31	99 17 43 48 76
45 17 75 65 57	28 40 19 72 12	25 12 74 75 67	60 40 60 81 19	24 62 01 61 16
96 76 28 12 54	22 01 11 94 25	71 96 16 16 88	68 64 36 74 45	19 59 50 88 92
43 31 67 72 30	24 02 94 08 63	38 32 36 66 02	69 36 38 25 39	48 03 45 15 22
50 44 66 44 21	66 06 58 05 62	68 15 54 35 02	42 35 48 96 32	14 52 41 52 48
22 66 22 15 86	26 63 75 41 99	58 42 36 72 24	58 37 52 18 51	03 37 18 39 11
96 24 40 14 51	23 22 30 88 57	95 67 47 29 83	94 69 40 06 07	18 16 36 78 86
31 73 91 61 19	60 20 72 98 48	98 57 07 23 69	65 95 39 69 58	56 80 30 19 44
78 60 73 99 84	43 89 94 36 45	56 69 47 07 41	90 22 91 07 12	78 35 34 08 72
84 37 90 61 56	70 10 23 98 05	85 11 34 76 60	76 48 45 34 60	01 64 18 39 96
36 67 10 08 23	98 93 35 08 86	99 29 76 29 81	33 34 91 58 93	63 14 52 32 52
07 28 59 07 48	89 64 58 89 75	83 85 62 27 89	30 14 78 56 27	86 63 59 80 02
10 15 83 87 60	79 24 31 66 56	21 48 24 06 93	91 98 94 05 49	01 47 59 38 00
55 19 68 97 65	03 73 52 16 56	00 53 55 90 27	33 42 29 38 87	22 13 88 83 34
53 81 29 13 39	35 01 20 71 34	62 33 74 82 14	53 73 19 09 03	56 54 29 56 93
51 86 32 68 92	33 98 74 66 99	40 14 71 94 58	45 94 19 38 81	14 44 99 81 07
35 91 70 29 13	80 03 54 07 27	96 94 78 32 66	50 95 52 74 33	13 80 55 62 54
37 71 67 95 13	20 02 44 95 94	64 85 04 05 72	01 32 90 76 14	53 89 74 60 41
93 66 13 83 27	92 79 64 64 72	28 54 96 53 84	48 14 52 98 94	56 07 93 89 30
02 96 08 45 65	13 05 00 41 84	93 07 54 72 59	21 45 57 09 77	19 48 56 27 44
49 83 43 48 35	82 88 33 69 96	72 36 04 19 76	47 45 15 18 60	82 11 08 95 97
84 60 71 62 46	40 80 81 30 37	34 39 23 05 38	25 15 35 71 30	88 12 57 21 77
18 71 30 88 71	44 91 14 88 47	89 23 30 63 15	56 34 20 47 89	99 82 93 24 98
79 69 10 61 78	71 32 76 95 62	87 00 22 58 40	92 54 01 75 25	43 11 71 99 31
75 93 36 57 83	56 20 14 82 11	74 21 97 90 65	96 42 68 63 86	74 54 13 26 94
38 30 92 29 03	06 28 81 39 38	62 25 06 84 63	61 29 08 93 67	04 32 92 08 09
51 29 50 10 34	31 57 75 95 80	51 97 02 74 77	76 15 48 49 44	18 55 63 77 09
21 31 38 86 24	37 79 81 53 74	73 24 16 10 33	52 83 90 94 76	70 47 14 54 36
29 01 23 87 88	58 02 39 37 67	42 10 14 20 92	16 55 23 42 45	54 96 09 11 06
95 33 95 22 00	18 74 72 00 18	38 79 58 69 32	81 76 80 26 92	82 80 84 25 39
90 84 60 79 80	24 36 59 87 38	82 07 53 89 35	96 35 23 79 18	05 98 90 07 35
46 40 62 98 82	54 97 20 56 95	15 74 80 08 32	16 46 70 50 80	67 72 16 42 79
20 31 89 03 43	38 46 82 68 72	32 14 82 99 70	80 60 47 18 97	63 49 30 21 30
71 59 73 05 50	08 22 23 71 77	91 01 93 20 49	82 96 59 26 94	66 39 67 98 60

Random Numbers (*Contd.*)

22 17 66 65 84	68 95 23 92 35	87 02 22 57 51	61 09 43 95 06	58 24 82 03 47
19 36 27 59 46	13 79 93 37 55	39 77 32 77 09	85 52 05 30 62	47 83 51 62 74
16 77 23 02 77	09 61 87 25 21	28 06 24 25 93	16 71 13 59 78	23 05 47 47 25
78 43 76 71 61	20 44 90 32 64	97 67 63 99 61	46 38 03 93 22	69 81 21 99 21
03 28 28 26 08	73 37 32 04 05	69 30 16 09 05	88 69 58 28 99	35 07 44 75 47
93 22 53 64 39	07 10 63 76 35	87 03 04 79 88	08 13 13 85 51	55 34 57 72 69
78 76 58 54 74	92 38 70 96 92	52 06 79 79 45	82 63 18 27 44	69 66 92 19 09
23 68 35 26 00	99 53 93 61 28	52 70 05 48 34	56 65 05 61 86	90 92 10 70 80
15 39 25 70 99	93 86 52 77 65	15 33 59 05 28	22 87 26 07 47	86 96 98 29 06
58 71 96 30 24	18 46 23 34 27	85 13 99 24 44	49 18 09 79 49	74 16 32 23 02
57 35 27 33 72	24 53 63 94 09	41 10 76 47 91	44 04 95 49 66	39 60 04 59 81
48 50 86 54 48	22 06 34 72 52	82 21 15 65 20	33 29 94 71 11	15 91 29 12 03
61 96 48 95 03	07 16 39 33 66	98 56 10 56 79	77 21 30 27 12	90 49 22 23 62
36 93 89 41 26	29 70 83 63 51	99 74 20 52 36	87 09 41 15 09	98 60 16 03 03
18 87 00 42 31	57 90 12 02 07	23 47 37 17 31	54 08 01 88 63	39 41 88 92 10
88 56 53 27 59	33 35 72 67 47	77 34 55 45 70	08 18 27 38 90	16 95 86 70 75
09 72 95 84 29	49 41 31 06 70	42 38 06 45 18	64 84 73 31 65	52 53 37 97 15
12 96 88 17 31	65 19 69 02 83	60 75 86 90 68	24 64 19 35 51	56 61 87 39 12
85 94 57 24 16	92 09 84 38 76	22 00 27 69 85	29 81 94 78 70	21 94 47 90 12
38 64 43 59 98	98 77 87 68 07	91 51 67 62 44	40 98 05 93 78	23 32 65 41 18
53 44 09 42 72	00 41 85 79 79	68 47 22 00 20	35 55 31 51 51	00 83 63 22 55
40 76 66 26 84	57 99 99 90 37	36 63 32 08 58	37 40 13 68 97	87 64 81 07 83
02 17 79 18 05	12 59 52 57 02	22 07 90 47 03	28 14 11 30 79	20 69 22 40 98
95 17 82 06 53	31 51 10 96 46	92 06 88 07 77	56 11 50 81 69	40 23 72 51 39
35 76 22 42 92	96 11 83 44 80	34 68 35 48 77	33 42 40 90 60	73 96 53 97 86
26 29 13 56 41	85 47 04 66 08	34 72 57 59 13	82 43 80 46 15	38 26 61 70 04
77 80 20 75 82	72 82 32 99 90	63 95 73 76 63	89 73 44 99 05	48 67 26 43 18
46 40 66 44 52	91 36 74 43 53	30 82 13 54 00	78 45 63 98 35	55 03 36 67 68
37 56 08 18 09	77 53 84 46 47	31 91 18 95 58	24 16 74 11 53	44 10 13 85 57
61 65 61 68 66	37 27 47 39 19	84 83 70 07 48	53 21 40 06 71	95 06 79 88 54
93 43 69 64 07	34 18 04 52 35	56 27 09 24 86	61 85 53 83 45	19 90 70 99 00
21 96 60 12 99	11 20 99 45 18	48 13 93 55 34	18 37 79 49 90	65 97 38 20 46
95 20 47 97 97	27 37 83 28 71	00 06 41 41 74	45 89 09 39 84	51 67 11 52 49
97 86 21 78 73	10 65 81 92 59	58 76 17 14 97	04 76 62 16 17	17 95 70 45 80
69 92 06 34 13	59 71 74 17 32	27 55 10 24 19	23 71 82 13 74	63 52 52 01 41
04 31 17 21 56	33 73 99 19 87	26 72 39 27 67	53 77 57 68 93	60 61 97 22 61
61 06 98 03 91	87 14 77 43 96	43 00 65 98 50	45 60 33 01 07	98 99 46 50 47
85 93 85 86 88	72 87 08 62 40	16 06 10 89 20	23 21 34 74 97	76 38 03 29 63
21 74 32 47 45	73 96 07 94 52	09 65 90 77 47	25 76 16 19 33	53 05 70 53 30
15 69 53 82 80	79 96 23 53 10	65 39 07 16 29	45 33 02 43 70	02 87 40 41 45
02 89 08 04 49	20 21 14 68 86	87 63 93 95 17	11 29 01 95 80	35 14 97 35 33
87 18 15 89 79	85 43 01 72 73	08 61 74 51 69	89 74 39 82 15	94 51 33 41 67
98 83 71 94 22	59 97 50 99 52	08 52 85 08 40	87 80 61 65 31	91 51 80 32 44
10 08 58 21 66	72 68 49 29 31	89 85 84 46 06	59 73 19 85 23	65 09 29 75 63
47 90 56 10 08	88 02 84 27 83	42 29 72 23 19	66 56 45 65 79	20 71 53 20 25
22 85 61 68 90	49 64 92 85 44	16 60 12 89 88	50 14 49 81 06	01 82 77 45 12
67 80 43 79 33	12 83 11 41 16	25 58 19 68 70	77 02 54 00 52	53 43 37 15 26
27 62 50 96 72	79 44 61 40 15	14 53 40 65 39	27 31 58 50 28	11 39 03 34 25
33 78 80 87 15	38 30 06 38 21	14 47 47 07 26	54 96 87 53 32	40 36 40 96 76
13 13 92 66 99	47 24 49 57 74	32 25 43 62 17	10 97 11 69 84	99 63 22 32 98

Random Numbers (*Contd.*)

10 27 53 96 23	71 50 54 36 23	54 31 04 82 98	04 14 12 15 09	26 78 25 47 47
28 41 50 61 88	64 85 27 20 18	83 36 36 05 56	39 71 65 09 62	94 76 62 11 89
34 21 42 57 02	59 19 18 97 48	80 30 03 30 98	05 24 67 70 07	84 97 50 87 46
61 81 77 23 23	82 82 1154 08	53 28 70 58 96	44 07 39 55 43	42 34 43 39 28
61 15 18 13 54	16 86 20 26 88	90 74 80 55 09	14 53 90 51 17	52 01 63 01 59
91 76 21 64 64	44 91 13 32 97	75 31 62 66 54	84 80 32 75 77	56 08 25 70 29
00 97 79 08 06	37 30 28 59 85	53 56 68 53 40	01 74 39 59 73	30 19 99 85 48
36 46 18 34 94	75 20 80 27 77	78 91 69 16 00	08 43 18 73 68	67 69 61 34 25
88 98 99 60 50	65 95 79 42 94	93 62 40 89 96	43 56 47 71 66	46 76 29 67 02
04 37 59 87 21	05 02 03 24 17	47 97 81 56 51	92 34 86 01 82	55 51 33 12 91
63 62 06 34 41	94 21 78 55 09	72 76 45 16 94	29 95 81 83 83	79 88 01 97 30
78 47 23 53 90	34 41 92 45 71	09 23 70 70 07	12 38 92 79 43	14 85 11 47 23
87 68 62 15 43	53 14 36 59 25	54 47 33 70 15	59 24 48 40 35	50 03 42 99 36
47 60 92 10 77	88 59 53 11 52	66 25 69 07 04	48 68 64 71 06	61 65 70 22 12
56 88 87 59 41	65 28 04 67 53	95 79 88 37 31	50 41 06 94 76	81 83 17 16 33
02 57 45 86 67	73 43 07 34 48	44 26 87 93 29	77 09 61 67 84	06 69 44 77 75
31 54 14 13 17	48 62 11 90 60	68 12 93 64 28	46 24 79 16 76	14 60 25 51 01
28 50 16 43 36	28 97 85 58 99	67 22 52 76 23	24 70 36 54 54	59 28 61 71 96
63 29 62 66 50	02 63 45 52 38	67 63 47 54 75	83 24 78 43 20	92 63 13 47 48
45 65 58 26 51	76 96 59 38 72	86 57 45 71 46	44 67 76 14 55	44 88 01 62 12
39 65 36 63 70	77 45 85 50 51	74 13 39 35 22	30 53 36 02 95	49 34 88 73 61
73 71 98 16 04	29 18 94 51 23	76 51 94 84 86	79 93 96 38 63	08 58 25 58 94
72 20 56 20 11	72 65 71 08 86	79 57 95 13 91	97 48 72 66 48	09 71 17 24 89
75 17 26 99 76	89 37 20 70 01	77 31 61 95 46	26 97 05 73 51	53 33 18 72 87
37 48 69 82 29	81 30 15 39 14	48 38 75 93 29	06 87 37 78 48	45 56 00 84 47
68 08 02 80 72	83 71 46 30 49	89 17 95 88 29	02 39 56 03 46	97 74 06 56 17
14 23 98 61 67	70 52 85 01 50	01 84 02 78 43	10 62 98 19 41	18 83 99 47 99
49 08 96 21 44	25 27 99 41 28	07 41 08 34 66	19 42 74 39 91	41 96 53 78 72
78 37 06 08 43	63 61 62 42 29	39 68 95 10 96	09 24 23 00 62	56 12 80 73 16
37 21 34 17 68	68 96 83 23 56	32 84 60 15 31	44 73 67 34 77	91 15 79 74 58
14 29 09 34 04	87 83 07 55 07	76 58 30 83 64	87 29 25 58 84	86 50 60 00 25
58 43 28 06 36	49 52 83 51 14	47 56 92 29 34	05 87 31 06 95	12 45 57 09 09
10 43 67 29 70	80 62 80 03 42	10 80 21 38 84	90 56 35 03 09	43 12 74 49 14
44 38 88 39 54	86 97 37 44 22	00 95 01 31 76	17 16 29 56 63	38 78 94 49 81
90 69 59 19 51	85 39 52 85 13	07 28 37 07 61	11 16 36 27 03	78 86 72 04 95
41 47 10 25 62	97 05 31 03 61	20 26 36 31 62	68 69 86 95 44	84 95 48 46 45
91 94 14 63 19	75 89 11 47 11	31 56 34 19 09	79 57 92 36 59	14 93 87 81 40
80 06 54 18 66	09 18 94 06 19	98 40 07 17 81	22 45 44 84 11	24 62 20 42 31
67 72 77 63 48	84 08 31 55 58	24 33 45 77 58	80 45 67 93 82	75 70 16 08 24
59 40 24 13 27	79 26 88 86 30	01 31 60 10 39	53 58 47 70 93	85 81 56 39 38
05 90 35 89 95	01 61 16 96 94	50 78 13 69 36	37 68 53 37 31	71 26 35 03 71
44 43 80 69 98	46 68 05 14 82	90 78 50 05 62	77 79 13 57 44	59 60 10 39 66
61 81 31 96 82	00 57 25 60 59	46 72 60 18 77	55 66 12 62 11	08 99 55 64 57
42 88 07 10 05	24 98 65 63 21	47 21 61 88 32	27 80 30 21 60	10 92 35 36 12
77 94 30 05 39	28 10 99 00 27	12 73 73 99 12	49 99 57 94 82	96 88 57 17 91
78 83 19 76 16	94 11 68 84 26	23 54 20 86 85	23 86 66 99 07	36 37 34 92 09
87 76 59 61 81	43 63 64 61 61	65 76 36 95 90	18 48 27 45 68	27 23 65 30 72
91 43 05 96 47	55 78 99 95 24	37 55 85 78 78	0148 41 19 10	35 19 54 07 73
84 97 77 72 73	09 62 06 65 72	87 12 49 03 60	41 15 20 76 27	50 47 02 29 16
87 41 60 76 83	44 88 96 07 80	83 05 83 38 96	73 70 66 81 90	30 56 10 48 59

Random Numbers (*Contd.*)

28 89 65 87 08	13 50 63 04 23	25 47 57 91 31	52 62 24 19 94	91 67 48 57 10
30 29 43 65 42	78 66 28 55 80	47 46 41 90 08	55 98 78 10 70	49 92 05 12 07
95 74 62 60 53	51 57 32 22 27	12 72 72 27 77	44 67 32 23 13	67 95 07 76 30
01 85 54 96 72	66 86 65 64 60	56 59 75 36 75	46 44 33 63 71	54 50 06 44 75
10 91 46 96 86	19 83 52 47 53	65 00 51 93 51	30 80 05 19 29	56 23 27 19 03
05 33 18 08 51	51 78 57 26 17	34 87 96 23 95	89 99 93 39 79	11 28 94 15 52
04 43 13 37 00	79 68 96 26 60	70 39 83 66 56	62 03 55 86 57	77 55 33 62 02
05 85 40 25 24	73 52 93 70 50	48 21 47 74 63	17 27 27 51 26	35 96 29 00 45
84 90 90 65 77	63 99 25 69 02	09 04 03 35 78	19 79 95 07 21	02 84 48 51 97
28 55 53 09 48	86 28 30 02 35	71 30 32 06 47	93 74 21 86 33	49 90 21 69 74
89 83 40 69 80	97 96 47 59 97	56 33 24 87 36	17 18 16 90 46	75 27 28 52 13
73 20 96 05 68	93 41 69 96 07	97 50 81 79 59	42 37 13 81 83	82 42 85 04 31
10 89 07 76 21	40 24 74 36 42	40 33 04 46 24	35 63 02 31 61	34 59 43 36 96
91 50 27 78 37	06 06 16 25 98	17 78 80 36 85	26 41 77 63 37	71 63 94 94 33
03 45 44 66 88	97 81 26 03 89	39 46 67 21 17	98 10 39 33 15	61 63 00 25 92
89 41 58 91 63	65 99 59 97 84	90 14 79 61 55	56 16 88 87 60	32 15 99 67 43
13 43 00 97 26	16 91 21 32 41	60 22 66 72 17	31 85 33 69 07	68 49 20 43 29
71 71 00 51 72	62 03 89 26 32	35 27 99 18 25	78 12 03 09 70	50 93 19 35 56
19 28 15 00 41	92 27 73 40 38	37 11 05 75 16	98 81 99 37 29	92 20 32 39 67
56 38 30 92 30	45 51 94 69 04	00 84 14 36 37	95 66 39 01 09	21 68 40 95 79
39 27 52 89 11	00 81 06 28 48	12 08 05 75 26	03 35 63 05 77	13 81 20 67 58
73 13 28 58 01	05 06 42 24 07	60 60 29 99 93	72 93 78 04 36	25 76 01 54 03
81 60 84 51 57	12 68 46 55 89	60 09 71 87 89	70 81 10 95 91	83 79 68 20 66
05 62 98 07 85	07 79 26 69 61	67 85 72 37 41	85 79 76 84 23	61 58 87 08 05
62 97 16 29 18	52 16 16 23 56	62 95 80 97 63	32 25 34 03 36	48 84 60 37 65
31 13 63 21 08	16 01 92 58 21	48 79 74 73 72	08 64 80 91 38	07 28 66 61 59
97 38 35 34 19	89 84 05 34 47	88 09 31 54 88	97 96 86 01 69	46 13 95 65 96
32 11 78 33 82	51 99 98 44 39	12 75 10 60 36	80 66 39 94 97	42 36 31 16 59
81 99 13 37 05	08 12 60 39 23	61 73 84 89 18	26 02 04 37 95	96 18 69 06 30
45 74 00 03 05	69 99 47 26 52	48 06 30 00 18	03 30 28 55 59	66 10 71 44 05
11 84 13 69 01	88 91 28 79 50	71 42 14 96 55	98 59 96 01 36	88 77 90 45 59
14 66 12 87 22	59 45 27 08 51	85 64 23 85 41	64 72 08 59 44	67 98 56 65 56
40 25 67 87 82	84 27 17 30 37	48 69 49 02 58	98 02 50 58 11	95 39 06 35 63
44 48 97 49 43	65 45 53 41 07	14 83 46 74 11	76 66 63 60 08	90 54 33 65 84
41 94 54 06 57	48 28 01 83 84	09 11 21 91 73	97 28 44 74 06	22 30 95 69 72
07 12 15 58 84	93 18 31 83 45	54 52 62 29 91	53 58 54 66 05	47 19 63 92 75
64 27 90 43 52	18 26 32 96 83	50 58 45 27 57	14 96 39 64 85	73 87 96 76 23
80 71 86 41 03	45 62 63 40 88	35 69 34 10 94	32 22 52 04 74	69 63 21 83 41
27 06 08 09 92	26 22 59 28 27	38 58 22 14 79	24 32 12 38 42	33 56 90 92 57
54 68 97 20 54	33 26 74 03 30	74 22 19 13 48	30 28 01 92 49	58 61 52 27 03
02 92 65 68 99	05 53 15 26 70	04 69 22 64 07	04 73 25 74 82	78 35 22 21 88
83 52 57 78 62	98 61 70 48 22	68 50 64 55 75	42 70 32 09 60	58 70 61 43 97
82 82 76 31 33	85 13 41 38 10	16 47 61 43 77	83 27 19 70 41	34 78 77 60 25
38 61 34 09 49	04 41 66 09 76	20 50 73 40 95	24 77 95 73 20	47 42 80 61 03
01 01 11 88 38	03 10 16 82 24	39 58 20 12 39	82 77 02 18 88	33 11 49 15 16
21 66 14 38 28	54 08 18 07 04	92 17 63 36 75	33 14 11 11 78	97 30 53 62 38
32 29 30 69 59	68 50 33 31 47	15 64 88 75 27	04 51 41 61 96	86 62 93 66 71
04 59 21 65 47	39 90 89 86 77	46 86 86 88 86	50 09 13 24 91	54 80 67 78 66
38 64 50 07 36	56 50 45 94 25	48 28 48 30 51	60 73 73 03 87	68 47 37 10 84
48 33 50 83 53	59 77 64 59 90	58 92 62 50 18	93 09 45 89 06	13 26 98 86 29

RANDOM STANDARDISED NORMAL DEVIATES (Z VALUES)

An indefinitely large random sample of these 'preferred' z-values may be taken with replacement in conjunction with a table of random numbers. Take a three-digit random number, using the first two digits to select a row and the third to select a column; the intersection gives a random standardised normal deviate. For example, if the random number is 861, the remainder after dividing by 500 is 361 and so the corresponding normal deviate is + 0.539. (The device of mapping the random numbers 500 to 999 on to those from 000 to 499 saves 'wasting' random numbers.)

The mean of the 500 tabulated values is 0.00 and the variance is 1.00.

	0	1	2	3	4	5	6	7	8	9
00	−0.179	−0.399	−0.235	−0.098	−0.465	+1.563	−1.085	+0.860	+0.388	+0.710
01	+0.421	+1.454	+0.904	+0.437	−2.120	+1.085	−0.277	−0.170	+0.018	−0.722
02	+0.210	−0.556	+0.465	−1.812	−2.748	−0.345	−0.251	+0.622	−1.015	+0.762
03	−1.598	+0.919	−0.266	−0.999	+0.308	−0.592	+0.817	−0.454	+1.598	+0.240
04	+1.717	+1.514	−0.012	−0.852	+0.118	+0.399	−0.123	+0.432	−0.470	+0.776
05	−0.308	+0.867	−0.372	+0.697	−1.787	+0.568	−0.002	−0.133	+0.545	−0.824
06	−0.421	+0.516	−0.038	+1.200	+0.063	−0.377	−1.007	−0.334	+1.299	+0.038
07	−0.776	+0.874	−1.265	−0.580	+0.377	−0.697	−2.226	−1.299	−0.796	−0.628
08	+0.640	−0.522	+0.023	−0.393	−1.412	−2.457	−1.580	+1.160	+0.008	+0.487
09	−0.319	+0.889	+1.180	−0.404	+1.322	+0.410	+1.468	+0.235	−0.810	−1.131
10	+0.610	−0.383	+1.812	+0.729	+0.204	−0.225	+0.169	−0.729	−0.432	+0.634
11	−0.174	−0.154	+0.098	+0.393	−3.090	+1.762	+1.530	+0.028	+0.950	−0.935
12	+2.576	−0.684	−1.200	+0.002	+0.261	−0.415	+0.598	−0.769	−0.169	−1.498
13	−1.103	+1.398	−0.653	+1.739	+0.476	+0.510	+0.782	−0.634	+0.562	−0.053
14	+1.635	+0.448	−1.530	−0.043	+2.290	−0.063	−1.695	+0.199	+1.211	−1.360
15	−0.068	−0.860	−0.194	−1.616	+0.334	+0.189	+0.927	−1.454	+0.958	+0.404
16	−1.960	+1.076	−0.671	−0.103	+1.041	+2.226	+1.838	−0.510	−1.322	+2.366
17	+0.443	−0.912	+0.251	−0.574	+1.131	−0.204	−0.324	−0.487	−1.287	+0.522
18	+1.360	+0.533	+1.094	+0.671	+0.852	−2.576	−0.539	−0.568	+0.225	−0.545
19	+0.810	+0.319	−1.514	+0.556	+1.112	−0.210	+0.292	+0.749	+0.882	+0.033
20	+0.616	+1.347	−1.866	−0.755	+0.329	+0.148	−0.058	−0.199	+0.048	+1.546
21	−0.598	−2.366	−0.831	+0.454	−0.118	−1.762	+0.493	+1.103	+0.361	+0.113
22	+0.426	+1.580	−1.112	+0.550	−1.254	−0.033	+0.143	−1.141	+0.366	−0.073
23	+0.831	−0.516	−1.717	−0.340	+1.655	+0.194	−0.388	−0.942	−1.243	−0.292
24	−0.640	−0.128	+1.276	−1.838	−0.410	+0.646	+2.075	−0.159	+1.695	+0.527
25	−0.927	+0.838	−1.546	+0.246	−0.742	−0.143	+2.457	+0.043	−1.058	−0.867
26	+1.232	+2.170	+0.088	−0.803	+0.574	+0.058	+0.282	+0.356	+0.350	−1.927
27	+0.935	+0.665	+2.034	−1.995	+0.703	−0.083	−1.468	+0.078	−0.966	−0.303
28	−1.739	−0.622	−1.563	+0.313	+0.220	−0.586	+0.272	+0.789	−1.335	+1.440
29	+0.990	−1.483	+0.154	−1.372	−1.896	+1.385	−1.041	+0.974	+0.482	−1.211
30	−0.189	−0.240	+0.133	−2.290	−0.616	−0.437	+0.459	−0.499	+0.845	+0.383
31	+1.866	−1.398	+0.068	+0.053	−2.034	+1.426	+1.254	+1.067	+0.592	+0.174
32	−0.018	+0.628	+0.230	+0.659	−0.298	+1.927	−0.282	+0.769	−0.690	+1.675
33	−0.646	−0.350	+0.324	−1.675	+1.190	−1.076	+1.287	−1.426	+0.345	−0.215
34	−1.150	−0.220	−0.533	+0.912	−0.710	−0.904	−0.817	−1.160	−0.919	−0.659
35	+0.103	−0.361	+1.024	−0.604	+0.966	−1.122	+0.604	−0.845	+0.736	−0.882
36	+1.243	+0.539	+0.684	−0.716	−0.482	−0.562	+0.277	−1.440	−0.366	−0.256
37	−0.093	−1.190	+0.580	−1.276	+0.563	−0.048	+0.742	−1.170	+1.960	+2.120
38	−0.261	−0.194	+0.303	+0.340	+1.498	−1.232	−0.078	−0.443	+1.141	+1.787
39	−0.230	−0.550	+0.266	−1.655	+0.999	−1.067	+1.058	+0.796	+0.415	+1.995
40	−0.148	+0.504	−0.028	+0.083	+0.824	−1.024	+1.412	−0.164	+1.150	−0.272
41	+1.122	+0.896	−0.789	+0.215	−0.426	−1.049	−0.974	+0.586	+1.311	−0.736
42	+0.499	−1.032	+0.159	+0.123	+2.748	−0.749	−0.665	−1.221	−1.180	+1.049
43	+0.678	−0.782	+0.470	+0.256	+0.298	−0.990	+0.287	+0.942	+0.128	+1.372
44	−1.347	+3.090	−0.896	+0.138	−0.838	+0.690	+1.007	+0.184	+0.164	+0.179
45	−1.094	−0.610	−0.287	+0.755	−0.459	−1.635	−0.108	−0.246	+1.032	−0.527
46	−0.088	−0.889	+0.803	−1.311	−0.703	+1.170	−0.113	+0.108	−0.874	+0.372
47	+0.093	−0.476	+1.265	−0.448	+1.015	−0.313	−0.958	+0.716	+1.483	+0.722
48	−0.950	−0.008	+0.012	+0.073	−0.762	−0.493	+1.896	+0.982	+1.616	+1.221
49	−0.329	−0.138	−0.504	−0.678	+1.335	−2.075	−1.385	−0.023	−0.356	−0.982

PRESENT VALUE FACTORS

The following table gives the present value of a single payment received *n* years in the future discounted at *x*% per year.

x / *Year*	1%	2%	3%	4%	5%	6%	7%	8%	9%	10%
1	.9901	.9804	.9709	.9615	.9524	.9434	.9346	.9259	.9174	.9091
2	.9803	.9612	.9426	.9246	.9070	.8900	.8734	.8573	.8417	.8264
3	.9706	.9423	.9151	.8890	.8638	.8396	.8163	.7938	.7722	.7513
4	.9610	.9238	.8885	.8548	.8227	.7921	.7629	.7350	.7084	.6830
5	.9515	.9057	.8626	.8219	.7835	.7473	.7130	.6806	.6499	.6209
6	.9420	.8880	.8375	.7903	.7462	.7050	.6663	.6302	.5963	.5645
7	.9327	.8706	.8131	.7599	.7107	.6651	.6227	.5835	.5470	.5132
8	.9235	.8535	.7894	.7307	.6768	.6274	.5820	.5403	.5019	4665
9	.9143	.8368	.7664	.7026	.6446	.5919	.5439	.5002	.4604	.4241
10	.9053	.8203	.7441	.6756	.6139	.5584	.5083	.4632	.4224	.3855
11	.8963	.8043	.7224	.6496	.5847	.5268	.4751	.4289	.3875	.3505
12	.8874	.7885	.7014	.6246	.5568	.4970	.4440	.3971	.3555	.3186
13	.8787	.7730	.6810	.6006	.5303	.4688	4150	.3677	.3262	.2897
14	.8700	.7579	.6611	.5775	.5051	.4423	.3878	.3405	.2992	.2633
15	.8613	.7430	.6419	.5553	.4810	.4173	.3624	.3152	.2745	.2394
16	.8528	.7284	.6232	.5339	.4581	.3936	.3387	.2919	.2519	.2176
17	.8444	.7142	.6050	.5134	.4363	.3714	.3166	.2703	.2311	.1978
18	.8360	.7002	.5874	.4936	.4155	.3503	.2959	.2502	.2120	.1799
19	.8277	.6864	.5703	.4746	.3957	.3305	.2765	.2317	.1945	.1635
20	.8195	.6730	.5537	.4564	.3769	.3118	.2584	.2145	.1784	.1486
21	.8114	.6598	.5375	.4388	.3589	.2942	.2415	.1987	.1637	.1351
22	.8034	.6468	.5219	.4220	.3418	.2775	.2257	.1839	.1502	.1228
23	.7954	.6342	.5067	.4057	.3256	.2618	.2109	.1703	.1378	.1117
24	.7876	.6217	.4919	.3901	.3101	.2470	.1971	.1577	.1264	.1015
25	.7798	.6095	.4776	.3751	.2953	.2330	.1842	.1460	.1160	.0923
26	.7720	.5976	.4637	.3607	.2812	.2198	.1722	.1352	.1064	.0839
27	.7644	.5859	.4502	.3468	.2678	.2074	.1609	.1252	.0976	.0763
28	.7568	.5744	.4371	.3335	.2551	.1956	.1504	.1159	.0895	.0693
29	.7493	.5631	.4243	.3207	.2429	.1846	.1406	.1073	.0822	.0630
30	.7419	.5521	.4120	.3083	.2314	.1741	.1314	.0994	.0754	.0573
31	.7346	.5412	.4000	.2965	.2204	.1643	.1228	.0920	.0691	.0521
32	.7273	.5306	.3883	.2851	.2099	.1550	.1147	.0852	.0634	.0474
33	.7201	.5202	.3770	.2741	.1999	.1462	.1072	.0789	.0582	.0431
34	.7130	.5100	.3660	.2636	.1904	.1379	.1002	.0730	.0534	.0391
35	.7059	.5000	.3554	.2534	.1813	.1301	.0937	.0676	.0490	.0356
36	.6989	.4902	.3450	.2437	.1727	.1227	.0875	.0626	.0449	.0323
37	.6920	.4806	.3350	.2343	.1644	.1158	.0818	.0580	.0412	.0294
38	.6852	.4712	.3252	.2253	.1566	.1092	.0765	.0537	.0378	.0267
39	.6784	.4619	.3158	.2166	.1491	.1031	.0715	.0497	.0347	.0243
40	.6717	.4529	.3066	.2083	.1420	.0972	.0668	.0460	.0318	.0221
41	.6650	.4440	.2976	.2003	.1353	.0917	.0624	.0426	.0292	.0201
42	.6584	.4353	.2890	.1926	.1288	.0865	.0583	.0395	.0268	.0183
43	.6519	.4268	.2805	.1852	.1227	.0816	.0545	.0365	.0246	.0166
44	.6454	.4184	.2724	.1780	.1169	.0770	.0509	.0338	.0226	.0151
45	.6391	.4102	.2644	.1712	.1113	.0727	.0476	.0313	.0207	.0137
46	.6327	.4022	.2567	.1646	.1060	.0685	.0445	.0290	.0190	.0125
47	.6265	.3943	.2493	.1583	.1009	.0647	.0416	.0269	.0174	.0113
48	.6203	.3865	.2420	.1522	.0961	.0610	.0389	.0249	.0160	.0103
49	.6141	.3790	.2350	.1463	.0916	.0575	.0363	.0230	.0147	.0094
50	.6080	.3715	.2281	.1407	.0872	.0543	.0339	.0213	.0134	.0085

Present Value Factors (*Contd.*)

Discount range 11% to 20%

Year \ *x*	11%	12%	13%	14%	15%	16%	17%	18%	19%	20%
1	.9009	.8929	.8850	.8772	.8696	.8621	.8547	.8475	.8403	.8333
2	.8116	.7972	.7831	.7695	.7561	.7432	.7305	.7182	.7062	.6944
3	.7312	.7118	,6931	.6750	.6575	.6407	.6244	.6086	.5934	.5787
4	.6587	.6355	.6133	.5921	.5718	.5523	.5337	.5158	.4987	.4823
5	.5935	.5674	.5428	.5194	.4972	.4761	.4561	.4371	.4190	.4019
6	.5346	.5066	.4803	.4556	.4323	.4104	.3898	.3704	.3521	.3349
7	.4817	.4523	.4251	.3996	.3759	.3538	.3332	.3139	.2959	.2791
8	.4339	.4039	.3762	.3506	.3269	.3050	.2848	.2660	.2487	.2326
9	.3909	.3606	.3329	.3075	.2843	.2630	.2434	.2255	.2090	.1938
10	.3522	.3220	.2946	.2697	.2472	.2267	.2080	.1911	.1756	.1615
11	.3173	.2875	.2607	.2366	.2149	.1954	.1778	.1619	.1476	.1346
12	.2858	.2567	.2307	.2076	.1869	.1685	.1520	.1372	.1240	.1122
13	.2575	.2292	.2042	.1821	.1625	.1452	.1299	.1163	.1042	.0935
14	.2320	.2046	.1807	.1597	.1413	.1252	.1110	.0985	.0876	.0779
15	.2090	.1827	.1599	.1401	.1229	.1079	.0949	.0835	.0736	.0649
16	.1883	.1631	.1415	.1229	.1069	.0930	.0811	.0708	.0618	.0541
17	.1696	.1456	.1252	.1078	.0929	.0802	.0693	.0600	.0520	.0451
18	.1528	.1300	.1108	.0946	.0808	.0691	.0592	.0508	.0437	.0376
19	.1377	.1161	.0981	.0829	.0703	.0596	.0506	.0431	.0367	.0313
20	.1240	.1037	.0868	.0728	.0611	.0514	.0433	.0365	.0308	.0261
21	.1117	.0926	.0768	.0638	.0531	.0443	.0370	.0309	.0259	.0217
22	.1007	.0826	.0680	.0560	.0462	.0382	.0316	.0262	.0218	.0181
23	.0907	.0738	.0601	.0491	.0402	.0329	.0270	.0222	.0183	.0151
24	.0817	.0659	.0532	.0431	.0349	.0284	.0231	.0188	.0154	.0126
25	.0736	.0588	.0471	.0378	.0304	.0245	.0197	.0160	.0129	.0105
26	.0663	.0525	.0417	.0331	.0264	.0211	.0169	.0135	.0109	.0087
27	.0597	.0469	.0369	.0291	.0230	.0182	.0144	.0115	.0091	.0073
28	.0538	.0419	.0326	.0255	.0200	.0157	.0123	.0097	.0077	.0061
29	.0485	.0374	.0289	.0224	.0174	.0135	.0105	.0082	.0064	.0051
30	.0437	.0334	.0256	.0196	.0151	.0116	.0090	.0070	.0054	.0042
31	.0394	.0298	.0226	.0172	.0131	.0100	.0077	.0059	.0046	.0035
32	.0355	.0266	.0200	.0151	.0114	.0087	.0066	.0050	.0038	.0029
33	.0319	.0238	.0177	.0132	.0099	.0075	.0056	.0042	.0032	.0024
34	.0288	.0212	.0157	.0116	.0086	.0064	.0048	.0036	.0027	.0020
35	.0259	.0189	.0139	.0102	.0075	.0055	.0041	.0030	.0023	.0017
36	.0234	.0169	.0123	.0089	.0065	.0048	.0035	.0026	.0019	.0014
37	.0210	.0151	.0109	.0078	.0057	.0041	.0030	.0022	.0016	.0012
38	.0190	.0135	.0096	.0069	.0049	.0036	.0026	.0019	.0013	.0010
39	.0171	.0120	.0085	.0060	.0043	.0031	.0022	.0016	.0011	.0008
40	.0154	.0107	.0075	.0053	.0037	.0026	.0019	.0013	.0010	.0007
41	.0139	.0096	.0067	.0046	.0032	.0023	.0016	.0011	.0008	.0006
42	.0125	.0086	.0059	.0041	.0028	.0020	.0014	.0010	.0007	.0005
43	.0112	.0076	.0052	.0036	.0025	.0017	.0012	.0008	.0006	.0004
44	.0101	.0068	.0046	.0031	.0021	.0015	.0010	.0007	.0005	.0003
45	.0091	.0061	.0041	.0027	.0019	.0013	.0009	.0006	.0004	.0003
46	.0082	.0054	.0036	.0024	.0016	.0011	.0007	.0005	.0003	.0002
47	.0074	.0049	.0032	.0021	.0014	.0009	.0006	.0004	.0003	.0002
48	.0067	.0043	.0028	.0019	.0012	.0008	.0005	.0004	.0002	.0002
49	.0060	.0039	.0025	.0016	.0011	.0007	.0005	.0003	.0002	.0001
50	.0054	.0035	.0022	.0014	.0009	.0006	.0004	.0003	.0002	.0001

Present Value Factors (*Contd.*)

Discount range 21% to 30%

x / *Year*	21%	22%	23%	24%	25%	26%	27%	28%	29%	30%
1	0.8264	0.8197	0.8130	0.8065	0.8000	0.7937	0.7874	0.7813	0.7752	0.7692
2	0.6830	0.6719	0.6610	0.6504	0.6400	0.6299	0.6200	0.6104	0.6009	0.5917
3	0.5645	0.5507	0.5374	0.5245	0.5120	0.4999	0.4882	0.4768	0.4658	0.4552
4	0.4665	0.4514	0.4369	0.4230	0.4096	0.3968	0.3844	0.3725	0.3611	0.3501
5	0.3855	0.3700	0.3552	0.3411	0.3277	0.3149	0.3027	0.2910	0.2799	0.2693
6	0.3186	0.3033	0.2888	0.2751	0.2621	0.2499	0.2383	0.2274	0.2170	0.2072
7	0.2633	0.2486	0.2348	0.2218	0.2097	0.1983	0.1877	0.1776	0.1682	0.1594
8	0.2176	0.2038	0.1909	0.1789	0.1678	0.1574	0.1478	0.1388	0.1304	0.1226
9	0.1799	0.1670	0.1552	0.1443	0.1342	0.1249	0.1164	0.1084	0.1011	0.0943
10	0.1486	0.1369	0.1262	0.1164	0.1074	0.0992	0.0916	0.0847	0.0784	0.0725
11	0.1228	0.1122	0.1026	0.0938	0.0859	0.0787	0.0721	0.0662	0.0607	0.0558
12	0.1015	0.0920	0.0834	0.0757	0.0687	0.0625	0.0568	0.0517	0.0471	0.0429
13	0.0839	0.0754	0.0678	0.0610	0.0550	0.0496	0.0447	0.0404	0.0365	0.0330
14	0.0693	0.0618	0.0551	0.0492	0.0440	0.0393	0.0352	0.0316	0.0283	0.0254
15	0.0573	0.0507	0.0448	0.0397	0.0352	0.0312	0.0277	0.0247	0.0219	0.0195
16	0.0474	0.0415	0.0364	0.0320	0.0281	0.0248	0.0218	0.0193	0.0170	0.0150
17	0.0391	0.0340	0.0296	0.0258	0.0225	0.0197	0.0172	0.0150	0.0132	0.0116
18	0.0323	0.0279	0.0241	0.0208	0.0180	0.0156	0.0135	0.0118	0.0102	0.0089
19	0.0267	0.0229	0.0196	0.0168	0.0144	0.0124	0.0107	0.0092	0.0079	0.0068
20	0.0221	0.0187	0.0159	0.0135	0.0115	0.0098	0.0084	0.0072	0.0061	0.0053
21	0.0183	0.0154	0.0129	0.0109	0.0092	0.0078	0.0066	0.0056	0.0048	0.0040
22	0.0151	0.0126	0.0105	0.0088	0.0074	0.0062	0.0052	0.0044	0.0037	0.0031
23	0.0125	0.0103	0.0086	0.0071	0.0059	0.0049	0.0041	0.0034	0.0029	0.0024
24	0.0103	0.0085	0.0070	0.0057	0.0047	0.0039	0.0032	0.0027	0.0022	0.0018
25	0.0085	0.0069	0.0057	0.0046	0.0038	0.0031	0.0025	0.0021	0.0017	0.0014
26	0.0070	0.0057	0.0046	0.0037	0.0030	0.0025	0.0020	0.0016	0.0013	0.0011
27	0.0058	0.0047	0.0037	0.0030	0.0024	0.0019	0.0016	0.0013	0.0010	0.0008
28	0.0048	0.0038	0.0030	0.0024	0.0019	0.0015	0.0012	0.0010	0.0008	0.0006
29	0.0040	0.0031	0.0025	0.0020	0.0015	0.0012	0.0010	0.0008	0.0006	0.0005
30	0.0033	0.0026	0.0020	0.0016	0.0012	0.0010	0.0008	0.0006	0.0005	0.0004
31	0.0027	0.0021	0.0016	0.0013	0.0010	0.0008	0.0006	0.0005	0.0004	0.0003
32	0.0022	0.0017	0.0013	0.0010	0.0008	0.0006	0.0005	0.0004	0.0003	0.0002
33	0.0019	0.0014	0.0011	0.0008	0.0006	0.0005	0.0004	0.0003	0.0002	0.0002
34	0.0015	0.0012	0.0009	0.0007	0.0005	0.0004	0.0003	0.0002	0.0002	0.0001
35	0.0013	0.0009	0.0007	0.0005	0.0004	0.0003	0.0002	0.0002	0.0001	0.0001
36	0.0010	0.0008	0.0006	0.0004	0.0003	0.0002	0.0002	0.0001	0.0001	0.0001
37	0.0009	0.0006	0.0005	0.0003	0.0003	0.0002	0.0001	0.0001	0.0001	0.0001
38	0.0007	0.0005	0.0004	0.0003	0.0002	0.0002	0.0001	0.0001	0.0001	
39	0.0006	0.0004	0.0003	0.0002	0.0002	0.0001	0.0001	0.0001		
40	0.0005	0.0004	0.0003	0.0002	0.0001	0.0001	0.0001	0.0001		
41	0.0004	0.0003	0.0002	0.0001	0.0001	0.0001	0.0001			
42	0.0003	0.0002	0.0002	0.0001	0.0001	0.0001				
43	0.0003	0.0002	0.0001	0.0001	0.0001					
44	0.0002	0.0002	0.0001	0.0001	0.0001					
45	0.0002	0.0001	0.0001	0.0001						
46	0.0002	0.0001	0.0001	0.0001						
47	0.0001	0.0001	0.0001							
48	0.0001	0.0001								
49	0.0001	0.0001								
50	0.0001									

Present Value Factors (*Contd.*)

Discount range 21% to 30%

Year \ *x*	31%	32%	33%	34%	35%	36%	37%	38%	39%	40%
1	0.7634	0.7576	0.7519	0.7463	0.7407	0.7353	0.7299	0.7246	0.7194	0.7143
2	0.5827	0.5739	0.5653	0.5569	0.5487	0.5407	0.5328	0.5251	0.5176	0.5102
3	0.4448	0.4348	0.4251	0.4156	0.4064	0.3975	0.3889	0.3805	0.3724	0.3644
4	0.3396	0.3294	0.3196	0.3102	0.3011	0.2923	0.2839	0.2757	0.2679	0.2603
5	0.2592	0.2495	0.2403	0.2315	0.2230	0.2149	0.2072	0.1998	0.1927	0.1859
6	0.1979	0.1890	0.1807	0.1727	0.1652	0.1580	0.1512	0.1448	0.1386	0.1328
7	0.1510	0.1432	0.1358	0.1289	0.1224	0.1162	0.1104	0.1049	0.0997	0.0949
8	0.1153	0.1085	0.1021	0.0962	0.0906	0.0854	0.0806	0.0760	0.0718	0.0678
9	0.0880	0.0822	0.0768	0.0718	0.0671	0.0628	0.0588	0.0551	0.0516	0.0484
10	0.0672	0.0623	0.0577	0.0536	0.0497	0.0462	0.0429	0.0399	0.0371	0.0346
11	0.0513	0.0472	0.0434	0.0400	0.0368	0.0340	0.0313	0.0289	0.0267	0.0247
12	0.0392	0.0357	0.0326	0.0298	0.0273	0.0250	0.0229	0.0210	0.0192	0.0176
13	0.0299	0.0271	0.0245	0.0223	0.0202	0.0184	0.0167	0.0152	0.0138	0.0126
14	0.0228	0.0205	0.0185	0.0166	0.0150	0.0135	0.0122	0.0110	0.0099	0.0090
15	0.0174	0.0155	0.0139	0.0124	0.0111	0.0099	0.0089	0.0080	0.0072	0.0064
16	0.0133	0.0118	0.0104	0.0093	0.0082	0.0073	0.0065	0.0058	0.0051	0.0046
17	0.0101	0.0089	0.0078	0.0069	0.0061	0.0054	0.0047	0.0042	0.0037	0.0033
18	0.0077	0.0068	0.0059	0.0052	0.0045	0.0039	0.0035	0.0030	0.0027	0.0023
19	0.0059	0.0051	0.0044	0.0038	0.0033	0.0029	0.0025	0.0022	0.0019	0.0017
20	0.0045	0.0039	0.0033	0.0029	0.0025	0.0021	0.0018	0.0016	0.0014	0.0012
21	0.0034	0.0029	0.0025	0.0021	0.0018	0.0016	0.0013	0.0012	0.0010	0.0009
22	0.0026	0.0022	0.0019	0.0016	0.0014	0.0012	0.0010	0.0008	0.0007	0.0006
23	0.0020	0.0017	0.0014	0.0012	0.0010	0.0008	0.0007	0.0006	0.0005	0.0004
24	0.0015	0.0013	0.0011	0.0009	0.0007	0.0006	0.0005	0.0004	0.0004	0.0003
25	0.0012	0.0010	0.0008	0.0007	0.0006	0.0005	0.0004	0.0003	0.0003	0.0002
26	0.0009	0.0007	0.0006	0.0005	0.0004	0.0003	0.0003	0.0002	0.0002	0.0002
27	0.0007	0.0006	0.0005	0.0004	0.0003	0.0002	0.0002	0.0002	0.0001	0.0001
28	0.0005	0.0004	0.0003	0.0003	0.0002	0.0002	0.0001	0.0001	0.0001	0.0001
29	0.0004	0.0003	0.0003	0.0002	0.0002	0.0001	0.0001	0.0001	0.0001	0.0001
30	0.0003	0.0002	0.0002	0.0002	0.0001	0.0001	0.0001	0.0001	0.0001	
31	0.0002	0.0002	0.0001	0.0001	0.0001	0.0001	0.0001			
32	0.0002	0.0001	0.0001	0.0001	0.0001	0.0001				
33	0.0001	0.0001	0.0001	0.0001	0.0001					
34	0.0001	0.0001	0.0001							
35	0.0001	0.0001								
36	0.0001									
37										
38										
39										
40										
41										
42										
43										
44										
45										
46										
47										
48										
49										
50										

CUMULATIVE PRESENT VALUE FACTORS

The following table gives the present value of n annual payments of ₹1 received at the end of each of the next n years with a constant discount rate of $x\%$.

Year \ x	1%	2%	3%	4%	5%	6%	7%	8%	9%	10%
1	0.990	0.980	0.971	0.962	0.952	0.943	0.935	0.926	0.917	0.909
2	1.970	1.942	1.914	1.886	1.859	1.833	1.808	1.783	1.759	1.736
3	2.941	2.884	2.829	2.775	2.723	2.673	2.624	2.577	2.531	2.487
4	3.902	3.808	3.717	3.630	3.546	3.465	3.387	3.312	3.240	3.170
5	4.854	4.713	4.580	4.452	4.329	4.212	4.100	3.993	3.890	3.791
6	5.796	5.601	5.417	5.242	5.076	4.917	4.767	4.623	4.486	4.355
7	6.728	6.472	6.230	6.002	5.786	5.583	5.389	5.206	5.033	4.868
8	7.652	7.326	7.020	6.733	6.463	6.210	5.971	5.747	5.535	5.335
9	8.566	8.162	7.786	7.435	7.108	6.802	6.515	6.247	5.995	5.759
10	9.471	8.983	8.530	8.111	7.722	7.360	7.023	6.710	6.418	6.145
11	10.368	9.787	9.253	8.761	8.306	7.887	7.499	7.139	6.805	6.495
12	11.255	10.575	9.954	9.385	8.863	8.384	7.943	7.536	7.161	6.814
13	12.134	11.348	10.635	9.986	9.393	8.853	8.358	7.904	7.487	7.103
14	13.004	12.106	11.296	10.563	9.899	9.295	8.745	8.244	7.786	7.367
15	13.865	12.849	11.938	11.119	10.380	9.712	9.108	8.559	8.061	7.606
16	14.718	13.578	12.561	11.652	10.838	10.106	9.446	8.851	8.312	7.824
17	15.562	14.292	13.166	12.166	11.274	10.477	9.763	9.122	8.544	8.021
18	16.398	14.992	13.754	12.659	11.689	10.828	10.059	9.372	8.756	8.201
19	17.226	15.679	14.324	13.134	12.085	11.158	10.335	9.604	8.950	8.365
20	18.045	16.352	14.878	13.590	12.462	11.470	10.594	9.818	9.128	8.513
21	18.857	17.011	15.415	14.029	12.821	11.764	10.835	10.017	9.292	8.649
22	19.660	17.658	15.937	14.451	13.163	12.042	11.061	10.201	9.442	8.771
23	20.456	18.292	16.444	14.857	13.488	12.304	11.272	10.371	9.580	8.883
24	21.243	18.914	16.936	15.247	13.798	12.551	11.469	10.529	9.707	8.985
25	22.023	19.524	17.413	15.622.	14.094	12.784	11.653	10.675	9.823	9.077
26	22.795	20.121	17.877	15.983	14.375	13.003	11.825	10.810	9.929	9.161
27	23.559	20.707	18.327	16.330	14.643	13.211	11.986	10.935	10.027	9.237
28	24.316	21.281	18.764	16.663	14.898	13.406	12.137	11.051	10.116	9.306
29	25.066	21.845	19.189	16.984	15.141	13.591	12.277	11.158	10.198	9.369
30	25.807	22.397	19.601	17.292	15.372	13.765	12.409	11.258	10.274	9.427
31	26.542	22.938	20.001	17.589	15.593	13.929	12.532	11.350	10.343	9.479
32	27.269	23.468	20.389	17.874	15.802	14.084	12.646	11.435	10.406	9.526
33	27.989	23.989	20.766	18.148	16.002	14.231	12.753	11.514	10.464	9.569
34	28.702	24.499	21.132	18.411	16.193	14.368	12.854	11.587	10.518	9.608
35	29.408	24.999	21.487	18.665	16.374	14.499	12.947	11.654	10.567	9.644
36	30.107	25.489	21.832	18.909	16.547	14.621	13.035	11.717	10.612	9.676
37	30.799	25.969	22.167	19.143	16.711	14.737	13.117	11.775	10.653	9.706
38	31.484	26.441	22.492	19.368	16.868	14.846	13.193	11.829	10.691	9.732
39	32.163	26.903	22.808	19.585	17.017	14.949	13.265	11.878	10.725	9.757
40	32.835	27.355	23.115	19.793	17.159	15.047	13.331	11.924	10.757	9.779
41	33.500	27.799	23.412	19.993	17.294	15.138	13.394	11.967	10.786	9.799
42	34.158	28.235	23.701	20.186	17.423	15.225	13.452	12.006	10.813	9.817
43	34.810	28.662	23.982	20.371	17.546	15.306	13.507	12.043	10.838	9.834
44	35.455	29.080	24.254	20.549	17.663	15.383	13.558	12.077	10.860	9.849
45	36.094	29.490	24.519	20.720	17.774	15.456	13.605	12.108	10.881	9.863
46	36.727	29.892	24.775	20.885	17.880	15.525	13.650	12.137	10.900	9.875
47	37.354	30.287	25.025	21.043	17.981	15.589	13.691	12.164	10.917	9.886
48	37.974	30.673	25.267	21.195	18.077	15.650	13.730	12.189	10.933	9.897
49	38.588	31.052	25.502	21.342	18.168	15.708	13.766	12.212	10.948	9.906
50	39.196	31.424	25.730	21.482	18.256	15.762	13.800	12.233	10.962	9.915

Cumulative Present Value Factors (*Contd.*)

Discount range 11% to 20%

x / *Year*	11%	12%	13%	14%	15%	16%	17%	18%	19%	20%
1	0.901	0.893	0.885	0.877	0.870	0.862	0.855	0.848	0.840	0.833
2	1.713	1.690	1.668	1.647	1.626	1.605	1.585	1.566	1.547	1.528
3	2.444	2.402	2.361	2.322	2.283	2.246	2.210	2.174	2.140	2.106
4	3.102	3.037	2.975	2.914	2.855	2.798	2.743	2.690	2.639	2.589
5	3.696	3.605	3.517	3.433	3.352	3.274	3.199	3.127	3.058	2.991
6	4.231	4.111	3.998	3.889	3.785	3.685	3.589	3.498	3.410	3.326
7	4.712	4.564	4.423	4.288	4.160	4.039	3.922	3.812	3.706	3.605
8	5.146	4.968	4.799	4.639	4.487	4.344	4.207	4.078	3.954	3.837
9	5.537	5.328	5.132	4.947	4.772	4.607	4.451	4.303	4.163	4.031
10	5.889	5.650	5.426	5.216	5.019	4.833	4.659	4.494	4.339	4.193
11	6.207	5.938	5.687	5.453	5.234	5.029	4.836	4.656	4.487	4.327
12	6.492	6.194	5.918	5.660	5.421	5.197	4.988	4.793	4.611	4.439
13	6.750	6.424	6.122	5.843	5.583	5.342	5.118	4.910	4.715	4.533
14	6.982	6.628	6.303	6.002	5.724	5.468	5.229	5.008	4.802	4.611
15	7.191	6.811	6.463	6.142	5.847	5.576	5.324	5.092	4.876	4.676
16	7.379	6.974	6.604	6.265	5.954	5.669	5.405	5.162	4.938	4.730
17	7.549	7.120	6.729	6.373	6.047	5.749	5.475	5.222	4.990	4.775
18	7.702	7.250	6.840	6.468	6.128	5.818	5.534	5.273	5.033	4.812
19	7.839	7.366	6.938	6.551	6.198	5.877	5.584	5.316	5.070	4.844
20	7.963	7.469	7.025	6.623	6.259	5.929	5.628	5.353	5.101	4.870
21	8.075	7.562	7.102	6.687	6.312	5.973	5.665	5.384	5.127	4.892
22	8.176	7.645	7.170	6.743	6.359	6.011	5.696	5.410	5.149	4.910
23	8.266	7.718	7.230	6.792	6.399	6.044	5.723	5.432	5.167	4.925
24	8.348	7.784	7.283	6.835	6.434	6.073	5.746	5.451	5.182	4.937
25	8.422	7.843	7.330	6.873	6.464	6.097	5.766	5.467	5.195	4.948
26	8.488	7.896	7.372	6.906	6.491	6.118	5.783	5.480	5.206	4.957
27	8.548	7.943	7.409	6.935	6.514	6.136	5.797	5.492	5.215	4.964
28	8.601	7.984	7.441	6.961	6.534	6.152	5.810	5.502	5.223	4.970
29	8.650	8.022	7.470	6.983	6.551	6.166	5.820	5.510	5.229	4.975
30	8.694	8.055	7.496	7.003	6.566	6.177	5.829	5.517	5.235	4.979
31	8.733	8.085	7.519	7.020	6.579	6.187	5.837	5.523	5.239	4.983
32	8.769	8.112	7.539	7.035	6.591	6.196	5.844	5.528	5.243	4.986
33	8.800	8.135	7.556	7.048	6.600	6.203	5.849	5.532	5.246	4.988
34	8.829	8.157	7.572	7.060	6.609	6.210	5.854	5.535	5.249	4.990
35	8.855	8.176	7.586	7.070	6.617	6.215	5.858	5.538	5.251	4.992
36	8.879	8.192	7.598	7.079	6.623	6.220	5.862	5.541	5.253	4.993
37	8.900	8.208	7.609	7.087	6.629	6.224	5.865	5.543	5.255	4.994
38	8.919	8.221	7.619	7.094	6.634	6.228	5.867	5.545	5.256	4.995
39	8.936	8.233	7.627	7.100	6.638	6.231	5.869	5.547	5.257	4.996
40	8.951	8.244	7.635	7.105	6.642	6.234	5.871	5.548	5.258	4.997
41	8.965	8.253	7.641	7.110	6.645	6.236	5.873	5.549	5.259	4.997
42	8.977	8.262	7.647	7.114	6.648	6.238	5.874	5.550	5.260	4.998
43	8.989	8.270	7.652	7.117	6.650	6.240	5.875	5.551	5.260	4.998
44	8.999	8.276	7.657	7.120	6.652	6.241	5.876	5.552	5.261	4.999
45	9.008	8.282	7.661	7.123	6.654	6.242	5.877	5.552	5.261	4.999
46	9.016	8.288	7.665	7.126	6.656	6.243	5.878	5.553	5.262	4.999
47	9.023	8.293	7.668	7.128	6.657	6.244	5.879	5.553	5.262	4.999
48	9.030	8.297	7.671	7.130	6.658	6.245	5.879	5.554	5.262	5.000
49	9.036	8.301	7.673	7.131	6.659	6.246	5.880	5.554	5.262	5.000
50	9.042	8.304	7.675	7.133	6.660	6.246	5.880	5.554	5.262	5.000

Cumulative Present Value Factors (*Contd.*)

Discount range 21% to 30%

x / *Year*	21%	22%	23%	24%	25%	26%	27%	28%	29%	30%
1	0.826	0.820	0.813	0.807	0.800	0.794	0.787	0.781	0.775	0.769
2	1.509	1.492	1.474	1.457	1.440	1.424	1.407	1.392	1.376	1.361
3	2.074	2.042	2.011	1.981	1.952	1.924	1.896	1.869	1.842	1.816
4	2.540	2.494	2.448	2.404	2.362	2.320	2.280	2.241	2.203	2.166
5	2.926	2.864	2.804	2.746	2.689	2.635	2.583	2.532	2.483	2.436
6	3.245	3.167	3.092	3.021	2.951	2.885	2.821	2.759	2.700	2.643
7	3.508	3.416	3.327	3.242	3.161	3.083	3.009	2.937	2.868	2.802
8	3.725	3.619	3.518	3.421	3.329	3.241	3.157	3.076	2.999	2.925
9	3.905	3.786	3.673	3.566	3.463	3.366	3.273	3.184	3.100	3.019
10	4.054	3.923	3.799	3.682	3.571	3.465	3.365	3.269	3.178	3.092
11	4.177	4.036	3.902	3.776	3.656	3.544	3.437	3.335	3.239	3.147
12	4.278	4.128	3.985	3.852	3.725	3.606	3.493	3.387	3.286	3.190
13	4.362	4.203	4.053	3.913	3.780	3.656	3.538	3.427	3.322	3.223
14	4.431	4.265	4.108	3.962	3.824	3.695	3.573	3.459	3.351	3.249
15	4.489	4.315	4.153	4.001	3.859	3.726	3.601	3.484	3.373	3.268
16	4.536	4.357	4.190	4.033	3.887	3.751	3.623	3.503	3.390	3.283
17	4.575	4.391	4.219	4.059	3.910	3.771	3.640	3.518	3.403	3.295
18	4.608	4.419	4.243	4.080	3.928	3.786	3.654	3.530	3.413	3.304
19	4.634	4.442	4.263	4.097	3.942	3.799	3.664	3.539	3.421	3.310
20	4.656	4.460	4.279	4.110	3.954	3.809	3.673	3.546	3.427	3.316
21	4.675	4.476	4.292	4.121	3.963	3.816	3.679	3.552	3.432	3.320
22	4.690	4.488	4.302	4.130	3.970	3.823	3.684	3.556	3.435	3.323
23	4.702	4.499	4.311	4.137	3.976	3.827	3.689	3.559	3.438	3.325
24	4.713	4.507	4.318	4.143	3.981	3.831	3.692	3.562	3.441	3.327
25	4.721	4.514	4.323	4.147	3.985	3.834	3.694	3.564	3.442	3.328
26	4.728	4.520	4.328	4.151	3.988	3.837	3.696	3.566	3.444	3.330
27	4.734	4.525	4.332	4.154	3.990	3.839	3.698	3.567	3.445	3.330
28	4.739	4.528	4.335	4.157	3.992	3.840	3.699	3.568	3.445	3.331
29	4.743	4.531	4.337	4.159	3.994	3.842	3.700	3.569	3.446	3.331
30	4.746	4.534	4.339	4.160	3.995	3.843	3.701	3.570	3.446	3.332
31	4.749	4.536	4.341	4.161	3.996	3.843	3.701	3.570	3.447	3.332
32	4.751	4.538	4.342	4.162	3.997	3.844	3.702	3.570	3.447	3.332
33	4.753	4.539	4.343	4.163	3.997	3.844	3.702	3.571	3.447	3.333
34	4.754	4.540	4.344	4.164	3.998	3.845	3.703	3.571	3.448	3.333
35	4.756	4.541	4.345	4.164	3.998	3.845	3.703	3.571	3.448	3.333
36	4.757	4.542	4.345	4.165	3.998	3.845	3.703	3.571	3.448	3.333
37	4.757	4.543	4.346	4.165	3.999	3.846	3.703	3.571	3.448	3.333
38	4.758	4.543	4.346	4.165	3.999	3.846	3.703	3.571	3.448	3.333
39	4.759	4.544	4.347	4.166	3.999	3.846	3.703	3.572	3.448	3.333
40	4.759	4.544	4.347	4.166	3.999	3.846	3.703	3.572	3.448	3.333
41	4.760	4.544	4.347	4.166	3.999	3.846	3.704	3.572	3.448	3.333
42	4.760	4.545	4.347	4.166	3.999	3.846	3.704	3.572	3.448	3.333
43	4.760	4.545	4.347	4.166	4.000	3.846	3.704	3.572	3.448	3.333
44	4.760	4.545	4.348	4.166	4.000	3.846	3.704	3.572	3.448	3.333
45	4.761	.4.545	4.348	4.166	4.000	3.846	3.704	3.572	3.448	3.333
46	4.761	4.545	4.348	4.166	4.000	3.846	3.704	3.572	3.448	3.333
47	4.761	4.545	4.348	4.166	4.000	3.846	3.704	3.572	3.448	3.333
48	4.761	4.545	4.348	4.166	4.000	3.846	3.704	3.572	3.448	3.333
49	4.761	4.545	4.348	4.166	4.000	3.846	3.704	3.572	3.448	3.333
50	4.761	4.545	4.348	4.166	4.000	3.846	3.704	3.572	3.448	3.333

Cumulative Present Value Factors (*Contd.*)

Discount range 31% to 40%

Year \ *x*	31%	32%	33%	34%	35%	36%	37%	38%	39%	40%
1	0.763	0.758	0.752	0.746	0.741	0.735	0.730	0.725	0.719	0.714
2	1.346	1.332	1.317	1.303	1.289	1.276	1.263	1.250	1.237	1.225
3	1.791	1.766	1.742	1.719	1.696	1.674	1.652	1.630	1.609	1.589
4	2.131	2.096	2.062	2.029	1.997	1.966	1.936	1.906	1.877	1.849
5	2.390	2.345	2.302	2.261	2.220	2.181	2.143	2.106	2.070	2.035
6	2.588	2.534	2.483	2.433	2.385	2.339	2.294	2.251	2.209	2.168
7	2.739	2.677	2.619	2.562	2.508	2.455	2.404	2.355	2.308	2.263
8	2.854	2.786	2.721	2.658	2.598	2.540	2.485	2.431	2.380	2.331
9	2.942	2.868	2.798	2.730	2.665	2.603	2.544	2.487	2.432	2.379
10	3.009	2.930	2.855	2.784	2.715	2.649	2.587	2.526	2.469	2.414
11	3.060	2.978	2.899	2.824	2.752	2.683	2.618	2.555	2.496	2.438
12	3.100	3.013	2.931	2.854	2.779	2.708	2.641	2.576	2.515	2.456
13	3.130	3.040	2.956	2.876	2.799	2.727	2.658	2.592	2.529	2.469
14	3.152	3.061	2.974	2.892	2.814	2.740	2.670	2.603	2.538	2.478
15	3.170	3.076	2.988	2.905	2.825	2.750	2.679	2.611	2.546	2.484
16	3.183	3.088	2.999	2.914	2.834	2.757	2.685	2.616	2.551	2.489
17	3.193	3.097	3.006	2.921	2.840	2.763	2.690	2.621	2.554	2.492
18	3.201	3.104	3.012	2.926	2.844	2.767	2.693	2.624	2.557	2.494
19	3.207	3.109	3.017	2.930	2.847	2.770	2.696	2.626	2.559	2.496
20	3.211	3.113	3.020	2.933	2.850	2.772	2.698	2.627	2.560	2.497
21	3.215	3.116	3.023	2.935	2.852	2.773	2.699	2.629	2.561	2.498
22	3.217	3.118	3.024	2.937	2.853	2.775	2.700	2.629	2.562	2.499
23	3.219	3.120	3.026	2.938	2.854	2.775	2.701	2.630	2.563	2.499
24	3.221	3.121	3.027	2.939	2.855	2.776	2.701	2.630	2.563	2.499
25	3.222	3.122	3.028	2.939	2.855	2.776	2.702	2.631	2.563	2.499
26	3.223	3.123	3.028	2.940	2.856	2.777	2.702	2.631	2.564	2.500
27	3.224	3.123	3.029	2.940	2.856	2.777	2.702	2.631	2.564	2.500
28	3.224	3.124	3.029	2.941	2.856	2.777	2.702	2.631	2.564	2.500
29	3.224	3.124	3.029	2.941	2.857	2.777	2.702	2.631	2.564	2.500
30	3.225	3.124	3.030	2.941	2.857	2.777	2.702	2.631	2.564	2.500
31	3.225	3.124	3.030	2.941	2.857	2.777	2.702	2.631	2.564	2.500
32	3.225	3.125	3.030	2.941	2.857	2.778	2.702	2.631	2.564	2.500
33	3.225	3.125	3.030	2.941	2.857	2.778	2.702	2.631	2.564	2.500
34	3.225	3.125	3.030	2.941	2.857	2.778	2.702	2.631	2.564	2.500
35	3.225	3.125	3.030	2.941	2.857	2.778	2.702	2.631	2.564	2.500
36	3.226	3.125	3.030	2.941	2.857	2.778	2.702	2.631	2.564	2.500
37	3.226	3.125	3.030	2.941	2.857	2.778	2.702	2.631	2.564	2.500
38	3.226	3.125	3.030	2.941	2.857	2.778	2.702	2.631	2.564	2.500
39	3.226	3.125	3.030	2.941	2.857	2.778	2.702	2.631	2.564	2.500
40	3.226	3.125	3.030	2.941	2.857	2.778	2.702	2.631	2.564	2.500
41	3.226	3.125	3.030	2.941	2.857	2.778	2.702	2.631	2.564	2.500
42	3.226	3.125	3.030	2.941	2.857	2.778	2.702	2.631	2.564	2.500
43	3.226	3.125	3.030	2.941	2.857	2.778	2.702	2.631	2.564	2.500
44	3.226	3.125	3.030	2.941	2.857	2.778	2.702	2.631	2.564	2.500
45	3.226	3.125	3.030	2.941	2.857	2.778	2.702	2.631	2.564	2.500
46	3.226	3.125	3.030	2.941	2.857	2.778	2.702	2.631	2.564	2.500
47	3.226	3.125	3.030	2.941	2.857	2.778	2.702	2.631	2.564	2.500
48	3.226	3.125	3.030	2.941	2.857	2.778	2.702	2.631	2.564	2.500
49	3.226	3.125	3.030	2.941	2.857	2.778	2.702	2.631	2.564	2.500
50	3.226	3.125	3.030	2.941	2.857	2.778	2.702	2.631	2.564	2.500

CAPITAL RECOVERY FACTORS

The following table gives the equal annual payment to be made for *n* years in the future to repay loan principal and interest with interest at *x*% per year.

x / *Year*	1%	2%	3%	4%	5%	6%	7%	8%	9%	10%
1	1.0100	1.0200	1.0300	1.0400	1.0500	1.0600	1.0700	1.0800	1.0900	1.1000
2	0.5075	0.5150	0.5226	0.5302	0.5378	0.5454	0.5531	0.5608	0.5685	0.5762
3	0.3400	0.3468	0.3535	0.3603	0.3672	0.3741	0.3811	0.3880	0.3951	0.4021
4	0.2563	0.2626	0.2690	0.2755	0.2820	0.2886	0.2952	0.3019	0.3087	0.3155
5	0.2060	0.2122	0.2184	0.2246	0.2310	0.2374	0.2439	0.2505	0.2571	0.2638
6	0.1725	0.1785	0.1846	0.1908	0.1970	0.2034	0.2098	0.2163	0.2229	0.2296
7	0.1486	0.1545	0.1605	0.1666	0.1728	0.1791	0.1856	0.1921	0.1987	0.2054
8	0.1307	0.1365	0.1425	0.1485	0.1547	0.1610	0.1675	0.1740	0.1807	0.1874
9	0.1167	0.1225	0.1284	0.1345	0.1407	0.1470	0.1535	0.1601	0.1668	0.1736
10	0.1056	0.1113	0.1172	0.1233	0.1295	0.1359	0.1424	0.1490	0.1558	0.1627
11	0.0965	0.1022	0.1081	0.1141	0.1204	0.1268	0.1334	0.1401	0.1469	0.1540
12	0.0888	0.0946	0.1005	0.1066	0.1128	0.1193	0.1259	0.1327	0.1397	0.1468
13	0.0824	0.0881	0.0940	0.1001	0.1065	0.1130	0.1197	0.1265	0.1336	0.1408
14	0.0769	0.0826	0.0885	0.0947	0.1010	0.1076	0.1143	0.1213	0.1284	0.1357
15	0.0721	0.0778	0.0838	0.0899	0.0963	0.1030	0.1098	0.1168	0.1241	0.1315
16	0.0679	0.0737	0.0796	0.0858	0.0923	0.0990	0.1059	0.1130	0.1203	0.1278
17	0.0643	0.0700	0.0760	0.0822	0.0887	0.0954	0.1024	0.1096	0.1170	0.1247
18	0.0610	0.0667	0.0727	0.0790	0.0855	0.0924	0.0994	0.1067	0.1142	0.1219
19	0.0581	0.0638	0.0698	0.0761	0.0827	0.0896	0.0968	0.1041	0.1117	0.1195
20	0.0554	0.0612	0.0672	0.0736	0.0802	0.0872	0.0944	0.1019	0.1095	0.1175
21	0.0530	0.0588	0.0649	0.0713	0.0780	0.0850	0.0923	0.0998	0.1076	0.1156
22	0.0509	0.0566	0.0627	0.0692	0.0760	0.0830	0.0904	0.0980	0.1059	0.1140
23	0.0489	0.0547	0.0608	0.0673	0.0741	0.0813	0.0887	0.0964	0.1044	0.1126
24	0.0471	0.0529	0.0590	0.0656	0.0725	0.0797	0.0872	0.0950	0.1030	0.1113
25	0.0454	0.0512	0.0574	0.0640	0.0710	0.0782	0.0858	0.0937	0.1018	0.1102
26	0.0439	0.0497	0.0559	0.0626	0.0696	0.0769	0.0846	0.0925	0.1007	0.1092
27	0.0424	0.0483	0.0546	0.0612	0.0683	0.0757	0.0834	0.0914	0.0997	0.1083
28	0.0411	0.0470	0.0533	0.0600	0.0671	0.0746	0.0824	0.0905	0.0989	0.1075
29	0.0399	0.0458	0.0521	0.0589	0.0660	0.0736	0.0814	0.0896	0.0981	0.1067
30	0.0387	0.0446	0.0510	0.0578	0.0651	0.0726	0.0806	0.0888	0.0973	0.1061
31	0.0377	0.0436	0.0500	0.0569	0.0641	0.0718	0.0798	0.0881	0.0967	0.1055
32	0.0367	0.0426	0.0490	0.0559	0.0633	0.0710	0.0791	0.0875	0.0961	0.1050
33	0.0357	0.0417	0.0482	0.0551	0.0625	0.0703	0.0784	0.0869	0.0956	0.1045
34	0.0348	0.0408	0.0473	0.0543	0.0618	0.0696	0.0778	0.0863	0.0951	0.1041
35	0.0340	0.0400	0.0465	0.0536	0.0611	0.0690	0.0772	0.0858	0.0946	0.1037
36	0.0332	0.0392	0.0458	0.0529	0.0604	0.0684	0.0767	0.0853	0.0942	0.1033
37	0.0325	0.0385	0.0451	0.0522	0.0598	0.0679	0.0762	0.0849	0.0939	0.1030
38	0.0318	0.0378	0.0445	0.0516	0.0593	0.0674	0.0758	0.0845	0.0935	0.1027
39	0.0311	0.0372	0.0438	0.0511	0.0588	0.0669	0.0754	0.0842	0.0932	0.1025
40	0.0305	0.0366	0.0433	0.0505	0.0583	0.0665	0.0750	0.0839	0.0930	0.1023
41	0.0299	0.0360	0.0427	0.0500	0.0578	0.0661	0.0747	0.0836	0.0927	0.1020
42	0.0293	0.0354	0.0422	0.0495	0.0574	0.0657	0.0743	0.0833	0.0925	0.1019
43	0.0287	0.0349	0.0417	0.0491	0.0570	0.0653	0.0740	0.0830	0.0923	0.1017
44	0.0282	0.0344	0.0412	0.0487	0.0566	0.0650	0.0738	0.0828	0.0921	0.1015
45	0.0277	0.0339	0.0408	0.0483	0.0563	0.0647	0.0735	0.0826	0.0919	0.1014
46	0.0272	0.0335	0.0404	0.0479	0.0559	0.0644	0.0733	0.0824	0.0917	0.1013
47	0.0268	0.0330	0.0400	0.0475	0.0556	0.0641	0.0730	0.0822	0.0916	0.1011
48	0.0263	0.0326	0.0396	0.0472	0.0553	0.0639	0.0728	0.0820	0.0915	0.1010
49	0.0259	0.0322	0.0392	0.0469	0.0550	0.0637	0.0726	0.0819	0.0913	0.1009
50	0 0255	00318	0.0389	0.0466	0.0548	0.0634	0.0725	0.0817	0.0912	0.1009

Capital Recovery Factors (*Contd.*)

Discount range 11% to 20%

Year / x	11%	12%	13%	14%	15%	16%	17%	18%	19%	20%
1	1.1100	1.1200	1.1300	1.1400	1.1500	1.1600	1.1700	1.1800	1.1900	1.2000
2	0.5839	0.5917	0.5995	0.6073	0.6151	0.6230	0.6308	0.6387	0.6466	0.6545
3	0.4092	0.4163	0.4235	0.4307	0.4380	0.4453	0.4526	0.4599	0.4673	0.4747
4	0.3223	0.3292	0.3362	0.3432	0.3503	0.3574	0.3645	0.3717	0.3790	0.3863
5	0.2706	0.2774	0.2843	0.2913	0.2983	0.3054	0.3126	0.3198	0.3271	0.3344
6	0.2364	0.2432	0.2502	0.2572	0.2642	0.2714	0.2786	0.2859	0.2933	0.3007
7	0.2122	0.2191	0.2261	0.2332	0.2404	0.2476	0.2549	0.2624	0.2699	0.2774
8	0.1943	0.2013	0.2084	0.2156	0.2229	0.2302	0.2377	0.2452	0.2529	0.2606
9	0.1806	0.1877	0.1949	0.2022	0.2096	0.2171	0.2247	0.2324	0.2402	0.2481
10	0.1698	0.1770	0.1843	0.1917	0.1993	0.2069	0.2147	0.2225	0.2305	0.2385
11	0.1611	0.1684	0.1758	0.1834	0.1911	0.1989	0.2068	0.2148	0.2229	0.2311
12	0.1540	0.1614	0.1690	0.1767	0.1845	0.1924	0.2005	0.2086	0.2169	0.2253
13	0.1482	0.1557	0.1634	0.1712	0.1791	0.1872	0.1954	0.2037	0.2121	0.2206
14	0.1432	0.1509	0.1587	0.1666	0.1747	0.1829	0.1912	0.1997	0.2082	0.2169
15	0.1391	0.1468	0.1547	0.1628	0.1710	0.1794	0.1878	0.1964	0.2051	0.2139
16	0.1355	0.1434	0.1514	0.1596	0.1679	0.1764	0.1850	0.1937	0.2025	0.2114
17	0.1325	0.1405	0.1486	0.1569	0.1654	0.1740	0.1827	0.1915	0.2004	0.2094
18	0.1298	0.1379	0.1462	0.1546	0.1632	0.1719	0.1807	0.1896	0.1987	0.2078
19	0.1276	0.1358	0.1441	0.1527	0.1613	0.1701	0.1791	0.1881	0.1972	0.2065
20	0.1256	0.1339	0.1424	0.1510	0.1598	0.1687	0.1777	0.1868	0.1960	0.2054
21	0.1238	0.1322	0.1408	0.1495	0.1584	0.1674	0.1765	0.1857	0.1951	0.2044
22	0.1223	0.1308	0.1395	0.1483	0.1573	0.1664	0.1756	0.1848	0.1942	0.2037
23	0.1210	0.1296	0.1383	0.1472	0.1563	0.1654	0.1747	0.1841	0.1935	0.2031
24	0.1198	0.1285	0.1373	0.1463	0.1554	0.1647	0.1740	0.1835	0.1930	0.2025
25	0.1187	0.1275	0.1364	0.1455	0.1547	0.1640	0.1734	0.1829	0.1925	0.2021
26	0.1178	0.1267	0.1357	0.1448	0.1541	0.1634	0.1729	0.1825	0.1921	0.2018
27	0.1170	0.1259	0.1350	0.1442	0.1535	0.1630	0.1725	0.1821	0.1917	0.2015
28	0.1163	0.1252	0.1344	0.1437	0.1531	0.1625	0.1721	0.1818	0.1915	0.2012
29	0.1156	0.1247	0.1339	0.1432	0.1527	0.1622	0.1718	0.1815	0.1912	0.2010
30	0.1150	0.1241	0.1334	0.1428	0.1523	0.1619	0.1715	0.1813	0.1910	0.2008
31	0.1145	0.1237	0.1330	0.1425	0.1520	0.1616	0.1713	0.1811	0.1909	0.2007
32	0.1140	0.1233	0.1327	0.1421	0.1517	0.1614	0.1711	0.1809	0.1907	0.2006
33	0.1136	0.1229	0.1323	0.1419	0.1515	0.1612	0.1710	0.1808	0.1906	0.2005
34	0.1133	0.1226	0.1321	0.1416	0.1513	0.1610	0.1708	0.1806	0.1905	0.2004
35	0.1129	0.1223	0.1318	0.1414	0.1511	0.1609	0.1707	0.1806	0.1904	0.2003
36	0.1126	0.1221	0.1316	0.1413	0.1510	0.1608	0.1706	0.1805	0.1904	0.2003
37	0.1124	0.1218	0.1314	0.1411	0.1509	.0.1607	0.1705	0.1804	0.1903	0.2002
38	0.1121	0.1216	0.1313	0.1410	0.1507	0.1606	0.1704	0.1803	0.1903	0.2002
39	0.1119	0.1215	0.1311	0.1409	0.1506	0.1605	0.1704	0.1803	0.1902	0.2002
40	0.1117	0.1213	0.1310	0.1407	0.1506	0.1604	0.1703	0.1802	0.1902	0.2001
41	0.1115	0.1212	0.1309	0.1407	0.1505	0.1604	0.1703	0.1802	0.1902	0.2001
42	0.1114	0.1210	0.1308	0.1406	0.1504	0.1603	0.1702	0.1802	0.1901	0.2001
43	0.1113	0.1209	0.1307	0.1405	0.1504	0.1603	0.1702	0.1801	0.1901	0.2001
44	0.1111	0.1208	0.1306	0.1404	0.1503	0.1602	0.1702	0.1801	0.1901	0.2001
45	0.1110	0.1207	0.1305	0.1404	0.1503	0.1602	0.1701	0.1801	0.1901	0.2001
46	0.1109	0.1207	0.1305	0.1403	0.1502	0.1602	0.1701	0.1801	0.1901	0.2000
47	0.1108	0.1206	0.1304	0.1403	0.1502	0.1601	0.1701	0.1801	0.1901	0.2000
48	0.1107	0.1205	0.1304	0.1403	0.1502	0.1601	0.1701	0.1801	0.1900	0.2000
49	0.1107	0.1205	0.1303	0.1402	0.1502	0.1601	0.1701	0.1801	0.1900	0.2000
50	0.1106	0.1204	0.1303	0.1402	0.1501	0.1601	0.1701	0.1800	0.1900	0.2000

Capital Recovery Factors (*Contd.*)

Discount range 21% to 30%

x / *Year*	21%	22%	23%	24%	25%	26%	27%	28%	29%	30%
1	1.2100	1.2200	1.2300	1.2400	1.2500	1.2600	1.2700	1.2800	1.2900	1.3000
2	0.6625	0.6705	0.6784	0.6864	0.6944	0.7025	0.7105	0.7186	0.7267	0.7348
3	0.4822	0.4897	0.4972	0.5047	0.5123	0.5199	0.5275	0.5352	0.5429	0.5506
4	0.3936	0.4010	0.4085	0.4159	0.4234	0.4310	0.4386	0.4462	0.4539	0.4616
5	0.3418	0.3492	0,3567	0.3642	0.3718	0.3795	0.3872	0.3949	0.4027	0.4106
6	0.3082	0.3158	0.3234	0.3311	0.3388	0.3466	0.3545	0.3624	0.3704	0.3784
7	0.2851	0.2928	0.3006	0.3084	0.3163	0.3243	0.3324	0.3405	0.3486	0.3569
8	0.2684	0.2763	0.2843	0.2923	0.3004	0.3086	0.3168	0.3251	0.3335	0.3419
9	0.2561	0.2641	0.2722	0.2805	0.2888	0.2971	0.3056	0.3140	0.3226	0.3312
10	0.2467	0.2549	0.2632	0.2716	0.2801	0.2886	0.2972	0.3059	0.3147	0.3235
11	0.2394	0.2478	0.2563	0.2649	0.2735	0.2822	0.2910	0.2998	0.3088	0.3177
12	0.2337	0.2423	0.2509	0.2596	0.2684	0.2773	0.2863	0.2953	0.3043	0.3135
13	0.2292	0.2379	0.2467	0.2556	0.2645	0.2736	0.2826	0.2918	0.3010	0.3102
14	0.2256	0.2345	0.2434	0.2524	0.2615	0.2706	0.2799	0.2891	0.2984	0.3078
15	0.2228	0.2317	0.2408	0.2499	0.2591	0.2684	0.2777	0.2871	0.2965	0.3060
16	0.2204	0.2295	0.2387	0.2479	0.2572	0.2666	0.2760	0.2855	0.2950	0.3046
17	0.2186	0.2278	0.2370	0.2464	0.2558	0.2652	0.2747	0.2843	0.2939	0.3035
18	0.2170	0.2263	0.2357	0.2451	0.2546	0.2641	0.2737	0.2833	0.2930	0.3027
19	0.2158	0.2251	0.2346	0.2441	0.2537	0.2633	0.2729	0.2826	0.2923	0.3021
20	0.2147	0.2242	0.2337	0.2433	0.2529	0.2626	0.2723	0.2820	0.2918	0.3016
21	0.2139	0.2234	0.2330	0.2426	0.2523	0.2620	0.2718	0.2816	0.2914	0.3012
22	0.2132	0.2228	0.2324	0.2421	0.2519	0.2616	0.2714	0.2812	0.2911	0.3009
23	0.2127	0.2223	0.2320	0.2417	0.2515	0.2613	0.2711	0.2810	0.2908	0.3007
24	0.2122	0.2219	0.2316	0.2414	0.2512	0.2610	0.2709	0.2808	0.2906	0.3006
25	0.2118	0.2215	0.2313	0.2411	0.2509	0.2608	0.2707	0.2806	0.2905	0.3004
26	0.2115	0.2213	0.2311	0.2409	0.2508	0.2606	0.2705	0.2805	0.2904	0.3003
27	0.2112	0.2210	0.2309	0.2407	0.2506	0.2605	0.2704	0.2804	0.2903	0.3003
28	0.2110	0.2208	0.2307	0.2406	0.2505	0.2604	0.2703	0.2803	0.2902	0.3002
29	0.2108	0.2207	0.2306	0.2405	0.2504	0.2603	0.2703	0.2802	0.2902	0.3001
30	0.2107	0.2206	0.2305	0.2404	0.2503	0.2603	0.2702	0.2802	0.2901	0.3001
31	0.2106	0.2205	0.2304	0.2403	0.2502	0.2602	0.2702	0.2801	0.2901	0.3001
32	0.2105	0.2204	0.2303	0.2402	0.2502	0.2602	0.2701	0.2801	0.2901	0.3001
33	0.2104	0.2203	0.2302	0.2402	0.2502	0.2601	0.2701	0.2801	0.2901	0.3001
34	0.2103	0.2203	0.2302	0.2402	0.2501	0.2601	0.2701	0.2801	0.2901	0.3000
35	0.2103	0.2202	0.2302	0.2401	0.2501	0.2601	0.2701	0.2800	0.2900	0.3000
36	0.2102	0.2202	0.2301	0.2401	0.2501	0.2601	0.2700	0.2800	0.2900	0.3000
37	0.2102	0.2201	0.2301	0.2401	0.2501	0.2601	0.2700	0.2800	0.2900	0.3000
38	0.2102	0.2201	0.2301	0.2401	0.2501	0.2600	0.2700	0.2800	0.2900	0.3000
39	0.2101	0.2201	0.2301	0.2401	0.2500	0.2600	0.2700	0.2800	0.2900	0.3000
40	0.2101	0.2201	0.2301	0.2400	0.2500	0.2600	0.2700	0.2800	0.2900	0.3000
41	0.2101	0.2201	0.2300	0.2400	0.2500	0.2600	0.2700	0.2800	0.2900	0.3000
42	0.2101	0.2201	0.2300	0.2400	0.2500	0.2600	0.2700	0.2800	0.2900	0.3000
43	0.2101	0.2200	0.2300	0.2400	0.2500	0.2600	0.2700	0.2800	0.2900	0.3000
44	0.2100	0.2200	0.2300	0.2400	0.2500	0.2600	0.2700	0.2800	0.2900	0.3000
45	0.2100	0.2200	0.2300	0.2400	0.2500	0.2600	0.2700	0.2800	0.2900	0.3000
46	0.2100	0.2200	0.2300	0.2400	0.2500	0.2600	0.2700	0.2800	0.2900	0.3000
47	0.2100	0.2200	0.2300	0.2400	0.2500	0.2600	0.2700	0.2800	0.2900	0.3000
48	0.2100	0.2200	0.2300	0.2400	0.2500	0.2600	0.2700	0.2800	0.2900	0.3000
49	0.2100	0.2200	0.2300	0.2400	0.2500	0.2600	0.2700	0.2800	0.2900	0.3000
50	0.2100	0.2200	0.2300	0.2400	0.2500	0.2600	0.2700	0.2800	0.2900	0.3000

Capital Recovery Factors (*Contd.*)

Discount range 31% to 40%

x / *Year*	31%	32%	33%	34%	35%	36%	37%	38%	39%	40%
1	1.3100	1.3200	1.3300	1.3400	1.3500	1.3600	1.3700	1.3800	1.3900	1.4000
2	0.7429	0.7510	0.7592	0.7674	0.7755	0.7837	0.7919	0.8002	0.8084	0.8167
3	0.5584	0.5662	0.5740	0.5818	0.5897	0.5976	0.6055	0.6134	0.6214	0.6294
4	0.4694	0.4772	0.4850	0.4929	0.5008	0.5087	0.5167	0.5247	0.5327	0.5408
5	0.4185	0.4264	0.4344	0.4424	0.4505	0.4586	0.4667	0.4749	0.4831	0.4914
6	0.3865	0.3946	0.4028	0.4110	0.4193	0.4276	0.4359	0.4443	0.4528	0.4613
7	0.3652	0.3735	0.3819	0.3903	0.3988	0.4073	0.4159	0.4245	0.4332	0.4419
8	0.3504	0.3589	0.3675	0.3762	0.3849	0.3936	0.4024	0.4113	0.4201	0.4291
9	0.3399	0.3487	0.3575	0.3663	0.3752	0.3841	0.3931	0.4022	0.4112	0.4203
10	0.3323	0.3412	0.3502	0.3592	0.3683	0.3774	0.3866	0.3958	0.4050	0.4143
11	0.3268	0.3358	0.3450	0.3542	0.3634	0.3727	0.3820	0.3913	0.4007	0.4101
12	0.3226	0.3319	0.3411	0.3505	0.3598	0.3692	0.3787	0.3881	0.3976	0.4072
13	0.3196	0.3289	0.3383	0.3477	0.3572	0.3667	0.3763	0.3859	0.3955	0.4051
14	0.3172	0.3267	0.3362	0.3457	0.3553	0.3649	0.3746	0.3842	0.3939	0.4036
15	0.3155	0.3251	0.3346	0.3443	0.3539	0.3636	0.3733	0.3831	0.3928	0.4026
16	0.3142	0.3238	0.3335	0.3432	0.3529	0.3626	0.3724	0.3822	0.3920	0.4018
17	0.3132	0.3229	0.3326	0.3424	0.3521	0.3619	0.3718	0.3816	0.3915	0.4013
18	0.3124	0.3222	0.3320	0.3418	0.3516	0.3614	0.3713	0.3812	0.3910	0.4009
19	0.3118	0.3216	0.3315	0.3413	0.3512	0.3610	0.3709	0.3808	0.3907	0.4007
20	0.3114	0.3212	0.3311	0.3410	0.3509	0.3608	0.3707	0.3806	0.3905	0.4005
21	0.3111	0.3209	0.3308	0.3407	0.3506	0.3606	0.3705	0.3804	0.3904	0.4003
22	0.3108	0.3207	0.3306	0.3405	0.3505	0.3604	0.3704	0.3803	0.3903	0.4002
23	0.3106	0.3205	0.3305	0.3404	0.3504	0.3603	0.3703	0.3802	0.3902	0.4002
24	0.3105	0.3204	0.3304	0.3403	0.3503	0.3602	0.3702	0.3802	0.3901	0.4001
25	0.3104	0.3203	0.3303	0.3402	0.3502	0.3602	0.3701	0.3801	0.3901	0.4001
26	0.3103	0.3202	0.3302	0.3402	0.3501	0.3601	0.3701	0.3801	0.3901	0.4001
27	0.3102	0.3202	0.3301	0.3401	0.3501	0.3601	0.3701	0.3801	0.3901	0.4000
28	0.3102	0.3201	0.3301	0.3401	0.3501	0.3601	0.3701	0.3800	0.3900	0.4000
29	0.3101	0.3201	0.3301	0.3401	0.3501	0.3600	0.3700	0.3800	0.3900	0.4000
30	0.3101	0.3201	0.3301	0.3401	0.3500	0.3600	0.3700	0.3800	0.3900	0.4000
31	0.3101	0.3201	0.3300	0.3400	0.3500	0.3600	0.3700	0.3800	0.3900	0.4000
32	0.3101	0.3200	0.3300	0.3400	0.3500	0.3600	0.3700	0.3800	0.3900	0.4000
33	0.3100	0.3200	0.3300	0.3400	0.3500	0.3600	0.3700	0.3800	0.3900	0.4000
34	0.3100	0.3200	0.3300	0.3400	0.3500	0.3600	0.3700	0.3800	0.3900	0.4000
35	0.3100	0.3200	0.3300	0.3400	0.3500	0.3600	0.3700	0.3800	0.3900	0.4000
36	0.3100	0.3200	0.3300	0.3400	0.3500	0.3600	0.3700	0.3800	0.3900	0.4000
37	0.3100	0.3200	0.3300	0.3400	0.3500	0.3600	0.3700	0.3800	0.3900	0.4000
38	0.3100	0.3200	0.3300	0.3400	0.3500	0.3600	0.3700	0.3800	0.3900	0.4000
39	0.3100	0.3200	0.3300	0.3400	0.3500	0.3600	0.3700	0.3800	0.3900	0.4000
40	0.3100	0.3200	0.3300	0.3400	0.3500	0.3600	0.3700	0.3800	0.3900	0.4000
41	0.3100	0.3200	0.3300	0.3400	0.3500	0.3600	0.3700	0.3800	0.3900	0.4000
42	0.3100	0.3200	0.3300	0.3400	0.3500	0.3600	0.3700	0.3800	0.3900	0.4000
43	0.3100	0.3200	0.3300	0.3400	0.3500	0.3600	0.3700	0.3800	0.3900	0.4000
44	0.3100	0.3200	0.3300	0.3400	0.3500	0.3600	0.3700	0.3800	0.3900	0.4000
45	0.3100	0,3200	0.3300	0.3400	0.3500	0.3600	0.3700	0.3800	0.3900	0.4000
46	0.3100	0.3200	0.3300	0.3400	0.3500	0.3600	0.3700	0.3800	0.3900	0.4000
47	0.3100	0.3200	0.3300	0.3400	0.3500	0.3600	0.3700	0.3800	0.3900	0.4000
48	0.3100	0.3200	0.3300	0.3400	0.3500	0.3600	0.3700	0.3800	0.3900	0.4000
49	0.3100	0.3200	0.3300	0.3400	0.3500	0.3600	0.3700	0.3800	0.3900	0.4000
50	0.3100	0.3200	0.3300	0.3400	0.3500	0.3600	0.3700	0.3800	0.3900	0.4000

TRANSFORMATION BETWEEN *r* AND *z*

r	*z*	*r*	*z*	*r*	*z*	*r*	*z*	*r*	*z*
0.000	0.000	0.096	0.096	0.222	0.226	0.318	0.329	0.414	0.440
0.002	0.002	0.098	0.098	0.224	0.228	0.320	0.332	0.416	0.443
0.004	0.004	0.100	0.100	0.226	0.230	0.322	0.334	0.418	0.445
0.006	0.006	0.102	0.102	0.228	0.232	0.324	0.336	0.420	0.448
0.008	0.008	0.104	0.104	0.230	0.234	0.326	0.338	0.422	0.450
0.010	0.010	0.106	0.106	0.232	0.236	0.328	0.341	0.424	0.453
0.012	0.012	0.108	0.108	0.234	0.238	0.330	0.343	0.426	0.455
0.014	0.014	0.110	0.110	0.236	0.241	0.332	0.345	0.428	0.457
0.016	0.016	0.112	0.112	0.238	0.243	0.334	0.347	0.430	0.460
0.018	0.018	0.144	0.145	0.240	0.245	0.336	0.350	0.432	0.462
0.020	0.020	0.146	0.147	0.242	0.247	0.338	0.352	0.434	0.465
0.022	0.022	0.148	0.149	0.244	0.249	0.340	0.354	0.436	0.467
0.024	0.024	0.150	0.151	0.246	0.251	0.342	0.356	0.438	0.470
0.026	0.026	0.152	0.153	0.248	0.253	0.344	0.359	0.440	0.472
0.028	0.028	0.154	0.155	0.250	0.255	0.346	0.361	0.442	0.475
0.030	0.030	0.156	0.157	0.252	0.258	0.348	0.363	0.444	0.477
0.032	0.032	0.158	0.159	0.254	0.260	0.350	0.365	0.446	0.480
0.034	0.034	0.160	0.161	0.256	0.262	0.352	0.368	0.448	0.482
0.036	0.036	0.162	0.163	0.258	0.264	0.354	0.370	0.450	0.485
0.038	0.038	0.164	0.165	0.260	0.266	0.356	0.372	0.452	0.487
0.040	0.040	0.166	0.168	0.262	0.268	0.358	0.375	0.454	0.490
0.042	0.042	0.168	0.170	0.264	0.270	0.360	0.377	0.456	0.492
0.044	0.044	0.170	0.172	0.266	0.273	0.362	0.379	0.458	0.495
0.046	0.046	0.172	0.174	0.268	0.275	0.364	0.381	0.460	0.497
0.048	0.048	0.174	0.176	0.270	0.277	0.366	0.384	0.462	0.500
0.050	0.050	0.176	0.178	0.272	0.279	0.368	0.386	0.464	0.502
0.052	0.052	0.178	0.180	0.274	0.281	0.370	0.388	0.466	0.505
0.054	0.054	0.180	0.182	0.276	0.283	0.372	0.391	0.468	0.508
0.056	0.056	0.182	0.184	0.278	0.286	0.374	0.393	0.470	0.510
0.058	0.058	0.184	0.186	0.280	0.288	0.376	0.395	0.472	0.513
0.060	0.060	0.186	0.188	0.282	0.290	0.378	0.398	0.474	0.515
0.062	0.062	0.188	0.190	0.284	0.292	0.380	0.400	0.476	0.518
0.064	0.064	0.190	0.192	0.286	0.294	0.382	0.402	0.478	0.520
0.066	0.066	0.192	0.194	0.288	0.296	0.384	0.405	0.480	0.523
0.068	0.068	0.194	0.196	0.290	0.299	0.386	0.407	0.482	0.526
0.070	0.070	0.196	0.199	0.292	0.301	0.388	0.409	0.484	0.528
0.072	0.072	0.198	0.201	0.294	0.303	0.390	0.412	0.486	0.531
0.074	0.074	0.200	0.203	0.296	0.305	0.392	0.414	0.488	0.533
0.076	0.076	0.202	0.205	0.298	0.307	0.394	0.417	0.490	0.536
0.078	0.078	0.204	0.207	0.300	0.310	0.396	0.419	0.492	0.539
0.080	0.080	0.206	0.209	0.302	0.312	0.398	0.421	0.494	0.541
0.082	0.082	0.208	0.211	0.304	0.314	0.400	0.424	0.496	0.544
0.084	0.084	0.210	0.213	0.306	0.316	0.402	0.426	0.498	0.547
0.086	0.086	0.212	0.215	0.308	0.318	0.404	0.428	0.500	0.549
0.088	0.088	0.214	0.217	0.310	0.321	0.406	0.431	0.502	0.552
0.090	0.090	0.216	0.219	0.312	0.323	0.408	0.433	0.504	0.555
0.092	0.092	0.218	0.222	0.314	0.325	0.410	0.436	0.506	0.557
0.094	0.094	0.220	0.224	0.316	0.327	0.412	0.438	0.508	0.560

Transformation Between *r* and *z* (*contd.*)

r	*z*	*r*	*z*	*r*	*z*	*r*	*z*	*r*	*z*
0.510	0.563	0.608	0.706	0.706	0.879	0.804	1.110	0.902	1.483
0.512	0.565	0.610	0.709	0.708	0.883	0.806	1.116	0.904	1.494
0.514	0.568	0.612	0.712	0.710	0.887	0.808	1.121	0.906	1.505
0.516	0.571	0.614	0.715	0.712	0.891	0.810	1.127	0.908	1.516
0.518	0.574	0.616	0.719	0.714	0.895	0.812	1.133	0.910	1.528
0.520	0.576	0.618	0.722	0.716	0.899	0.814	1.139	0.912	1.539
0.522	0.579	0.620	0.725	0.718	0.904	0.816	1.145	0.914	1.551
0.524	0.582	0.622	0.728	0.720	0.908	0.818	1.151	0.916	1.564
0.526	0.585	0.624	0.732	0.722	0.912	0.820	1.157	0.918	1.576
0.528	0.587	0.626	0.735	0.724	0.916	0.822	1.163	0.920	1.589
0.530	0.590	0.628	0.738	0.726	0.920	0.824	1.169	0.922	1.602
0.532	0.593	0.630	0.741	0.728	0.924	0.826	1.175	0.924	1.616
0.534	0.596	0.632	0.745	0.730	0.929	0.828	1.182	0.926	1.630
0.536	0.599	0.634	0.748	0.732	0.933	0.830	1.188	0.928	1.644
0.538	0.601	0.636	0.751	0.734	0.937	0.832	1.195	0.930	1.658
0.540	0.604	0.638	0.755	0.736	0.942	0.834	1.201	0.932	1.673
0.542	0.607	0.640	0.758	0.738	0.946	0.836	1.208	0.934	1.689
0.544	0.610	0.642	0.762	0.740	0.950	0.838	1.214	0.936	1.705
0.546	0.613	0.644	0.765	0.742	0.955	0.840	1.221	0.938	1.721
0.548	0.616	0.646	0.768	0.744	0.959	0.842	1.228	0.940	1.738
0.550	0.618	0.648	0.772	0.746	0.964	0.844	1.235	0.942	1.756
0.552	0.621	0.650	0.775	0.748	0.968	0.846	1.242	0.944	1.774
0.554	0.624	0.652	0.779	0.750	0.973	0.848	1.249	0.946	1.792
0.556	0.627	0.654	0.782	0.752	0.978	0.850	1.256	0.948	1.812
0.558	0.630	0.656	0.786	0.754	0.982	0.852	1.263	0.950	1.832
0.560	0.633	0.658	0.789	0.756	0.987	0.854	1.271	0.952	1.853
0.562	0.636	0.660	0.793	0.758	0.991	0.856	1.278	0.954	1.874
0.564	0.639	0.662	0.796	0.760	0.996	0.858	1.286	0.956	1.897
0.566	0.642	0.664	0.800	0.762	1.001	0.860	1.293	0.958	1.921
0.568	0.645	0.666	0.804	0.764	1.006	0.862	1.301	0.960	1.946
0.570	0.648	0.668	0.807	0.766	1.011	0.864	1.309	0.962	1.972
0.572	0.650	0.670	0.811	0.768	1.015	0.866	1.317	0.964	2.000
0.574	0.653	0.672	0.814	0.770	1.020	0.868	1.325	0.966	2.029
0.576	0.656	0.674	0.818	0.772	1.025	0.870	1.333	0.968	2.060
0.578	0.659	0.676	0.822	0.774	1.030	0.872	1.341	0.970	2.092
0.580	0.662	0.678	0.825	0.776	1.035	0.874	1.350	0.972	2.127
0.582	0.665	0.680	0.829	0.778	1.040	0.876	1.358	0.974	2.165
0.584	0.669	0.682	0.833	0.780	1.045	0.878	1.367	0.976	2.205
0.586	0.672	0.684	0.837	0.782	1.050	0.880	1.376	0.978	2.249
0.588	0.675	0.686	0.840	0.784	1.056	0.882	1.385	0.980	2.298
0.590	0.678	0.688	0.844	0.786	1.061	0.884	1.394	0.982	2.351
0.592	0.681	0.690	0.848	0.788	1.066	0.886	1.403	0.984	2.410
0.594	0.684	0.692	0.852	0.790	1.071	0.888	1.412	0.986	2.477
0.596	0.687	0.694	0.856	0.792	1.077	0.890	1.422	0.988	2.555
0.598	0.690	0.696	0.860	0.794	1.082	0.892	1.432	0.990	2.647
0.600	0.693	0.698	0.863	0.796	1.088	0.894	1.442	0.992	2.759
0.602	0.696	0.700	0.867	0.798	1.093	0.896	1.452	0.994	2.903
0.604	0.699	0.702	0.871	0.800	1.099	0.898	1.462	0.996	3.106
0.606	0.703	0.704	0.875	0.802	1.104	0.900	1.472	0.998	3.453

RECIPROCALS OF NUMBERS

From 1 to 10

[Numbers in different columns to be subtracted, not added]

	0	1	2	3	4	5	6	7	8	9	Mean Differences								
											1	2	3	4	5	6	7	8	9
1.0	1.000	9901	9804	9709	9615	9524	9434	9346	9259	9174									
1.1	.9091	9009	8929	8850	8772	8696	8621	8547	8475	8403									
1.2	.8333	8264	8197	8130	8065	8000	7937	7874	7813	7752									
1.3	.7692	7634	7576	7519	7463	7407	7353	7299	7246	7194									
1.4	.7143	7092	7042	6993	6944	6897	6849	6803	6757	6711	5	10	14	19	24	29	33	38	43
1.5	.6667	6623	6579	6536	6494	6452	6410	6369	6369	6289	4	8	13	17	21	25	29	33	38
1.6	.0250	6211	6173	6135	6098	6061	6024	5988	5952	5917	4	7	11	15	18	22	26	29	33
1.7	.5882	5848	5814	5780	5747	5714	5682	5650	5618	5587	3	6	10	13	16	20	23	26	29
1.8	.5556	5525	5495	5464	5435	5405	53^6	5348	5319	5291	3	6	9	12	15	17	20	23	26
1.9	.5263	5236	5208	5181	5155	5128	5102	5076	5051	5025	3	5	8	11	13	16	18	21	24
2.0	.5000	4975	4950	4926	4902	4878	4854	4831	4808	4785	2	5	7	10	12	14	17	19	21
2.1	.4762	4739	4717	4695	4673	4651	4630	4608	4587	4566	2	4	7	9	11	13	15	17	20
2.2	.4545	4525	4505	4484	4464	4444	4425	4405	4386	4367	2	4	6	8	10	12	14	16	18
2.3	.4348	4329	4310	4292	4274	4255	4237	4219	4202	4184	2	4	5	7	9	11	13	14	16
2.4	.4167	4149	4132	4115	4098	4082	4065	4049	4032	4016	2	3	5	7	8	10	12	13	15
2.5	.4000	3984	3968	3953	3937	3922	3906	3891	3876	3876	2	3	5	6	8	9	11	12	14
2.6	.3846	3831	3817	3802	3788	3774	3759	3745	3731	3717	1	3	4	6	7	8	10	11	13
2.7	.3704	3690	3676	3663	3650	3636	3623	3610	3597	3584	1	3	4	5	7	8	9	11	12
2.8	.3571	3559	3546	3534	3521	3509	3497	3484	3472	3460	1	2	4	5	6	7	9	10	11
2.9	.3448	3436	3425	3413	3401	3390	3378	3367	3356	3344	1	2	3	5	6	7	8	9	10
3.0	.3333	3322	3311	3300	3289	3279	3268	3257	3247	3236	1	2	3	4	5	6	7	9	10
3.1	.3226	3215	3205	3195	3185	3175	3165	3155	3145	3135	1	2	3	4	5	6	7	8	9
3.2	.3125	3115	3106	3096	3086	3077	3067	3058	3049	3040	1	2	3	4	5	6	7	8	9
3.3	.3030	3021	3012	3003	2994	2985	2976	2967	2959	2950	1	2	3	4	4	5	6	7	8
3.4	.2941	2933	2924	2915	2907	2899	2890	2882	2874	2865	1	2	3	3	4	5	6	7	8
3.5	.2857	2849	2841	2833	2825	2817	2809	2801	2793	2786	1	2	2	3	4	5	6	6	7
3.6	.2778	2770	2762	2755	2747	2740	2732	2725	2717	2710	1	2	2	3	4	5	5	6	7
3.7	.2703	2695	2688	2681	2674	2667	2660	2653	2646	2639	1	1	2	3	4	4	5	6	6
3.8	.2632	2625	2618	2611	2604	2597	2591	2584	2577	2571	1	1	2	3	3	4	5	5	6
3.9	.2564	2558	2551	2545	2538	2532	2525	2519	2513	2506	1	1	2	3	3	4	4	5	6
4.0	.2500	2494	2488	2481	2475	2469	2463	2457	2451	2445	1	1	2	2	2	4	4	5	5
4.1	.2439	2433	2427	2421	2415	2410	2404	2398	2392	2387	1	1	2	2	3	3	4	5	5
4.2	.2381	2375	2370	2364	2358	2353	2347	2342	2336	2331	1	1	2	2	3	3	4	4	5
4.3	.2326	2320	2315	2309	2304	2299	2294	2288	2283	2278	1	1	2	2	3	3	4	4	5
4.4	.2273	2268	2262	2257	2252	2247	2242	2237	2232	2227	1	1	2	2	3	3	4	4	5
4.5	.2222	2217	2212	2208	2203	2198	2193	2188	2183	2179	0	1	1	2	2	3	3	4	4
4.6	.2174	2169	2165	2160	2155	2151	2146	2141	2137	2132	0	1	1	2	2	3	3	4	4
4.7	.2128	2123	2119	2114	2110	2105	2101	2096	2092	2088	0	1	1	2	2	3	3	4	4
4.8	.2083	2079	2075	2070	2066	2062	2058	2053	2049	2045	0	1	1	2	2	3	3	3	4
4.9	2041	2037	2033	2028	2024	2020	2016	2012	2008	2004	0	1	1	2	2	2	3	3	4
5.0	.2000	1996	1992	1988	1984	1980	1976	1972	1969	1965	0	1	1	2	2	2	3	3	4
5.1	.1961	1957	1953	1949	1946	1942	1938	1934	1931	1927	0	1	1	2	2	2	3	3	3
5.2	.1923	1919	1916	1912	1908	1905	1901	1898	1894	1890	0	1	1	1	2	2	3	3	3
5.3	.1887	1883	1880	1876	1873	1869	1866	1862	1859	1855	0	1	1	1	2	2	2	3	3
5.4	.1852	1848	1845	1842	1838	1835	1832	1828	1825	1821	0	1	1	1	2	2	2	3	3

Reciprocals of Numbers (*Contd.*)

	0	1	2	3	4	5	6	7	8	9	Mean Differences								
											1	2	3	4	5	6	7	8	9
5.5	.1818	1815	1812	1808	1805	1802	1799	1795	1792	1789	0	1	1	1	2	2	2	3	3
5.6	.1786	1783	1779	1776	1773	1770	1767	1764	1761	1757	0	1	1	1	2	2	2	3	3
5.7	.1754	1751	1748	1745	1742	1739	1736	1733	1730	1727	0	1	1	1	1	2	2	2	3
5.8	.1724	1721	1718	1715	1712	1709	1706	1704	1701	1698	0	1	1	1	1	2	2	2	3
5.9	.1695	1692	1689	1686	1684	1681	1678	1675	1672	1669	0	1	1	1	1	2	2	2	3
6.0	.1667	1664	1661	1658	1656	1653	1650	1647	1645	1642	0	1	1	1	1	2	2	2	3
6.1	.1639	1637	1634	1631	1629	1626	1623	1621	1618	1616	0	1	1	1	1	2	2	2	2
6.2	.1613	1610	1608	1605	1603	1600	1597	1595	1592	1590	0	1	1	1	1	2	2	2	2
6.3	.1587	1585	1582	1580	1577	1575	1572	1570	1567	1565	0	0	1	1	1	1	2	2	2
6.4	.1562	1560	1558	1555	1553	1550	1548	1546	1543	1541	0	0	1	1	1	1	2	2	2
6.5	.1538	1536	1534	1531	1529	1527	1524	1522	1520	1517	0	0	1	1	1	1	2	2	2
6.6	.1515	1513	1511	1508	1506	1504	1502	1499	1497	1495	0	0	1	1	1	1	2	2	2
6.7	.1493	1490	1488	1486	1484	1481	1479	1477	1475	1473	0	0	1	1	1	1	2	2	2
6.8	.1471	1468	1466	1464	1462	1460	1458	1456	1453	1451	0	0	1	1	1	1	2	2	2
6.9	.1449	1447	1445	1443	1441	1439	1437	1435	1433	1431	0	0	1	1	1	1	2	2	2
7.0	.1429	1427	1425	1422	1420	1418	1416	1414	1412	1410	0	0	1	1	1	1	2	2	2
7.1	.1408	1406	1404	1403	1401	1399	1397	1395	1393	1391	0	0	1	1	1	1	1	2	2
7.2	.1389	1387	1385	1383	1381	1379	1377	1376	1374	1372	0	0	1	1	1	1	1	2	2
7.3	.1370	1368	1366	1364	1362	1361	1359	1357	1355	1353	0	0	1	1	1	1	1	2	2
7.4	.1351	1350	1348	1346	1344	1342	1340	1339	1337	1335	0	0	1	1	1	1	1	2	2
7.5	.1333	1332	1330	1328	1326	1325	1323	1321	1319	1318	0	0	1	1	1	1	1	1	2
7.6	.1316	1314	1312	1311	1309	1307	1305	1304	1302	1300	0	0	1	1	1	1	1	1	2
7.7	.1299	1297	1295	1294	1292	129C	1289	1287	1285	1284	0	0	0	1	1	1	1	1	1
7.8	.1282	1280	1279	1277	1276	1274	1272	1271	1269	1267	0	0	0	1	1	1	1	1	1
7.9	.1266	1264	1263	1261	1259	1258	1256	1255	1253	1252	0	0	0	1	1	1	1	1	1
8.0	.1250	1248	1247	1245	1244	1242	1241	1239	1238	1236	0	0	0	1	1	1	1	1	1
8.1	.1235	1233	1232	1230	1229	1227	1225	1224	1222	1221	0	0	0	1	1	1	1	1	1
8.2	.1220	1218	1217	1215	1214	1212	1211	1209	1208	1206	0	0	0	1	1	1	1	1	1
8.3	.1205	1203	1202	1200	1199	1198	1196	1195	1193	1192	0	0	0	1	1	1	1	1	1
8.4	.1190	1189	1188	1186	1185	1183	1182	1181	1179	1178	0	0	0	1	1	1	1	1	1
8.5	.1176	1175	1174	1172	1171	1170	1168	1167	1166	1164	0	0	0	1	1	1	1	1	1
8.6	.1163	1161	1160	1159	1157	1156	1155	1153	1152	1151	0	0	0	1	1	1	1	1	1
8.7	.1149	1148	1147	1145	1144	1143	1142	1140	1139	1138	0	0	0	1	1	1	1	1	1
8.8	.1136	1135	1134	1133	1131	1130	1129	1127	1126	1125	0	0	0	1	1	1	1	1	1
8.9	.1124	1122	1121	1120	1119	1117	1116	1115	1114	1112	0	0	0	1	1	1	1	1	1
9.0	.1111	1110	1109	1107	1106	1105	1104	1103	1101	1100	0	0	0	1	1	1	1	1	1
9.1	.1099	1098	1096	1095	1094	1093	1092	1090	1089	1088	0	0	0	0	1	1	1	1	1
9.2	.1087	1086	1085	1083	1082	1081	1080	1079	1078	1076	0	0	0	0	1	1	1	1	1
9.3	.1075	1074	1073	1072	1071	1070	1068	1067	1066	1065	0	0	0	0	1	1	1	1	1
9.4	.1064	1063	1062	1060	1059	1058	1057	1056	1055	1054	0	0	0	0	1	1	1	1	1
9.5	.1053	1052	1050	1049	1048	1047	1046	1045	1044	1043	0	0	0	0	1	1	1	1	1
9.6	.1042	1041	1039	1038	1037	1036	1035	1034	1033	1032	0	0	0	1	1	1	1	1	1
9.7	.1031	1030	1029	1028	1027	1026	1025	1024	1022	1021	0	0	0	0	1	1	1	1	1
9.8	.1020	1019	1018	1017	1016	1015	1014	1013	1012	1011	0	0	0	0	1	1	1	1	1
9.9	.1010	1009	1008	1007	1006	1005	1004	1003	1002	1001	0	0	0	0	0	1	1	1	1

POWERS, ROOTS AND RECIPROCALS

n	n^2	n^3	$\sqrt{n}$	$\sqrt[3]{n}$	$\frac{1}{n}$	$\sqrt{10n}$	$\frac{1}{\sqrt{n}}$	$\frac{1}{\sqrt{10n}}$
1	1	1	1.0000	1.0000	1.00000	3.1623	1.00000	0.31623
2	4	8	1.4142	1.2599	0.50000	4.4721	0.70711	0.22361
3	9	27	1.7321	1.4422	0.33333	5.4772	0.57735	0.18257
4	16	64	2.0000	1.5874	0.25000	6.3245	0.50000	0.15811
5	25	125	2.2361	1.7100	0.20000	7.0711	0.44721	0.14142
6	36	216	2.4495	1.8171	0.16667	7.7460	0.40825	0.12910
7	49	343	2.6458	1.9129	0.14286	8.3666	0.37796	0.11952
8	64	512	2.8284	2.0000	0.12500	8.9443	0.35355	0.11180
9	81	729	3.0000	2.0801	0.11111	9.4868	0.33333	0.10541
10	100	1000	3.1623	2.1544	0.10000	10.0000	0.31623	0.10000
11	121	1331	3.3166	2.2240	0.09091	10.4881	0.30151	0.09535
12	144	1728	3.4641	2.2894	0.08333	10.9545	0.28868	0.09129
13	169	2197	3.6056	2.3513	0.07692	11.4018	0.27735	0.08771
14	196	2744	3.7417	2.4101	0.07143	11.8322	0.26726	0.08452
15	225	3375	3.8730	2.4662	0.06667	12.2474	0.25820	0.08165
16	256	4096	4.0000	2.5198	0.06250	12.6491	0.25000	0.07906
17	289	4913	4.1231	2.5713	0.05882	13.0384	0.24253	0.07670
18	324	5832	4.2426	2.6207	0.05556	13.4164	0.23570	0.07454
19	361	6859	4.3589	2.6684	0.05263	13.7840	0.22942	0.07255
20	400	8000	4.4721	2.7144	0.05000	14.1421	0.22361	0.07071
21	441	9261	4.5826	2.7586	0.04762	14.4914	0.21822	0.06901
22	484	10648	4.6904	2.8020	0.04545	14.8324	0.21320	0.06742
23	529	12167	4.7958	2.8439	0.04348	15.1658	0.20851	0.06594
24	576	13824	4.8990	2.8845	0.04167	15.4919	0.20412	0.06455
25	625	15625	5.0000	2.9240	0.04000	15.8114	0.20000	0.06325
26	676	17576	5.0990	2.9625	0.03846	16.1245	0.19612	0.06202
27	729	19683	5.1962	3.0000	0.03704	16.4317	0.19245	0.06086
28	784	21952	5.2915	3.0366	0.03571	16.7332	0.13898	0.05976
29	841	24389	5.3852	3.0723	0.03448	17.0294	0.18570	0.05872
30	900	27000	5.4772	3.1072	0.03333	17.3205	0.18257	0.05774
31	961	29791	5.5678	3.1414	0.03226	17.6068	0.17961	0.05680
32	1024	32768	5.6569	3.1748	0.03125	17.8885	0.17678	0.05590
33	1089	35937	5.7446	3.2075	0.03030	18.1659	0.17408	0.05505
34	1156	39304	5.8310	3.2396	0.02941	18.4391	0.17150	0.05423
35	1225	42875	5.9161	3.2711	0.02857	18.7083	0.16903	0.05345
36	1296	46656	6.0000	3.3019	0.02778	18.9737	0.16667	0.05270
37	1369	50653	6.0828	3.3322	0.02703	19.2354	0.16440	0.05199
38	1444	54872	6.1644	3.3620	0.02632	19.4936	0.16222	0.05130
39	1521	59319	6.2450	3.3912	0.02564	19.7484	0.16013	0.05064
40	1600	64000	6.3245	3.4200	0.02500	20.0000	0.15811	0.05000
41	1681	68921	6.4031	3.4482	0.02439	20.2485	0.15617	0.04939
42	1764	74088	6.4807	3.4760	0.02381	20.4939	0.15430	0.04880
43	1849	79507	6.5574	3.5034	0.02326	20.7364	0.15250	0.04822
44	1936	85184	6.6332	3.5302	0.02273	20.9762	0.15076	0.04767
45	2025	91125	6.7082	3.5569	0.02222	21.2132	0.14907	0.04714
46	2116	97336	6.7823	3.5830	0.02174	21.4476	0.14744	0.04663
47	2209	103823	6.8557	3.6088	0.02128	21.6795	0.14586	0.04613
48	2304	110592	6.9282	3.6342	0.02083	21.9089	0.14434	0.04564
49	2401	117649	7.0000	3.6593	0.02041	22.1359	0.14286	0.04518
50	2500	125000	7.0711	3.6840	0.02000	22.3607	0.14142	0.04472

Powers, Roots and Reciprocals (*Contd.*)

n	n^2	n^3	$\sqrt{n}$	$\sqrt[3]{n}$	$\frac{1}{n}$	$\sqrt{10n}$	$\frac{1}{\sqrt{n}}$	$\frac{1}{\sqrt{10n}}$
51	2601	132651	7.1414	3.7084	0.01961	22.5832	0.14003	0.04428
52	2704	140608	7.2111	3.7325	0.01923	22.8035	0.13868	0.04385
53	2809	148877	7.2802	3.7563	0.01887	23.0217	0.13736	0.04344
54	2916	157464	7.3485	3.7798	0.01852	23.2379	0.13608	0.04303
55	3025	166375	7.4162	3.8030	0.01818	23.4521	0.13484	0.04264
56	3136	175616	7.4833	3.8259	0.01786	23.6643	0.13363	0.04226
57	3249	185193	7.5498	3.8485	0.01754	23.8747	0.13245	0.04189
58	3364	195112	7.6158	3.8709	0.01724	24.0832	0.13131	0.04152
59	3481	205379	7.6811	3.8930	0.01695	24.2899	0.13019	0.04117
60	3600	216000	7.7460	3.9149	0.01667	24.4949	0.12910	0.04082
61	3721	226981	7.8102	3.9365	0.01639	24.6982	0.12804	0.04049
62	3844	238328	7.8740	3.9579	0.01613	24.8998	0.12700	0.04016
63	3969	250047	7.9373	3.9791	0.01587	25.0998	0.12599	0.03984
64	4096	262144	8.0000	4.0000	0.01563	25.2982	0.12500	0.03953
65	4225	274625	8.0623	4.0207	0.01538	25.4951	0.12403	0.03922
66	4356	287496	8.1240	4.0412	0.01515	25.6905	0.12309	0.03892
67	4489	300763	8.1854	4.0615	0.01493	25.8844	0.12217	0.03863
68	4624	314432	8.2462	4.0817	0.01471	26.0768	0.12127	0.03835
69	4761	328509	8.3066	4.1016	0.01449	26.2679	0.12039	0.03807
70	4900	343000	8.3666	4.1213	0.01429	26.4575	0.11952	0.03780
71	5041	357911	8.4261	4.1408	0.01408	26.6458	0.11868	0.03753
72	5184	373248	8.4853	4.1602	0.01389	26.8328	0.11785	0.03727
73	5329	389017	8.5440	4.1793	0.01370	27.0185	0.11704	0.03701
74	5476	405224	8.6023	4.1983	0.01351	27.2029	0.11625	0.03676
75	5625	421875	8.6603	4.2172	0.01333	27.3861	0.11547	0.03651
76	5776	438976	8.7178	4,2358	0.01316	27.5681	0.11471	0.03627
77	5929	456533	8.7750	4.2543	0.01299	27.7489	0.11396	0.03604
78	6084	474552	8.8318	4.2727	0.01282	27.9285	0.11323	0.03581
79	6241	493039	8.8882	4.2908	0.01266	28.1069	0.11251	0.03558
80	6400	512000	8.9443	4.3089	0.01250	28.2843	0.11180	0.03536
81	6561	531441	9.0000	4.3267	0.01235	28.4601	0.11111	0.03514
82	6724	551368	9.0554	4.3445	0.01220	28.6356	0.11043	0.03492
83	6889	571787	9.1104	4.3621	0.01205	28.8097	0.10976	0.03471
84	7056	592704	9.1652	4.3795	0.01190	28.9828	0.10911	0.03450
85	7225	614125	9.2195	4.3968	0.01176	29.1548	0.10847	0.03430
86	7396	636056	9.2736	4.4140	0.01163	29.3258	0.10783	0.03410
87	7569	658503	9.3274	4.4310	0.01149	29.4958	0.10721	0.03390
88	7744	681472	9.3808	4.4480	0.01136	29.6648	0.10660	0.03371
89	7921	704969	9.4340	4.4647	0.01124	29.8329	0.10600	0.03352
90	8100	729000	9.4868	4.4814	0.01111	30.0000	0.10541	0.03333
91	8281	753571	9.5394	4.4979	0.01099	30.1662	0.10483	0.03315
92	8464	778688	9.5917	4.5144	0.01087	30.3315	0.10426	0.03297
93	8649	804357	9.6437	4.5307	0.01075	30.4959	0.10370	0.03279
94	8836	830584	9.6954	4.5468	0.01064	30.6594	0.10314	0.03262
95	9025	857375	9.7468	4.5629	0.01053	30.8221	0.10260	0.03244
96	9216	884736	9.7980	4.5789	0.01042	30.9839	0.10206	0.03227
97	9409	912673	9.8489	4.5947	0.01031	31.1448	0.10153	0.03211
98	9604	941192	9.8995	4.6104	0.01020	31.3050	0.10102	0.03194
99	9801	970299	9.9499	4.6261	0.01010	31.4643	0.10050	0.03178
100	10000	1000000	10.0000	4.6416	0.01000	31.6228	0.10000	0.03162

LOGARITHMS

	0	1	2	3	4	5	6	7	8	9	Mean Differences								
											1	2	3	4	5	6	7	8	9
10	0000	0043	0086	0128	0170	0212	0253	0294	0334	0374	4	8	12	17	21	25	29	33	37
11	0414	0453	0492	0531	0569	0607	0645	0682	0719	0775	4	8	11	15	19	23	26	30	34
12	0792	0828	0864	0899	0934	0969	1004	1038	1072	1106	3	7	10	14	17	21	24	28	31
13	1139	1173	1206	1239	1271	1303	1335	1367	1399	1430	3	6	10	13	16	19	23	26	29
14	1461	1492	1523	1553	1584	1614	1644	1673	1703	1732	3	6	9	12	15	18	21	24	27
15	1761	1790	1818	1847	1875	1903	1931	1959	1987	2014	3	6	8	11	14	17	20	22	25
16	2041	2068	2095	2122	2148	2175	2201	2227	2253	2279	3	5	8	11	13	16	18	21	24
17	2304	2330	2355	2380	2405	2430	2455	2480	2504	2529	2	5	7	10	12	15	17	20	22
18	2553	2577	2601	2625	2648	2672	2695	2718	2742	2765	2	5	7	9	12	14	16	19	21
19	2788	2810	2833	2856	2878	2900	2923	2945	2967	2989	2	4	7	9	11	13	16	18	20
20	3010	3032	3054	3075	3096	3118	3139	3160	3181	3201	2	4	6	8	11	13	15	17	19
21	3222	3243	3363	3284	3304	3324	3345	3365	3385	3404	2	4	6	8	10	12	14	16	18
22	3424	3444	3464	3483	3502	3522	3541	3560	3579	3598	2	4	6	8	10	12	14	15	17
23	3617	3636	3655	3464	3692	3711	3729	3747	3766	3784	2	4	6	7	9	11	13	15	17
24	3802	3820	3838	3856	3874	3892	3909	3927	3945	3962	2	4	5	7	9	11	12	14	16
25	3979	3997	4014	4031	4048	4065	4082	4099	4116	4133	2	3	5	7	9	10	12	14	15
26	4150	4166	4183	4200	4216	4232	4249	4265	4281	4298	2	3	5	7	8	10	11	13	15
27	4314	4330	4346	4362	4378	4393	4409	4425	4440	4456	2	3	5	6	8	9	11	13	14
28	4472	4487	4502	4518	4533	4548	4564	4579	4594	4609	2	3	5	6	8	9	11	12	14
29	4624	4639	4654	4669	4683	4698	4713	4728	4742	4757	1	3	4	6	7	9	10	12	13
30	4771	4786	4800	4814	4829	4843	4857	4871	4886	4900	1	3	4	6	7	9	10	11	13
31	4914	4928	4942	4955	4969	4983	4997	5011	5024	5038	1	3	4	6	7	8	10	11	12
32	5051	5065	5079	5092	5105	5119	5132	5145	5159	5172	1	3	4	5	7	8	10	11	12
33	5185	5198	5211	5224	5237	5250	5263	5276	5289	5302	1	3	4	5	6	8	9	10	12
34	5315	5328	5340	5353	5366	5378	5391	5403	5416	5428	1	3	4	5	6	8	9	10	11
35	5441	5453	5465	5478	5490	5502	5514	5527	5539	5551	1	2	4	5	6	7	9	10	11
36	5563	5575	5587	5599	5611	5623	5635	5647	5658	5670	1	2	4	5	6	7	8	10	11
37	5682	5694	5705	5717	5729	5740	5752	5763	5775	5786	1	2	3	5	6	7	8	9	10
38	5798	5809	5821	5832	5843	5855	5866	5877	5888	5899	1	2	3	5	6	7	8	9	10
39	5911	5922	5933	5944	5955	5966	5977	5988	5999	6010	1	2	3	4	5	7	8	9	10
40	6021	6031	6042	6053	6064	6075	6085	6096	6107	6117	1	2	3	4	5	6	8	9	10
41	6128	6138	6149	6160	6170	6180	6191	6201	6212	6222	1	2	3	4	5	6	7	8	9
42	6232	6243	6253	6263	6274	6284	6294	6304	6314	6325	1	2	3	4	5	6	7	8	9
43	6335	6345	6355	6365	6375	6385	6395	6405	6415	6425	1	2	3	4	5	6	7	8	9
44	6435	6444	6454	6464	6474	6484	6493	6503	6513	6522	1	2	3	4	5	6	7	8	9
45	6532	6542	6551	6561	6571	6580	6590	6599	6609	6618	1	2	3	4	5	6	7	8	9
46	6628	6637	6646	6656	6665	6675	6684	6693	6702	6712	1	2	3	4	5	6	7	7	8
47	6721	6730	6739	6749	6758	6767	6776	6785	6794	6803	1	2	3	4	5	6	6	7	8
48	6812	6821	6830	6839	6848	6857	6866	6875	6884	6893	1	2	3	4	5	5	6	7	8
49	6902	6911	6920	6928	6937	6946	6955	6964	6972	6981	1	2	3	4	4	5	6	7	8
50	6990	6998	7007	7016	7024	7033	7042	7050	7059	7067	1	2	3	4	5	6	6	7	8
51	7076	7084	7093	7101	7110	7118	7126	7135	7143	7152	1	2	3	3	4	5	6	7	8
52	7160	7168	7177	7185	7193	7202	7210	7218	7226	7235	1	2	2	3	4	5	6	7	7
53	7243	7251	7259	7267	7275	7284	7292	7300	7308	7316	1	2	2	3	4	5	6	6	7
54	7324	7332	7340	7348	7356	7364	7372	7380	7388	7396	1	2	2	3	4	5	6	6	7

Logarithms (*Contd.*)

	0	1	2	3	4	5	6	7	8	9	1	2	3	4	5	6	7	8	9
55	7404	7412	7419	7427	7435	7443	7451	7459	7466	7474	1	2	2	3	4	5	5	6	7
56	7482	7490	7497	7505	7513	7520	7528	7536	7543	7551	1	2	2	3	4	5	5	6	7
57	7559	7566	7574	7582	7589	7597	7604	7612	7619	7627	1	1	2	3	4	5	5	6	7
58	7634	7642	7649	7657	7664	7672	7679	7686	7694	7701	1	1	2	3	4	4	5	6	7
59	7709	7716	7723	7731	7738	7745	7752	7760	7767	7774	1	1	2	3	4	4	5	6	7
60	7782	7789	7796	7803	7810	7818	7825	7832	7839	7846	1	1	2	3	4	4	5	6	6
61	7853	7860	7868	7875	7882	7889	7896	7903	7910	7917	1	1	2	3	4	4	5	6	6
62	7924	7931	7938	7945	7952	7959	7966	7973	7980	7987	1	1	2	3	3	4	5	6	6
63	7993	8000	8007	8014	8021	8028	8035	8041	8048	8055	1	1	2	3	3	4	5	5	6
64	8062	8069	8075	8082	8089	8096	8102	8109	8116	8122	1	1	2	3	3	4	5	5	6
65	8129	8136	8142	8149	8156	8162	8169	8176	8182	8189	1	1	2	3	3	4	5	5	6
66	8195	8202	8209	8215	8222	8228	8235	8241	8248	8254	1	1	2	3	3	4	5	5	6
67	8261	8267	8274	8280	8287	8293	8299	8306	8312	8319	1	1	2	3	3	4	5	5	6
68	8325	8331	8338	8344	8351	8357	8363	8370	8376	8382	1	1	2	3	3	4	4	5	6
69	8388	8395	8401	8407	8414	8420	8426	8432	8439	8445	1	1	2	2	3	4	4	5	6
70	8451	8457	8463	8470	8476	8482	8488	8494	8500	8506	1	1	2	2	3	4	4	5	6
71	8513	8519	8525	8531	8537	8543	8549	8555	8561	8567	1	1	2	2	3	4	4	5	6
72	8573	8579	8585	8591	8597	8603	8609	8615	8621	8627	1	1	2	2	3	4	4	5	5
73	8633	8639	8645	8651	8657	8663	8669	8675	8681	8686	1	1	2	2	3	4	4	5	5
74	8692	8698	8704	8710	8716	8722	8727	8733	8739	8745	1	1	2	2	3	4	4	5	5
75	8751	8756	8762	8768	8774	8779	8785	8791	8797	8802	1	1	2	2	3	3	4	5	5
76	8808	8814	8820	8825	8831	8837	8842	8848	8854	8859	1	1	2	2	3	3	4	5	5
77	8865	8871	8876	8882	8887	8893	8899	8904	8910	8915	1	1	2	2	3	3	4	4	5
78	8921	8927	8932	8938	8943	8949	8954	8960	8965	8971	1	1	2	2	3	3	4	4	5
79	8976	8982	8987	8993	8998	9004	9009	9015	9020	9025	1	1	2	2	3	4	4	4	5
80	9031	9036	9042	9047	9053	9058	9063	9069	9074	9079	1	1	2	2	3	3	4	4	5
81	9085	9090	9096	9101	9106	9112	9117	9122	9128	9133	1	1	2	2	3	3	4	4	5
82	9138	9143	9149	9154	9159	9165	9170	9175	9180	9186	1	1	2	2	3	3	4	4	5
83	9191	9196	9201	9206	9212	9217	9222	9227	9232	9238	1	1	2	2	3	3	4	4	5
84	9243	9248	9253	9258	9263	9269	9274	9279	9284	9289	1	1	2	2	3	3	4	4	5
85	9294	9299	9304	9309	9315	9320	9325	9330	9335	9340	1	1	2	2	3	3	4	4	5
86	9345	9350	9355	9360	9365	9370	9375	9380	9385	9390	1	1	2	2	3	3	4	4	5
87	9395	9400	9405	9410	9415	9420	9425	9430	9435	9440	0	1	1	2	2	3	3	4	4
88	9445	9450	9455	9460	9465	9469	9474	9479	9484	9489	0	1	1	2	2	3	3	4	4
89	9494	9499	9504	9509	9513	9518	9523	9528	9533	9538	0	1	1	2	2	3	3	4	4
90	9542	9547	9552	9557	9562	9566	9571	9576	9581	9586	0	1	1	2	2	3	3	4	4
91	9590	9595	9600	9605	9609	9614	9619	9624	9628	9633	0	1	1	2	2	3	3	4	4
92	9638	9643	9647	9652	9657	9661	9666	9671	9675	9680	0	1	1	2	2	3	3	4	4
93	9685	9689	9694	9699	9703	9708	9713	9717	9722	9727	0	1	1	2	2	3	3	4	4
94	9731	9736	9741	9745	9750	9754	9759	9763	9768	9773	0	1	1	2	2	3	3	4	4
95	9777	9782	9786	9791	9795	9800	9805	9809	9814	9818	0	1	1	2	2	3	3	4	4
96	9823	9827	9832	9836	9841	9845	9850	9854	9859	9863	0	1	1	2	2	3	3	4	4
97	9868	9872	9877	9881	9886	9890	9894	9899	9903	9908	0	1	1	2	2	3	3	4	4
98	9912	9917	9921	9926	9930	9934	9939	9943	9948	9952	0	1	1	2	2	3	3	4	4
99	9956	9961	9965	9969	9974	9978	9983	9987	9991	9996	0	1	1	2	2	3	3	3	4

ANTILOGARITHMS

	0	1	2	3	4	5	6	7	8	9	Mean Differences								
											1	2	3	4	5	6	7	8	9
.00	1000	1002	1005	1007	1009	1012	1014	1016	1019	1021	0	0	1	1	1	1	2	2	2
.01	1023	1026	1028	1030	1033	1035	1038	1040	1042	1045	0	0	1	1	1	1	2	2	2
.02	1047	1050	1052	1054	1057	1059	1062	1064	1067	1069	0	0	1	1	1	1	2	2	2
.03	1072	1074	1076	1079	1081	1084	1086	1089	1091	1094	0	0	1	1	1	1	2	2	2
.04	1096	1099	1102	1104	1107	1109	1112	1114	1117	1119	0	1	1	1	1	2	2	2	2
.05	1122	1125	1127	1130	1132	1135	1138	1140	1143	1146	0	1	1	1	1	2	2	2	2
.06	1148	1151	1153	1156	1159	1161	1164	1167	1169	1172	0	1	1	1	1	2	2	2	2
.07	1175	1178	1180	1183	1186	1189	1191	1194	1197	1199	0	1	1	1	1	2	2	2	2
.08	1202	1205	1208	1211	1213	1216	1219	1222	1225	1227	0	1	1	1	2	2	2	2	3
.09	1230	1233	1236	1239	1242	1245	1247	1250	1253	1256	0	1	1	1	2	2	2	2	3
.10	1259	1262	1265	1268	1271	1274	1276	1279	1282	1285	0	1	1	1	2	2	2	2	3
.11	1288	1291	1294	1297	1300	1303	1306	1309	1312	1315	0	1	1	1	2	2	2	2	3
.12	1318	1321	1324	1327	1330	1334	1337	1340	1343	1346	0	1	1	1	2	2	2	2	3
.13	1349	1352	1355	1358	1361	1365	1368	1371	1374	1377	0	1	1	1	2	2	2	3	3
.14	1380	1384	1387	1390	1393	1396	1400	1403	1406	1409	0	1	1	1	2	2	2	3	3
.15	1413	1416	1419	1422	1426	1429	1432	1435	1439	1442	0	1	1	1	2	2	2	3	3
.16	1445	1449	1452	1455	1459	1462	1466	1469	1472	1476	0	1	1	1	2	2	2	3	3
.17	1479	1483	1486	1489	1493	1496	1500	1503	1507	1510	0	1	1	1	2	2	2	3	3
.18	1514	1517	1521	1524	1528	1531	1535	1538	1542	1545	0	1	1	1	2	2	2	3	3
.19	1549	1552	1556	1560	1563	1567	1570	1574	1578	1581	0	1	1	1	2	2	3	3	3
.20	1585	1589	1592	1596	1600	1603	1607	1611	1614	1618	0	1	1	1	2	2	3	3	3
.21	1622	1626	1629	1633	1637	1641	1644	1648	1652	1656	0	1	1	2	2	2	3	3	3
.22	1660	1663	1667	1671	1675	1679	1683	1687	1690	1694	0	1	1	2	2	2	3	3	3
.23	1698	1702	1706	1710	1714	1718	1722	1726	1730	1734	0	1	1	2	2	2	3	3	4
.24	1738	1742	1746	1750	1754	1758	1762	1766	1770	1774	0	1	1	2	2	2	3	3	4
.25	1778	1782	1786	1791	1795	1799	1803	1807	1811	1816	0	1	1	2	2	2	3	3	4
.26	1820	1824	1828	1832	1837	1841	1845	1849	1854	1858	0	1	1	2	2	3	3	3	4
.27	1862	1866	1871	1875	1879	1884	1888	1892	1897	1901	0	1	1	2	2	3	3	3	4
.28	1905	1910	1914	1919	1923	1928	1932	1936	1941	1945	0	1	1	2	2	3	3	4	4
.29	1950	1954	1959	1963	1968	1972	1977	1982	1986	1991	0	1	1	2	2	3	3	4	4
.30	1995	2000	2004	2009	2014	2018	2023	2028	2032	2037	0	1	1	2	2	3	3	4	4
.31	2042	2046	2051	2056	2061	2065	2070	2075	2080	2084	0	1	1	2	2	3	3	4	4
.32	2089	2094	2099	2104	2109	2113	2118	2123	2128	2133	0	1	1	2	2	3	3	4	4
.33	2138	2143	2148	2153	2158	2163	2168	2173	2178	2183	0	1	1	2	2	3	3	4	4
.34	2188	2193	2198	2203	2208	2213	2218	2223	2228	2234	1	1	2	2	3	3	4	4	5
.35	2239	2244	2249	2254	2259	2265	2270	2275	2280	2286	1	1	2	2	3	3	4	4	5
.36	2291	2296	2301	2307	2312	2317	2323	2328	2333	2339	1	1	2	2	3	3	4	4	5
.37	2344	2350	2355	2360	2366	2371	2377	2382	2388	2393	1	1	2	2	3	3	4	4	5
.38	2399	2404	2410	2415	2421	2427	2432	2438	2443	2449	1	1	2	2	3	3	4	4	5
.39	2455	2460	2466	2472	2477	2483	2489	2495	2500	2506	1	1	2	2	3	3	4	5	5
.40	2512	2518	2523	2529	2535	2541	2547	2553	2559	2564	1	1	2	2	3	4	4	5	5
.41	2570	2576	2582	2588	2594	2600	2606	2612	2618	2624	1	1	2	2	3	4	4	5	5
.42	2630	2636	2642	2649	2655	2661	2667	2673	2679	2685	1	1	2	2	3	4	4	5	6
.43	2692	2698	2704	2710	2716	2723	2729	2735	2742	2748	1	1	2	3	3	4	4	5	6
.44	2754	2761	2767	2773	2780	2786	2793	2799	2805	2812	1	1	2	3	3	4	4	5	6
.45	2818	2825	2831	2838	2844	2851	2858	2864	2871	2877	1	1	2	3	3	4	5	5	6
.46	2884	2891	2897	2904	2911	2917	2924	2931	2938	2944	1	1	2	3	3	4	5	5	6
.47	2951	2958	2965	2972	2979	2985	2992	2999	3006	3013	1	1	2	3	3	4	5	5	6
.48	3020	3027	3034	3041	3048	3055	3062	3069	3076	3083	1	1	2	3	4	4	5	6	6
.49	3090	3097	3105	3112	3119	3126	3133	3141	3148	3155	1	1	2	3	4	4	5	6	6

Antilogarithms (*Contd.*)

	0	1	2	3	4	5	6	7	8	9	Mean Differences 1	2	3	4	5	6	7	8	9
.50	3162	3170	3177	3184	3192	3199	3206	3214	3221	3228	1	1	2	3	4	5	5	6	7
.51	3236	3243	3251	3258	3266	3273	3281	3289	3296	3304	1	2	2	3	4	5	5	6	7
.52	3311	3319	3327	3334	3342	3350	3357	3365	3373	3381	1	2	2	3	4	5	5	6	7
.53	3388	3396	3404	3412	3420	3428	3436	3443	3451	3459	1	2	2	3	4	5	6	6	7
.54	3467	3475	3483	3491	3499	3508	3516	3524	3532	3540	1	2	2	3	4	5	6	6	7
.55	3548	3556	3565	3573	3581	3589	3597	3606	3614	3622	1	2	2	3	4	5	6	7	7
.56	3631	3639	3648	3656	3664	3673	3681	3690	3698	3707	1	2	3	3	4	5	6	7	8
.57	3715	3724	3733	3741	3750	3758	3767	3776	3784	3793	1	2	3	3	4	5	6	7	8
.58	3802	3811	3819	3828	3837	3846	3855	3864	3873	3882	1	2	3	4	4	5	6	7	8
.59	3890	3899	3908	3917	3926	3936	3945	3954	3963	3972	1	2	3	4	5	5	6	7	8
.60	3981	3990	3999	4009	4018	4027	4036	4046	4055	4064	1	2	3	4	5	6	6	7	8
.61	4074	4083	4093	4102	4111	4121	4130	4140	4150	4159	1	2	3	4	5	6	7	8	9
.62	4169	4178	4188	4198	4207	4217	4227	4236	4246	4256	1	2	3	4	5	6	7	8	9
.63	4266	4276	4285	4295	4305	4315	4325	4335	4345	4355	1	2	3	4	5	6	7	8	9
.64	4365	4375	4385	4395	4406	4416	4426	4436	4446	4457	1	2	3	4	5	6	7	8	9
.65	4467	4477	4487	4498	4508	4519	4529	4539	4550	4560	1	2	3	4	5	6	7	8	9
.66	4571	4581	4592	4603	4613	4624	4634	4645	4656	4667	1	2	3	4	5	6	7	9	10
.67	4677	4688	4699	4710	4721	4732	4742	4753	4764	4775	1	2	3	4	5	7	8	9	10
.68	4786	4797	4808	4819	4831	4842	4853	4864	4875	4887	1	2	3	4	6	7	8	9	10
.69	4898	4909	4920	4932	4943	4955	4966	4977	4989	5000	1	2	3	5	6	7	8	9	10
.70	5012	5023	5035	5047	5058	5070	5082	5093	5105	5117	1	2	4	5	6	7	8	9	11
.71	5129	5140	5152	5164	5176	5188	5200	5212	5224	5236	1	2	4	5	6	7	8	10	11
.72	5248	5260	5272	5284	5297	5309	5321	5333	5346	5358	1	2	4	5	6	7	9	10	11
.73	5370	5383	5395	5408	5420	5433	5445	5458	5470	5483	1	3	4	5	6	8	9	10	11
.74	5495	5508	5521	5534	5546	5559	5572	5585	5598	5610	1	3	4	5	7	8	9	10	11
.75	5623	5636	5649	5662	5675	5689	5702	5715	5728	5741	1	3	4	5	7	8	9	10	12
.76	5754	5768	5781	5794	5808	5821	5834	5848	5861	5875	1	3	4	5	7	8	9	11	12
.77	5888	5902	5916	5929	5943	5957	5970	5984	5998	6012	1	3	4	5	7	8	10	11	12
.78	6026	6039	6053	6067	6081	6095	6109	6124	6138	6152	1	3	4	6	7	8	10	11	13
.79	6166	6180	6194	6209	6223	6237	6252	6266	6281	6295	1	3	4	6	7	9	10	12	13
.80	6310	6324	6339	6353	6368	6383	6397	6412	6427	6442	1	3	4	6	7	9	10	12	13
.81	6457	6471	6486	6501	6516	6531	6546	6561	6577	6592	2	3	5	6	8	9	11	12	14
.82	6607	6622	6637	6653	6668	6683	6699	6714	6730	6745	2	3	5	6	8	9	11	12	14
.83	6761	6776	6792	6808	6823	6839	6855	6871	6887	6902	2	3	5	6	8	9	11	13	14
.84	6918	6934	6950	6966	6982	6998	7015	7031	7047	7063	2	3	5	6	8	10	12	13	15
.85	7079	7096	7112	7129	7145	7161	7178	7194	7211	7228	2	3	5	7	8	10	12	13	15
.86	7244	7261	7278	7295	7311	7328	7345	7362	7379	7396	2	3	5	7	8	10	12	13	15
.87	7413	7430	7447	7464	7482	7499	7516	7534	7551	7568	2	3	5	7	9	10	12	14	16
.88	7586	7603	7621	7638	7656	7674	7691	7709	7727	7745	2	4	5	7	9	11	12	14	16
.89	7762	7780	7798	7816	7834	7852	7870	7889	7907	7925	2	4	5	7	9	11	13	14	16
.90	7943	7962	7980	7998	8017	8035	8054	8072	8091	8110	2	4	6	7	9	11	13	15	17
.91	8128	8147	8166	8185	8204	8222	8241	8260	8279	8299	2	4	6	8	9	11	13	15	17
.92	8318	8337	8356	8375	8395	8414	8433	8453	8472	8492	2	4	6	8	10	12	14	15	17
.93	8511	8531	8551	8570	8590	8610	8630	8650	8670	8690	2	4	6	8	10	12	14	16	18
.94	8710	8730	8750	8770	8790	8810	8831	8851	8872	8892	2	4	6	8	10	12	14	16	18
.95	8913	8933	8954	8974	8995	9016	9036	9057	9078	9099	2	4	6	8	10	12	15	17	19
.96	9120	9141	9162	9183	9204	9226	9247	9628	9290	9311	2	4	6	8	11	13	15	17	19
.97	9333	9354	9376	9397	9419	9441	9462	9484	9506	9528	2	4	7	9	11	13	15	17	20
.98	9550	9572	9594	9616	9638	9661	9683	9705	9727	9750	2	4	7	9	11	13	16	18	20
.99	9772	9795	9817	9840	9863	9886	9908	9931	9954	9977	2	5	7	9	11	14	16	18	20